# WINNING WARS

# WINNING WARS

## The Enduring Nature and Changing Character of Victory from Antiquity to the 21st Century

*Editor*

MATTHIAS STROHN

CASEMATE
*Oxford & Philadelphia*

Published in Great Britain and the United States of America in 2020 by
CASEMATE PUBLISHERS
The Old Music Hall, 106–108 Cowley Road, Oxford OX4 1JE, UK
and
1950 Lawrence Road, Havertown, PA 19083, USA



Hardback Edition: ISBN 978-1-95271-500-6
Digital Edition: ISBN 978-1-95271-501-3

A CIP record for this book is available from the British Library



Printed and bound in the United States of America by Integrated Books International

Typeset by Versatile PreMedia Service (P) Ltd

For a complete list of Casemate titles, please contact:

CASEMATE PUBLISHERS (UK)
Telephone (01865) 241249
Email: casemate-uk@casematepublishers.co.uk
www.casematepublishers.co.uk

CASEMATE PUBLISHERS (US)
Telephone (610) 853-9131
Fax (610) 853-9146
Email: casemate@casematepublishers.com
www.casematepublishers.com

*Front cover image*: Special Forces soldiers from the 3rd Special Forces Group patrol a field in the Gulistan district of Farah, Afghanistan with Afghan National Army commandos from the 207th Kandak, 12 April 2009. (Photo by Spc. Joseph A. Wilson, US Army)

# Contents

# Foreword

*The Chief of the General Staff, General Sir Mark Carleton-Smith, KCB CBE ADC Gen*

We are all in the business of winning – it is in some ways the essence of life. And yet despite its fundamental nature, identifying what constitutes winning and how you might achieve it remains an abstract art. The concept of winning at war is no different – the factors are growing in complexity and the solutions appear ever more opaque. The character of warfare is changing at a rate that is perhaps as pronounced now as it has ever been; technology maintains its relentless and exponential progress, data and knowledge are increasingly ubiquitous, and the new domain of cyberspace is offering novel ways of interaction and subversion. These factors are combining to alter the balance of power both regionally and globally, which is manifested in friction and conflict along the boundaries and fault lines of states, religions, cultures and economies. And it is in this space that military and political leaders must seek to practice their trades for national advantage.

Forewarned is forearmed; anyone interested in international relations, military philosophy, politics, the utility of force, or the study of war, may find the following pages provide invaluable and enduring insights, accumulated over two millennia, into what has constituted 'winning' when the military instrument has been used.

# Editor's Acknowledgments

The idea for the book developed out of a research project that the Centre for Historical Analysis and Conflict Research (CHACR) carried out in support of the British Army's conceptual component of fighting power and the development of Army strategy in the 21st century. Turning Army in-house research, papers and briefings into a book is more complex and time consuming than one might think. I am grateful to the British Army and its commitment to CHACR that gave me the opportunity to produce this book. I am particularly grateful to the CHACR team and the director, Major-General (retired) Dr. Andrew Sharpe. The entire CHACR team was at some stage or other involved in the project and the open, friendly and academic atmosphere at CHACR has most definitely helped me when I was putting this book together.

Without the authors that contributed to the book, this opus would simply not have seen the light of day. They made my job as editor very easy indeed; their dedication to the project, the quality of the writings, and the ability to stick to sometimes rather tight timelines meant that being the editor was a pure joy. In particular I would like to thank the Chief of the General Staff of the British Army, General Sir Mark Carleton-Smith, who found the time to write the foreword to this book.

I would like to thank Casemate Publishers and the team that has worked with me on this project, in particular Ruth Sheppard and Isobel Fulton. I have worked with them on a number of projects and their professionalism has always impressed me.

The topic of the book is complex, perhaps even difficult. In many respects, you, the reader, will probably be left with more questions than answers once you have read it. This is the aim. To stay within the military realm, the conceptual component of fighting power can only be enhanced if we engage intellectually with the problems that present themselves to us. If this book makes you think, it has achieved its aim.

Matthias Strohn
Camberley, November 2020

CHACR

The Centre for Historical Analysis and Conflict Research (CHACR) is a 'think-tank' based in Robertson House, the former British Army Staff College, in Camberley. It was established on behalf of the British Army in 2016 to help to strengthen the conceptual component of fighting power and to offer independent and objective views to inform future force development. It is designed to appraise the Army of wider thinking in order to challenge convention and inform strategic decision-making. It is not, therefore, a vehicle to portray the Army's own thinking; on the contrary, it is a vehicle to bring varied views together to stimulate the Army's thinking.

# List of Contributors

**Kerry Brown** is Professor of Chinese Studies and Director of the Lau China Institute at King's College London. He is an Associate of the Asia Pacific Programme at Chatham House, London, an adjunct of the Australia New Zealand School of Government in Melbourne, and the co-editor of the *Journal of Current Chinese Affairs*, run from the German Institute for Global Affairs in Hamburg. From 2012 to 2015 he was Professor of Chinese Politics and Director of the China Studies Centre at the University of Sydney, Australia. Prior to this he worked at Chatham House from 2006 to 2012, as Senior Fellow and then Head of the Asia Programme. From 1998 to 2005 he worked at the British Foreign and Commonwealth Office, as First Secretary at the British Embassy in Beijing, and then as Head of the Indonesia, Philippine and East Timor Section. He lived in the Inner Mongolia region of China from 1994 to 1996. Professor Brown directed the Europe China Research and Advice Network (ECRAN) giving policy advice to the European External Action Service between 2011 and 2014. He is the author of 20 books on modern China.

**Aaron Edwards** is Senior Lecturer in Defence and International Affairs at the Royal Military Academy Sandhurst and Honorary Research Fellow in the School of History, Politics and International Relations at the University of Leicester. He obtained his PhD in Politics from Queen's University Belfast in 2006 and was appointed a Fellow of the Royal Historical Society in 2012. Aaron is the author of acclaimed books *Mad Mitch's Tribal Law: Aden and the End of Empire* (2014), *UVF: Behind the Mask* (2017) and *Agents of Influence: Britain's Secret Intelligence War Against the IRA* (2021).

**John France** is Professor Emeritus in the History Department of Swansea University and has held the Charles Boal Ewing Visiting Professorship at the United States Military Academy, West Point. His specialities are crusading and the history of medieval warfare. His recent books include *Hattin* (2015) and *Perilous Glory: The Rise of Western Military Power* (2011).

**Lothar Höbelt** is Associate Professor of Modern History and Lecturer at the Military Academy in Wiener Neustadt, a member of the Editorial Advisory Board of *War in History* and has published extensively on the Habsburg Monarchy, World War I and the post-war politics of Austria. He lives with his family in Bohemia.

**Rob Johnson** is Director of the Oxford Changing Character of War Research Centre, and Senior Research Fellow of Pembroke College at the University of Oxford. His primary research interests are strategic studies and decision-making, but he has written extensively on conflicts in Asia and the Middle East, the modalities of war from conventional operations to hybrid warfare. He is the author of *The Great War and the Middle East* (2016) and *Lawrence of Arabia on War* (2020), and he is the co-editor of *Military Strategy in the 21st Century* (2020), *The Conduct of War* (2021) and *World Information War* (2020). He is currently researching a major new study of Anglo-American wartime decision-making.

**Richard Kuno** is an EU national with extensive professional experience in the Arab world. He is fluent in Arabic and has lived and worked in numerous Arab countries, such as Syria, Lebanon, Tunisia, the UAE and Oman. His professional and scientific core interests are sources of conflict in the Middle East and geostrategy in the Gulf region. He has a PhD in Middle Eastern studies and a master's degree in history and he has been an affiliate scholar at two of the leading universities in the Arab world. Richard has been observing political and strategic developments in the Middle East and in the Gulf region for almost 20 years, and he shares his insights and experiences with a range of European and regional research institutions and consultancies. Richard Kuno is a pen name. He is currently living in Europe working on defence and security issues.

**Carter Malkasian** was the senior advisor for strategy to the Chairman of the Joint Chiefs of Staff, General Joseph Dunford (2015–19). He has extensive experience working in conflict zones, including nearly two years in Garmser district, Helmand province, Afghanistan, as a State Department political officer. His publications include *War Comes to Garmser: Thirty Years of Conflict on the Afghan Frontier* and *Illusions of Victory: The Anbar Awakening and the Rise of the Islamic State.*

**Daniel Marston** is Director of the Secretary of Defense Strategic Thinkers Program (STP) at the Paul H. Nitze School of Advanced International Studies (SAIS), Johns Hopkins University. He held a Professorship in Military Studies at the Australian National University. He has been a Visiting Fellow, on multiple occasions, with the Leverhulme Changing Character of War Program at the University of Oxford. He has been a special advisor, since 2006, in Iraq and Afghanistan with the US Army, USMC and British Army. He received a BA and MA from McGill University and a DPhil from Oxford University. He is a Fellow of the Royal Historical Society.

**Andrew Monaghan** is Director of the Russia Research Network, Ltd. Additionally he is a Non-Resident Associate Fellow of the NATO Defence College in Rome, where he is also the Commissioning Editor of the Russian Studies publication series, Senior Associate Fellow at the Royal United Services Institute in London, and Honorary Senior Research Fellow at the University of Birmingham. Previously employed by the Oxford Changing Character of War Centre, Chatham House, NATO and the UK Defence Academy, he has advised various governments and international organisations, as well as major companies. He has served as an expert witness to the UK's National Security Strategy Committee, the House of Commons Defence and Foreign Affairs Select Committees, and NATO's Parliamentary Assembly. He is widely published and the author of several books, including *Dealing with the Russians* (2019), *Power in Modern Russia* (2017) and *The New Politics of Russia* (2016).

**Ali Parchami** is Senior Lecturer in the Department of Defence and International Affairs at the Royal Military Academy Sandhurst. He received his BA, MPhil and DPhil from the University of Oxford and, between 2004 and 2007, was Stipendiary Lecturer at Exeter College, Oxford. His research interests encompass the fields of history, international relations and regional studies – with particular emphasis on Roman antiquity and strategic developments in the contemporary Middle East. This broad interdisciplinary interest is reflected in his publication record, which includes the monograph *Hegemonic Peace and Empire: The Pax Romana, Britannica and Americana* and peer-reviewed articles in imperial and commonwealth history, contemporary politics and Middle East journals. His latest projects focus on Sino-Iranian collaboration, Iran's expeditionary operations in the Middle East, and the evolution of the Islamic Republic's 'cascade strategy'.

**David Parrott** is Professor of Early Modern History and Fellow and Tutor at New College, University of Oxford. He has researched and written extensively on early modern military and political history. His previous publications include a study of the French army during the ministry of Cardinal Richelieu, and *The Business of War*, a broad study of military contracting and private enterprise in early modern Europe. His most recent book *1652: The Prince, the Cardinal and the Crisis of the Fronde*, offers a re-examination of the last and most devastating year of the French civil war of 1648–52, looking especially at the interaction of warfare, politics and personalities.

**Nicholas Rees** teaches and researches in international relations and international organisations, with research interests in the UN, EU and NATO and their roles in peace support operations. He has published widely on the EU and UN, focusing especially on the role of member states such as Ireland and Britain in these organisations. He was formerly Professor of International Relations and Dean of Humanities and Social Sciences at the University of Buckingham, where he is an Honorary Professorial Fellow in International Relations.

**Richard Reid** is Professor of African History in the Faculty of History and a Fellow of St Cross College at the University of Oxford. His work has focused particularly on the history of political culture, historical consciousness, warfare and militarism in Africa, notably eastern and northeast Africa, including Eritrea, Ethiopia, Uganda and Tanzania. He is the author of a number of books, including *A History of Modern Uganda* (2017) and *Warfare in African History* (2012), while a revised third edition of his *History of Modern Africa: from 1800 to the present* appeared in 2019. A former editor of the *Journal of African History*, most recently he has written a first-hand account of the social, political, and cultural impact of conflict in the Horn, *Shallow Graves: a memoir of the Ethiopia-Eritrea war* (2020).

**Jonathon Riley** served in the British Army for 40 years. He commanded on operations at every level from platoon to corps, in Northern Ireland, the Balkans, Sierra Leone, Iraq and Afghanistan. In the Balkans, he commanded the UN Task Force in the enclave of Gorazde and was later a NATO brigade commander. In Sierra Leone he commanded the Joint Task Force and was responsible for developing and implementing the campaign plan that led to the end of the ten-year war. He has since gone on to work for the Royal Jordanian Court and as an expert witness in international defence-related trials. Riley also holds three degrees including a PhD in modern history, he has published 25 books and is Visiting Professor in War Studies at King's College London.

**Andrew Sharpe** has commanded on operations in every rank from second lieutenant to brigadier during 34 years of military service. He also had a rich and varied career of staff appointments, notably including involvement in national and international strategic and campaign planning, and in training and education, including time as the Director of the Higher Command and Staff Course and as Deputy Commandant of the Staff College. His final appointment, as a major general, was as director of the MoD's independent think-tank the DCDC. For three years he ran the Chief of Defence Staff's strategic advisory panel. Since leaving the Army Doctor Sharpe has led a varied business, mentoring, speaking, writing and consultancy career, and is also the Director of the British Army's Centre for Historical Analysis and Conflict Research. He holds visiting fellowships at Exeter University and Kings College London, and a PhD in strategic leadership from Trinity College, Cambridge.

**Sir Hew Strachan**, FBA, FRSE, has been Wardlaw Professor of International Relations at the University of St Andrews since 2015. He is Life Fellow of Corpus Christi College, Cambridge, where he was successively Research Fellow and Fellow from 1975 to 1992, and Emeritus Fellow of All Souls College, Oxford. He was Professor of Modern History at Glasgow University 1992 to 2001, and Chichele Professor of the History of War at the University of Oxford 2002 to 2015. He was Commonwealth

War Graves Commissioner 2006–18, Trustee of the Imperial War Museum 2010–18, and a member of the national committees for the centenary of World War I of the United Kingdom, Scotland and France. His recent publications include *The First World War: To Arms* (2001), *The First World War: A New Illustrated History* (2003, based on his 10-part series for Channel 4) and *The Direction of War* (2013).

**Matthias Strohn**, MSt (Oxon), DPhil (Oxon), FRHistS, is Head of Historical Analysis at the Centre for Historical Analysis and Conflict Research, the British Army's strategic think tank, Visiting Professor of Military Studies at the University of Buckingham, and a member of the academic faculty at the Royal Military Academy Sandhurst. Matthias was educated at the universities of Münster (Germany) and Oxford. He holds a commission in the German Army and is a member of the military attaché reserve, having served on the defence attaché staffs in London, Paris and Madrid. Prior to this, he served as Military History Instructor at the German Staff College in Hamburg. He deployed to Iraq (with the British Army) and Afghanistan (with both the British Army and the German Bundeswehr). Matthias has published widely on 20th-century German and European military history; he has authored and edited 14 books and numerous articles.

**Christopher Tuck** is Reader in Strategic Studies at King's College London. He works in the Defence Studies Department at the United Kingdom Defence Academy. His publications include *Understanding Modern Warfare* (2016), *Understanding Land Warfare* (2014) and *Strategy, Confrontation and War Termination: Britain's Conflict with Indonesia, 1963–66* (2012).

# Introduction

## Sir Hew Strachan

'Attempts to solve major security issues by military means alone seldom succeed in the long term, even if initially enjoying apparent success.'[1] Those are the words of the fifth and most recent edition of *UK Defence Doctrine.* Published in November 2014, it reflects the bitter experience of the post-9/11 wars. In the Falklands in 1982, in the Persian Gulf in 1991, in Kosovo in 1999 and in Sierra Leone in 2000, military action had delivered political results in short order. War did indeed seem to be the continuation of policy by other means. However, after 2002 Britain and its allies fought wars which promised easy and quick success, but which persist to this day. Each of the interventions in Afghanistan, Iraq and Libya was characterised by an apparently easy victory, loudly and triumphantly proclaimed at the political level, but which then gave way to protracted conflict with – at best – indeterminate results.

*UK Defence Doctrine*'s equivocation about the utility of military force presents a persistent conundrum in the making of strategy. The doctrine applicable to war's next level down, *UK Joint Operations Doctrine,* was also published in a new edition in November 2014. At one level it reflected *UK Defence Doctrine* by eschewing words like 'winning' or 'victory', but in reality it deployed elements of both. It defined the 'military strategic end-state' as 'the successful completion of the military contribution to the desired outcome, reached when all the allocated military strategic objectives have been achieved.'[2] The words 'success' or 'successful' are used 21 times in *Joint Operations Doctrine*, and their pursuit is clearly – and many might feel properly – a benchmark for the British armed forces. But this determination to win evaporates when we revert once more to the higher level. The nearest that *UK Defence Doctrine* comes to defining what 'success' might actually look like in strategic, as opposed to operational, terms is to say that 'war is undertaken to maintain a position of advantage, establish a more advantageous situation, or influence the attitudes or behaviour of another party.'[3]

The note of qualification, the aspiration to relative rather than absolute success, contrasts with the expectation of a clear-cut and decisive victory derived from strategies of the past. In 1827 Carl von Clausewitz memorably defined wars as being of two types. He described one in geographical terms, as designed 'merely to occupy some of [the enemy's] frontier districts so that we can annex them or use them for bargaining at the peace negotiations.' This too was relative rather than absolute, but the objective of his other sort of war was much less ambiguous: 'To overthrow [the German is *niederwerfen*] the enemy, to annihilate [*vernichten*] him politically or render him defenceless [*wehrlos*], thus forcing him to sign whatever peace we please.'[4] No trace of this sort of winning is to be found in either of the two UK doctrines of November 2014.

At the tactical level, any suggestion that fighting is not intended to destroy the enemy results in logical absurdity. Infantry drills for fire and movement are designed to concentrate killing power in order to achieve an immediate effect or to secure a particular objective. In practice, not all tactics have destruction as their aim; they might be designed for reconnaissance or protection, to reassure a population or deter – not kill – a possible enemy. But success at the tactical level remains more quantifiable than at the strategic, and in some instances can be measured in ground gained or in enemy killed and captured. These victories give combat purpose and rationalise what soldiers do. They also give them moral purpose: to demand that they put their own lives at risk without a recognisable and justifiable reason for doing so would be unconscionable (which does not mean it does not happen).

The potential divergence between the tactics of fighting and the political context of war therefore creates a challenge which emerges most obviously at the operational level, not so much when it gives shape to tactics (its principal role in the 1980s, during the latter stages of the Cold War) as when it looks up to the political and strategic. This is where its gaze has been increasingly directed in the post-9/11 wars. It is here that the tactical imperative (and the military presumption that the aim is success, however defined) collides with the political objectives mediated at the strategic level.

In May 2006, Major General Richard Shirreff arrived in Iraq to conduct a reconnaissance before taking over in July as commander of the Multi-National Division in South-East Iraq (MND-(SE)). Since 2003 Basra, Iraq's principal port and its major city in the south, had been a British responsibility but by 2006 the bulk of British forces had been withdrawn from its streets. The city was increasingly in the hands of Jaish al-Mahdi (JAM), Shia militias who were backed by Iran. British efforts were focused on checking the flow of weapons across the Iran–Iraq border but in Basra itself were restricted to supporting the Iraqi police, who had themselves been infiltrated by JAM. Shirreff was worried that 'standing still amounted to defeat' and that Britain's objective had become 'extraction rather than necessarily achieving mission success.' He wanted to wrest back the initiative by focusing on Basra and the Shia militias, who were, he said, 'ingenious, capable and determined to inflict a humiliating defeat on us.' He wanted to hand back to the Iraqis a city which was in a governable state. If he did not, he feared the consequences not just for Iraq but also for the British army, whose credibility and reputation would be damaged both in its own eyes and in those of its American allies. Rather than an 'exit strategy', he wanted 'a winning strategy'. The plan for putting this intention into effect was initially called Operation *Salamanca* and then, in its reworked form, Operation *Sinbad*.[5]

Ultimately Operation *Sinbad* failed, and it might always have done so, but in August 2006 the challenges it faced came not just from within Iraq and Iran but also from London. As Justin Maciejewski, who commanded 2nd Battalion, The Rifles, in Basra in 2006–7 and so confronted the tactical consequences for what followed, put it:

> The British campaign in Iraq was designed to do just enough to underpin the credibility of the political and military relationship with the US, whilst minimising the domestic political consequences of financial and tactical over-exposure to events on the ground. From a British political viewpoint, the priorities for Operation Telic [the British commitment to the war in Iraq] after the invasion were damage limitation and economy of effort. It was this geostrategic equation that established the scale of military resources available in theatre, not the requirements of the campaign itself.[6]

At the UK's Permanent Joint Headquarters in Northwood, the chief of joint operations, Lieutenant General Nick Houghton, reflected the government line when he warned Shirreff to resist any 'rushes of testosterone'.[7] Britain was already shifting its main effort from Iraq to Afghanistan and would not resource what those on the ground in Iraq saw as essential to achieve success. Shirreff's plan to stabilise Basra after the military effort depended on the creation of a provincial reconstruction team, a model developed in Afghanistan, but the team which resulted had neither the money nor the personnel for the scale of the undertaking. The incoming chief of the general staff, General Sir Richard Dannatt, was worried by the strain on an army configured for only one war at a time, but now engaged in two, and counselled caution in Iraq while counting on Afghanistan to redeem Britain's military reputation. What little wind was left in Britain's war in Iraq was further reduced as rumours multiplied that Tony Blair, who had committed Britain to the invasion, was planning to stand down as prime minister in favour of Gordon Brown. In October Dannatt gave an interview to the *Daily Mail* in which he said that British troops needed to leave Iraq soon because they were worsening the security problem, not helping it. For him, the army's reputation was to be saved in Helmand, not in Basra; Shirreff said Dannatt's statement was a 'crass own goal'. When Blair visited Iraq in December 2006, Shirreff complained about the lack of cooperation in Basra between the Foreign and Commonwealth Office and the Department for International Development. What was even more striking was the lack of British cooperation with the Americans, who were now surging forces into Iraq, while the British were – in Shirreff's words to Blair – 'just dribbling'.[8]

The faults in this story were not all on one side. Military training stresses the need for action, the importance of seizing the initiative and the psychological value of the offensive. Regiments commemorate their battle honours and the history of war constructs its narratives around the notion of 'decisive victory'. These are the equivalents of the 'military end state' setting its own standards for declaring a win. They tell a story that is dominated by tactics and operations and only rarely shaped by politics and strategy. In practice battlefield success has frequently not translated into 'decisive victory', but instead proved transitory in its strategic and political outcomes. The Duke of Marlborough's greatest tactical and operational victory came at Blenheim in 1704, but the War of the Spanish Succession did not end for another ten years and was settled on terms which did not reflect the scale of Marlborough's success in this or subsequent battles. Frederick the Great 'decisively' defeated the French at Rossbach and the Austrians at Leuthen in successive months in 1757, one year after the outbreak of the Seven Years' War, but the war carried on until 1763 and by then Prussia was economically exhausted.

The long pedigree of the 'decisive' battle extends back to the Greeks' defeat of the Persians at Marathon in 490 BC. It was the earliest action described in Edward Creasy's massively successful *Fifteen Decisive Battles of the World* (1851), which had gone through 38 printings by 1891,[9] and it was where Jean Colin began *Les Grandes Batailles de l'Histoire*, published over 60 years later in 1913. More recently it has been a departure point for Victor Davis Hanson's arguments about hoplite warfare and the enduring success of what he calls the 'western way of war' in *Why the West Has Won: Carnage and Culture from Salamis to Vietnam* (2001) and of John Lynn's riposte, *Battle: A History of Combat and Culture from Ancient Greece to Modern America* (2003). The Greeks checked the Persian invasion at Marathon, but the Persians came back again ten years later, defeating the Spartans at Thermopylae and occupying Athens. This time the Greeks destroyed the Persians at sea, destroying their fleet at Salamis, and so breaking the Persian line of supply, and forcing the Persians to retreat once more. However, the unity which the Greek city states had achieved against the Persians collapsed in 432 BC, with the outbreak of the Peloponnesian War, during which the Persians became the power brokers within Greece and underpinned Sparta's eventual victory over Athens in 404 BC. By the following century both city states were backwaters and the balance of power had swung north to Macedonia. Marathon had looked decisive in 490 and had regained its lustre after Salamis confirmed its verdict, but any political and strategic benefits it had brought had evaporated thereafter: the verdict of history is conditional, dependent on the point in time when the judgement is being made.

For more modern historians, including both Creasy and Colin, before the outbreak of World War I the battle of Waterloo stood at the apex of this veneration of what was in essence a Napoleonic model of warfare – a short campaign (100 days in Waterloo's case) which reached its culmination in a single day's fighting (and not even a full day at Waterloo because the sodden state of the ground on 18 June 1815 prevented combat from starting until nearly lunchtime), and which produced a peace in Europe which lasted almost a century. There is of course a problem here, as the epithet 'Napoleonic' reveals. Napoleon lost this, his last battle. He had won many 'decisive battles', from Marengo to Austerlitz, but he lost the war. When he fled the battlefield in the gathering dark of 18 June 1815, he pressed on to Paris in order to rally support for the raising of fresh armies, but France was prostrated, not by one battle only but by 23 years of nearly unbroken warfare, and could summon neither the political will nor the resources for more. The cumulative effect of all France's battles and Napoleon's victories, as well as his defeats, determined the outcome of the war.

Clausewitz served in the Waterloo campaign, although he was not present at the battle itself. He opened book IV of *On War*, which he devoted to fighting (*Gefecht*), with a chapter on the nature of the modern battle (*Schlacht* in German, so we might call it major battle). His tone is ironic, suggesting that its decisiveness lay in the eye of the beholder more than in strictly measurable outcomes. He describes what contemporary commentators would call 'symmetrical warfare', in which 'contemporary armies have developed almost identically in military organisation and methods'. At the outset the two sides form up and a number of semi-independent fights develop. 'The battle smoulders away, like damp gunpowder.' When nobody can see any more, and 'darkness brings it to a halt', it is 'time to reckon up how much in the way of serviceable troops is left on either side' and to calculate the ground lost or gained. 'The results, along with personal impressions of the bravery and cowardice, intelligence or stupidity that one thinks one has observed in one's own troops and the enemy's, are then combined in an overall impression on which a decision is based.'[10]

Clausewitz made the outcome – of this battle at least – a matter of subjective judgment rather than objective facts. Paradoxically, however, he also argued that battle was the central act in war, since 'the concept of the engagement [*Gefecht*] lies at the root of all strategic action.'[11] Clausewitz regularly defined strategy as the use of the battle for the purpose of the war, but here he provided his successors with the arguments for inverting the sequence, for saying not that the battle served strategy but that strategy found its culmination in battle. By the end of the 19th century, most generals believed that their role, and the function of strategy, was to bring about battle on the most favourable terms, so as to deliver a decisive victory. Clausewitz, who had been on the receiving end of one of Napoleon's most convincing triumphs, the smashing of the Prussian army in the twin battles of Jena and Auerstedt in 1806, became for many commentators (and for one critic in particular, Basil Liddell Hart) the proponent of battle as the central act in war and as the direct route to victory.[12]

This understanding of battle was discredited by World War I. There were 'decisive' battles in its opening weeks, particularly at Tannenberg and on the Marne, but their strategic effects were defensive and negative. They protected France from the German invasion and East Prussia from the Russians, and they therefore prevented rather than enabled a quick victory. Moreover, they were not repeated. By 1916 what were called 'battles' on the Western Front, Verdun and the Somme, were in reality more akin to campaigns in both scale and length: the first lasted ten months and the second five. By their end the losses on both sides were comparable, the position of each had barely changed, and the same ground would be fought over again in subsequent years. At sea, the battle of Jutland (or Skagerrak for the Germans), fought on 31 May 1916, lasted a single day, but that was the only aspect in which it corresponded with Trafalgar, the model of decisive maritime victory so venerated by the Royal Navy. The two fleets were twice in contact for periods of about 20 minutes, and the strategic situation remained unchanged as a result. In 1918, the war ended, at least on the Western Front, without a climactic equivalent to Waterloo, and with the German army still intact and in enemy territory. Although there were major allied victories on other fronts – in Macedonia, at Megiddo in Palestine and at Vittorio Veneto in Italy – few historians linked them to the war's final outcome.[13]

Not dissimilar points can be made about World War II. The Germans won a sequence of dazzling victories between 1939 and 1941, overrunning much of Europe, but they still lost the war. The same point could be made about Japan. What were called 'battles' were not battles in the traditional sense: the battles of Britain and the Atlantic were campaigns, lasting months in the first case and years in the second, and so in many respects was Stalingrad, for many the 'turning point' in the land war. It has been argued that in the 18th century, soldier-kings accepted the verdict of the battlefield as conferring some sort of legal force on the resolution of disputes between states.[14] The picture is overdrawn: states did not necessarily seek terms after suffering a single defeat, but the victors did on the whole recognise the limits of what war could deliver and so proved readier to negotiate. That was not the case in the two world wars, whose scale – in terms of alliances, geographical reach, and levels of economic and social mobilisation – became incommensurate with what war might deliver unless they became wars of exhaustion. Air and sea power, the principal weapons of economic war, proved more important to the outcome of World War II than fighting on land. The differences in the output of equipment and munitions between the two sides proved decisive and the margin widened as much – if not more – through losses away from combat as in it, as the war's geographical scale wore out aeroplanes, ships and engines because of the distances they had to travel. Phillips O'Brien, who has argued in these terms, began his book on the subject by stating that, 'There were no decisive battles in World War II.'[15]

Since 1945 there have been short wars with decisive outcomes: the Arab–Israeli wars of 1966 and 1973 are cases in point, even if there has yet to be a lasting peace settlement in the Middle East. But most inter-state wars have been limited, and the most frequent forms of war have been irregular and insurgent. Their verdicts have come not on the battlefield, but over time and through eventual negotiation. Moreover, since the end of World War II, three major influences have served to undermine yet further the expectation of decisive victory in war.

The first of these was and is legal. In 1928 the Kellogg–Briand pact prohibited the use of war as an instrument of national policy except in cases of national self-defence. The pact failed to prevent war almost from the moment of its signature, but it still formed the foundation for the charge of waging aggressive war brought against Germans accused at the International Military Tribunal at Nuremberg. Moreover, in 1945 Article 51 of the charter of the United Nations incorporated the ideas of the Kellogg–Briand pact, appropriating to the Security Council the right to use force to maintain peace and security, and leaving only the right to self-defence to the individual state. In terms of international law, war lost its place as the final arbiter in inter-state relations. States are formally precluded from declaring war in most circumstances, and as a result normally don't do so. Obviously, *de facto* wars occur nonetheless, but one result has been a less clear demarcation between war and peace, with results that make it less certain when war can be deemed to have been won or lost. The 2014 *UK Defence Doctrine* reflected this ambiguity. 'Relationships between states, between groups and factions within a state, and between state and non-state actors are always competitive', it stated, 'and may be classified as those of cooperation, confrontation and conflict.' Note the refusal in this statement to use the word 'war'; note also the consequence that followed – 'Defence must be capable of operating across this spectrum' from cooperation through confrontation to conflict, or from peace to war.[16] If war and peace are on the same spectrum, and if cooperation, confrontation and conflict exist on a continuum, what is the legal difference between a state of peace and a state of *de facto* war?

The second influence was the impact of nuclear weapons. In 1946, Bernard Brodie asked in *The Absolute Weapon*, his edited volume written in response to the dropping of the two atomic bombs on Japan: 'If our cities can be wiped out in a day, if there is no good reason to expect the development of specific defenses against the bomb, if all the great powers are already within striking range of each other, if even substantial superiority in numbers of aircraft and bombs offers no real security, what possible avail can large armies and navies be?' Not all those conditions were already fulfilled by 1946 but by the 1950s they increasingly were, and by then too most opinion was persuaded of the conclusion which Brodie reached in 1946. 'Thus far the purpose of our military establishment has been to win wars. From now on its chief purpose must be to avert them. It can have almost no other useful purpose.'[17] Since the end of the Cold War, the salience of nuclear deterrence has diminished, but the notion that 'the purpose of our military establishment [is] to win wars' has not been revived.

That is in part the consequence of the third development, the growth of the idea of security, human, national and collective, as the overarching framework within which defence is nestled and where the spectrum from peace to war resides. Western democracies face threats to their security, from natural causes like pandemics and floods, as well as from human agency through cyberattacks and terrorism. In 2014,

*UK Defence Doctrine* began its chapter on strategy with a statement on national security and with a list of the external and internal threats which the nation confronted. It included disease and climatic events, whose impact policy might mitigate but whose onset it might struggle to avoid, alongside 'invasion, attack or blockade', whose prevention can be regarded as a primary function for defence traditionally understood. In general, most western democracies, especially within Europe, have come to discount the specific danger of war, not because its impact might not be as serious as a pandemic but because its probability is currently lower. This is war's standing with the UK's national security risk register. Moreover, if and when states go to war, as the United Kingdom has done three times since 9/11 (even if it has not declared it), they do so in pursuit not of winning, but of greater security. So states, despite living in much greater security than was the case for most of the 20th century, stress their insecurity by harping on about problems such as terrorism or Islamic radicalism. The latter present threats, but not major ones in terms of loss of life, and they are concepts, '-isms', not specific enemies except in a metaphorical or rhetorical sense, and not in a formally declared war. By making national security so all-encompassing and absolute, we heighten the awareness of insecurity, and by equating national security with strategy we are encouraged to apply strategy's concepts to the minimisation of the threats to security. However, this is the opposite of the approach required by war, which demands that the threat be confronted and overcome. Strategy in this context embraces risk as an opportunity for exploitation, not as a threat to be avoided. Almost by definition *UK Defence Doctrine*, in being so determined to avoid losing, eliminates the idea of winning.[18]

These three elements – the place of international law in conditioning the use of force; the presumption, however residual after the post-9/11 wars, that the purpose of that defence is to deter, not fight; and the open-endedness of security which fuses individual needs and rights with those of the state as a whole – have combined to foster a political and societal disengagement from the idea of victory. In the current catchphrase, soldiers have in the public's eyes become 'victims', not 'victors', misused by the state in wars of 'choice', not of necessity. In other words, they are killed or wounded in wars which do not have to be fought for existential reasons, and so do not have to be won.

For Clausewitz, living in the era of the French revolution and its veneration of the citizen soldier, an effective army which was politically aware because it was fighting for values to which the nation as a whole subscribed would also be an effective army – a nation in arms. In Iraq Richard Shirreff planned Operation *Sinbad* not as a citizen soldier but as an officer in a professional army concerned for the self-respect of the service to which he had dedicated his career and for its credibility within the army (also professional) of its principal ally. The war into which his plan fitted in 2006 was one to which the United States was still committed, and which its president, George W. Bush, believed could still be won, but that was no longer the view in the United Kingdom. The British people, having rallied round the flag when British troops were initially committed to the invasion of Iraq at the end of March 2003, had turned against the war by 2006, no longer believing it to be winnable and distrustful of its conduct by the United States. The military values which Shirreff embodied, although still widely respected within British society, no longer seemed apt for the war to which he wanted to apply them. British politicians, beginning with the prime minister, drifted because they could not see a way out of a situation for which they had to take a large measure of responsibility but which they did not have the ability to control or the determination to master.

At one level the societal response was more realistic than the military, and not just because it better reflected the actual resources that the British government was ready to put into the campaign. The more fundamental problem was that the post-9/11 wars were never winnable in the terms set by British (or even American) tactical and operational methods, however much they were bent into fresh shape in order to deal with the actual circumstances that confronted them. The vocabulary used at the political level to conceptualise the wars in Afghanistan and Iraq, 'the global war on terror' and later 'the long war', lacked utility for the purpose of framing strategy. Coherent strategy requires the prioritisation of objectives and the matching of means to ends: 'the global war on terror', with its limitless objectives and its under-resourced implementation, did neither. It misunderstood its enemies and it was not based on the sort of war its armed forces actually had to fight. But as a phrase it still said something important about what had happened to war after the end of the Cold War. The bounds of war had escaped the categorisation of armed conflicts inherited from the overarching threat of international war, the notional demarcation between 'major' and 'small' wars, between conventional and unconventional, regular and irregular, state and non-state, 'old' and 'new'. Instead war had become all these things in one, chaotic, protean and dynamic, constantly evolving as it eluded the clutches of states anxious to give it utility and to contain it by allocating it conceptual pigeon-holes. In the field, counterinsurgency provided an operational framework for military action, but was stripped of a strategic context or the political will to shape it for policy ends. Victory was hardly possible when there was so little idea of what winning would look like.

To put this point in the terminology of Thomas Hobbes' *Leviathan* (1651), 'warre', that perpetual conflict which Hobbes imagined as existing before the state secured the monopoly of force, had become joined with 'war', the use of armed conflict in the pursuit of the state's interests.[19] Hobbes' understanding of 'war' was that it was bounded, both by legal norms within war's own conduct and by the idea that its purpose ultimately was to resolve disputes between nations and so create peace. Those who wage war through insurgencies, civil wars and so-called 'new wars' with a view to winning are engaged in 'warre', and for them its persistence is itself a form of victory. Western democratic states, which responded with the tools of 'war', were handicapped because war had more forms than those for which they were prepared, either intellectually or in terms of equipment, even if they moved on from 'war' to 'warre'. Concepts like counterinsurgency and the 'comprehensive approach' were a form of catch-up, but still not the fundamental shift that was required to get ahead of the enemy. To do that would require western democracies to bend not just their armed forces out of shape, but also their states and their parent societies. This they lacked the will to do – the will to win.

Moreover, in adapting the concepts which they applied to 'war' to meet the problems posed by 'warre', they undermined the authority and inner certainty with which they had imbued those concepts in the first place. In 1977, the additional protocols to the Geneva conventions responded to the post-1945 colonial struggles for national liberation by retrospectively recognising their legitimacy and granting belligerent rights to 'freedom-fighters'. In doing so, they sundered the legal and ethical distinction between the authority to go to war, vested in the state, and the rights of those engaged in war, effectively legitimising revolution by giving belligerent status to those who fought as insurgents. The 1977 additional protocols were recognised as a sufficient threat to the state-dominated structure of the laws of war for the president of the United States, Ronald Reagan, to say that they did not apply to terrorists, precisely because they might.[20] Then, the adoption in 2005 by the United Nations of the 'responsibility to protect' peoples from the savagery and even genocidal tendencies of their own governments authorised outside interventions in civil wars and intra-state conflict, provided they had a Security Council mandate. The principle of state equality on which the United Nations rested was replaced by the categorisation of some countries as 'rogue' or 'failed' states, especially in the wake of the 9/11 attacks. The association of both law and war with the authority of the state was thereby eroded, with repercussions for both. Since 9/11 the legal regimes governing war have been in disarray, with the United States openly challenging them through processes like rendition and indefinite detention without trial, and the United Kingdom calling for a statute of limitation on accusations of war crimes against its own personnel and stripping British terrorists of their citizenship rather than bringing them to court.

'Democratic peace' theory, with its implication that because democracies do not normally fight each other they are inherently disposed to peace, muddies the waters yet further. The historical record since 1789 does not argue that democracies are incapable of fighting and winning wars. If anything, the reverse has been true. From the French revolution onwards, democracy has been a powerful agent for national mobilisation in time of war. In both world wars, democracies were ultimately able to prevail over more autocratic and totalitarian regimes. They may be reluctant to fight, but they are not pacifist. However, their armed forces cannot be expected to win when their very professionalism is interpreted as separating them from the society on whose behalf they fight and as denying them a civic voice in the framing of the policy objectives which they are expected to serve. As professionals, their operational concepts have consequences and sustained assumptions – like those of success and even of victory – which are not reflected in the national security strategies which govern their employment.

Going to war requires of those who do it a commitment to win – and a sense of what victory would look like. Without that, war lacks purpose and the strategy to shape it is robbed of coherence. Moreover, without either there is no obvious route to peace. During the course of the 20th century the path from peace to war proved to be a much easier and shorter one than the path from war to peace. World War I did not end with the armistices of 1918 but continued until the peace settlements were signed, the last of them – with the newly independent Turkish republic – in Lausanne in 1923. In the interim possibly four million people were killed in violent conflict across Europe and the Middle East, from Ireland to India and from Baltic to the Balkans, in addition to those killed in what we conventionally define as World War I.[21] In 1919 the statesmen in Paris did not appreciate both that the war was still ongoing and that the completion of the peace settlements therefore required the continued use of the military instrument and the application of the principles of strategy as much as of diplomacy. In the end the peace settlements did not deliver on the victories which the war had secured. Although there was a better understanding of these issues after World War II, it too did not end neatly in 1945: there were civil wars in Italy, Greece and China, and post-war struggles for national independence in South-East Asia and Africa. The colonial powers had forfeited their right to rule not least through their defeats at the hands of Germany and Japan. War and peace may exist on a continuum, but if there is no idea of their difference, war cannot deliver victory and success cannot be translated into peace.

## *Notes*

1 Ministry of Defence, *Joint Doctrine Publication 0-01: UK Defence Doctrine*, 5th edition (November 2014), 7, para 1.17. What follows is more a statement of opinion than a systematic discussion. For that some important departure points are: Gabriella Blum, 'The fog of victory', *European Journal of International Law*, Vol. 24, No. 1 (2013): 391–421; E. A. de Landmeier, 'What constitutes victory in modern war?', *Militaire Spectator* (20 March 2018); Robert Mandel, *The Meaning of Military Victory* (Boulder CO: Lynne Rienner Publishers, 2006); William C. Martel, *Victory in War: Foundations of Modern Military Policy* (Cambridge: Cambridge University Press, 2007; revised edn, 2011).

2 Ministry of Defence, *Joint Doctrine Publication 0.1: UK Joint Operations*, annex 1A, CDS planning directive, (November 2014), 25.

3 Ministry of Defence, *Joint Doctrine Publication 0-01: UK Defence Doctrine*, 18, para 1.54.

4 Carl von Clausewitz, *Vom Kriege*, Werner Hahlweg (ed) (Bonn: Duemmler, 1952), 77; this is my translation of the German; the English used in *On War*, translated by Michael Howard and Peter Paret (Princeton: Princeton University Press, 1976), 69, is more moderate than Clausewitz's German.

5 Justin Maciejewski, '"Best effort": Operation Sinbad and the Iraq campaign', in Jonathan Bailey, Richard Iron and Hew Strachan (eds), *British Generals in Blair's Wars* (Farnham: Ashgate Publishing

Ltd., 2013), 161; Jack Fairweather, *A War of Choice: the British in Iraq 2003–9* (London: Penguin Random House, 2013), 262

6 Maciejewski, 'Best effort', 157–8.

7 Fairweather, *War of Choice*, 261.

8 Tom Bower, *Broken Vows: Tony Blair: the Tragedy of Power* (London: Faber and Faber, 2016), 528, 531.

9 John Keegan, *The Face of Battle* (London: Penguin Random House, 1976), 57.

10 Clausewitz, *On War*, book IV, ch. 2, 'The nature of battle today', 226.

11 Ibid., 227.

12 Basil Liddell Hart, *The Ghost of Napoleon* (London: Faber and Faber, 1933).

13 I have developed these points more fully in Hew Strachan, 'La redéfinition de la bataille: Verdun et la Somme', in Alexandre Lafon (ed), *Les batailles de 1916* (Paris: Sorbonne Université Press, 2018), 15–41; Hew Strachan, 'La fin des batailles: stratèges et stratégies', in Bruno Cabannes (ed), *Une histoire de la guerre du XIXe siècle à nos jours* (Paris: SEUIL, 2018), 49–64.

14 James Q. Whitman, *The Verdict of Battle: The Law of Victory and Modern War* (Cambridge, MA: Harvard University Press, 2012).

15 Phillips Payson O'Brien, *How the War Was Won: Air-Sea Power and Allied Victory in World War II* (Cambridge: Cambridge University Press, 2015), 1.

16 Ministry of Defence, *Joint Doctrine Publication 0-01: UK Defence Doctrine,* 17, para 1.48.

17 Bernard Brodie, 'War in the atomic age', in Bernard Brodie (ed), *The Absolute Weapon* (New York: Harcourt, Brace, 1946); here quoted from Thomas G. Mahnken and Joseph A. Maiolo (eds), *Strategic Studies: A Reader* (London: Routledge, 2008), 203, 205.

18 Ministry of Defence, *Joint Doctrine Publication 0-01: UK Defence Doctrine*, 3–4, paras 1.1–1.4.

19 Thomas Hobbes, *Leviathan* (1651), part 1, ch. 13.

20 Paul Kennedy and George J. Andreopoulos, 'The laws of war: some concluding reflections', in Michael Howard, George J. Andreopoulos and Mark R. Shulman (eds), *The Laws of War: Constraints on Warfare in the Western World* (New Haven, CT: Yale University Press, 1994), 219; see also G. I. A. D. Draper, 'Wars of national liberation and war criminality', in Michael Howard, *Restraints on War: Studies in the Limitation of Armed Conflict* (Oxford: Oxford University Press, 1979), 147–8, 160.

21 Robert Gerwarth, *The Vanquished: Why the First World War Failed to End 1917–1923* (London: Farrar, Straus and Giroux, 2016).

CHAPTER I

# 'Winning' in Classical Antiquity and the Roman Conception of Victory

Ali Parchami

The notion of 'winning' in classical antiquity invokes, for a western audience, images of Alexander the Great or Julius Caesar. They may also be familiar with famous deeds from the period which, despite ending in defeat, still resonate of audacity and heroism: the last stand of King Leonidas of Sparta at Thermopylae; Hannibal's army crossing the Alps; the slave revolt led by Spartacus; or the military feats of the adventurer king, Pyrrhus of Epirus, who has bequeathed the term 'Pyrrhic Victory'. These examples remind us that the concept of victory can be fluid, subjective and sometimes understood and gauged only with the benefit of hindsight. They also serve as cautionary tales: what may ostensibly be a significant tactical or operational triumph could still end in strategic defeat. Pyrrhus, Hannibal and Spartacus were all visionary commanders who showed considerable ingenuity in the field of battle and yet all three were eventually overcome. Alexander's 'world conquest' proved transient and, as he lay dying, his empire was already being carved up by the *Diadochi* – his successor generals. Caesar's victorious military campaigns could neither safeguard his political career nor shield him from the daggers of assassins. Yet operational defeat can occasionally translate into victory: the small force of Greek allies led by Leonidas was resoundingly defeated by the Persian invaders but their demise has become part of a lore that is celebrated by posterity as a 'triumph in defeat'. The reception of defeat can, therefore, be an indicator of how different cultures view 'winning'.

While antiquity provides us with a richly varied range of successes in the face of adversity, it is Rome that western audiences will most likely associate with the idea of victory. Its enduring legacy notwithstanding, Rome ruled the Mediterranean basin for nearly a millennium: its formidable legions – with their unparalleled training, discipline and organisation – were to become a template for all professional military forces that have followed. In popular culture, Rome is often portrayed as a juggernaut which mercilessly crushed all opposition before it. This characterisation is not inaccurate, but we must remember that Rome's beginnings were that of a humble city-state – one of many such polities in the Italian peninsula. The rise of Rome was both incremental and fraught with numerous setbacks and defeats. So, what made the Romans one of the most fearsome powers in history? To attribute Rome's ascendency to the strength of its legions alone would be too simplistic and overlook what Virgil – the national poet – identified in the *Aeneid* as Rome's greatest strength: the character of its people supposedly moulded by the gods. Rome's complex social and political structures, and its cultural mores and religion, produced what could only be characterised as an irrepressible national will to dominate.

From the mid-Republic to the early decades of Empire, Rome was an expansionist state constantly at war. Though on occasions it was beaten by its enemies, it never acknowledged – much less could bring itself to accept – defeat. If a field army ever surrendered, the Senate could be expected to denounce the general, reject any agreed terms and declare for the continuation of the war. Even when they suffered military catastrophes, the Romans viewed these monumental defeats as a mere 'setback' on the path to victory. During the dark days of the Second Punic War, when Rome was engaged in a protracted and existential conflict with Carthage, it was not its legions that saved the city from the Hannibalic invasion but its remarkable resilience and determination to overcome and prevail. Nor was Rome ever willing to acquiesce to peace terms unless it involved the full and complete subjugation of the enemy. Tellingly, the Roman conception of peace – *pax* – is more akin to the English word 'pacification' than what is otherwise understood by our modern sensibilities. Rome could never make peace with a nemesis on equal terms because it viewed itself as an instrument of the gods. Its cultural rites entrenched a deep-seated patriotism that invited reverence for the state as the embodiment of public religion. Imbued with the idea that Rome was eternal and unstoppable, its citizen body was inculcated with the belief that, even if individual Romans could be conquered and Roman armies overwhelmed, the Roman people could never be defeated: for they were the personification of *invictus*.

To appreciate how 'winning' was understood in classical antiquity, we will first consider it in the broader context of ancient civilisations, particularly the Greek and the Hellenistic conceptions of victory, before turning our attention to Rome's unique interpretation.

## Winning in classical antiquity

Across the ancient world, the notion of victory was predicated on each civilisation's social-political traditions and was informed by their cultural outlook.[1] There were, nonetheless, commonalities: whether it was the Mesopotamian empires, the Old or New Kingdoms of Pharaonic Egypt, or the Carthaginians and the Greeks, religion played a significant role in how victory was viewed and received. Widely considered as a gift bestowed by the gods, each civilisation performed a set of religious rites to please their deities in the hope of securing victory in conflict or in a crisis. Just as important was the extent to which success in war determined the power and authority of rulers. Victory brought with it not only prestige but also legitimacy by signifying that rulers were favoured by the gods. This, in turn, underlined their role as the protectors of their subjects and helped project the image of effective leadership. Another by-product of a successful military campaign was riches in the form of booty, slaves and territories.[2] So victory came to be primarily associated with the favour of the gods, political power, military success and material gain.

To this mix we can add communal security, collective identity and individual ambitions. In an age when no international law existed to regulate the use of force, every polity had to undertake measures to defend itself against external attack. Defeat could be costly and entail heavy financial burdens or, worse, the massacre or the enslavement of an entire population. The cuneiform inscriptions of the Assyrian kings boastfully relate how they terrorised neighbouring peoples and, in some instances, altogether extinguished them. Similarly, the enslavement of the Israelites and their Babylonian captivity is a familiar tale from the Old Testament. Warding off adversaries and the ability to project power beyond one's realm was, therefore, a key measure of success. However, security is a relative term, and, for ancient communities, it was ephemeral and subject to inevitable challenge. For centuries the scourge of the Near East, the Assyrians were themselves eclipsed and conquered by an alliance of states who feared them. The Sumerians, the Hittites, the Phoenicians and the Medes are among the many illustrious ancient civilisations who briefly made their mark on the annals of history only to disappear.

As fundamental as physical security was to the conception of victory, for ancient Greeks this was a facet that was subsumed into a broader desire for the preservation of their Hellenic identity. Within the Greek world each *polis* – or city-state – jealously guarded its citizenship and was openly disdainful, even hostile, towards outsiders. The Greeks had a multi-layered view of identity: on the one hand, it was predicated on being a citizen of a sovereign city-state and, on the other hand, it was the recognition that every Greek belonged to the greater community of *Hellenes*. Serving the *polis* was a civic duty that, in principle, transcended all other considerations, including personal needs. Citizens were expected to protect the *polis* – not just its physicality but also its way of life – against internal and external threats; and, if necessary, to make the ultimate sacrifice in this endeavour. While fierce devotion to the *polis* predictably led to intra-Greek conflicts, it could also translate into the unification of the Hellenic people against 'barbarians'; or, at least, against their perceived corrupting influence. Whether it was in opposition to the Achaemenid Persians or the non-Hellenic tribes to the north, the fractious Greeks would willingly cast aside their differences in the cause of protecting their common heritage. The preservation of the collective Hellenic identity, in tandem with maintaining the distinct cultural and political identity of their individual *polis*, constituted for the Greeks the epitome of success and was synonymous with how they interpreted victory.

The importance they attached to the collective Hellenic identity is encapsulated by the Persian Wars: the conflict in which rival Greek city-states joined forces to repel the invading armies of the Achaemenid Kings. The Persians launched the campaign primarily to punish the 'haughty Greeks' with adding levy and territory to their realm representing secondary considerations. In contrast, the Greeks approached the war as an existential struggle: a choice between 'freedom and slavery', or, in the modern vernacular, the preservation of their culture against ignominious assimilation.[3] Although Athens was captured and burnt, the Persians were eventually defeated and forced to withdraw. Victory led to a flourishing that ushered in what the Victorians came to describe as the 'golden age' of Greece. Significantly, Victorian commentators celebrated the outcome of the war as a triumph of not only ancient Greece over 'Asiatic hordes', but also of western civilisation over the east. Were it not for the heroic Greek stance, they argued, western civilisation would have been destroyed in its infancy. The attitude of Victorian writers is a testament to how conceptions of victory can transmit through time, space and culture – with 19th- and 20th-century Europeans celebrating an ancient event, in a faraway land, by an ancient people, as an extension of their own identity.[4] Their outlook draws our attention to the role played by narrative and perspective: Herodotus, the chronicler of the conflict, was an Ionian Greek; and, as already noted, in spite of its depiction as an existential struggle, the conflict was a punitive campaign for the Persians along the periphery of their empire.

Whereas the Persian Wars centred on beating back an external antagonist, the Peloponnesian War was an intra-Greek affair fought for the future soul of the community. Once its ostensible causes are stripped away, it amounted to a clash of contrasting worldviews and rival coalitions competing for Hellenic supremacy. In one camp was the Delian League, led by a democratic, mercantile and vigorously imperialistic Athens. Against it stood the Peloponnesian alliance headed by an autocratic, insular and militaristic Sparta. Spanning a period of over two decades, and contested on the Greek mainland and in the Mediterranean as far as Sicily, the Peloponnesian War demonstrated the intensity by which each Greek city-state sought to protect and preserve its distinctive political identity and way of life. For much of the conflict Athens – a naval power – and Sparta – the pre-eminent power on land – were evenly matched. In the end, however, the Athenians overstretched themselves militarily, and with 'Persian gold' helping to tip the balance, Sparta succeeded in overthrowing the Athenian government and replacing it with a pseudo-puppet regime favourable to itself.

The story of the Peloponnesian War was recorded by the historian Thucydides, a former Athenian general who himself fought in the conflict. His celebrated annals, *The History of the Peloponnesian War*, provide us with valuable insights into contemporary Greek thinking on the character of war and victory. The Melian Dialogue, the most cited passage from the account, is well-known to the students of International Relations and is ubiquitously quoted in textbooks on Realpolitik. It relates how during the later stages of the war the Athenians offered the people of Melos an ultimatum to surrender or face destruction. The Melians attempted to reason with the Athenians and appealed to them on moral grounds. They received the stark response that 'the strong do what they can and the weak suffer what they must'.[5] Then, in an act that sent shockwaves across the Greek world for its brutality, the Athenians carried out their threat by massacring the Melian population and enslaving the survivors. Thus was borne, from the page of Thucydides, the expression 'might makes right' – the idea that the victor need not justify his actions nor be constrained by morality. Advocating that policymaking in war should be based on rationale calculations free of emotion, Thucydides may have eschewed morality as an unnecessary exigency but he also disapproved of terrorisation as a strategy when exercised by conquerors. He recognised that, as victory began to slip away from the Athenians, they became – in ever greater desperation – more ruthless and reckless in their methods. Instead of compromising or seeking a resolution, the Athenian leadership gradually succumbed to mob mentality and opted to pursue an all-or-nothing political and military strategy. Deeply critical of the binary mind-set of his countrymen, Thucydides viewed victory not in black and white terms but as a spectrum of colours with many different shades.

In Thucydides' judgement, the Peloponnesian War was nothing short of calamitous. He was not wrong. His Athenian homeland was occupied and was never to regain its former glory, while the losses and devastation suffered by the Greek world were proportionally comparable to the impact of World War I on Europe. Thucydides' bleak view, however, underscored an even greater tragedy: the bonds of Hellenic unity that had secured the victory of the Greeks in the Persian Wars were permanently shattered. As Thucydides sardonically remarked, what the Persians had failed to accomplish through invasion and force of arms they now realised by more insidious means.[6] The 'King's gold' inflamed inter-Greek enmities by stirring up tensions and the Persians selectively bribed and assisted rival Greek states with the aim of sowing divisions and weakening the Greeks. In Thucydides' view, a generation before the *Hellenes* had successfully repelled the Persians at the battles of Plataea and Salamis, but they had since grown complacent and their triumph was now turning into a strategic defeat.

Once again there is merit in Thucydides' assessment. In the final stages of the Peloponnesian War, Sparta – the hitherto stalwart bastion of Hellenism against foreign influence – had openly courted and had received Persian assistance against Athens and other fellow Greeks. Nor did the intervention of the Achaemenids stop with the end of the conflict. As Greek city-states continued to turn on one another, rival factions solicited financial and military support from the Great King and the Persians keenly and opportunistically complied. What had distinguished the Greeks from their contemporaries – and what had arguably constituted the bedrock of their interpretations of victory – was a fierce devotion to preserving the identity of their individual city-states and their categorical rejection of outside influence in favour of the common Hellenic heritage. But once they lost the will to unite as a community, and invited meddling from a foreign power, the Greeks were reduced to a vulnerable cluster of bickering city-states. Soon they were to find themselves at the mercy of an external foe to whom they lost not only their sovereignty as Hellenic peoples but also much of their distinctive identities as independent city-states.

The threat Thucydides had anticipated did not materialise from Persia but came from the emerging power of Macedon in the north. By the time the formidable Macedonian phalanx marched on mainland Greece, the fractured Hellenic *poleis* had become too weak and war-weary to put up any effective resistance. The wily Macedonian king, Philip II, portrayed himself not as a foreigner but as a *philhellene*; and in assuming the title of *Hegemon* he presented himself not as overarching monarch but as first among equals. Philip insisted that he was not a conqueror but merely the unifier of the Greeks against the threat posed by Persia. His clever distortions, however, could not disguise the reality that the Greeks were now mere subjects in the growing Macedonian Empire. Any remaining doubt was removed once the Greek city-states rebelled against Philip's son and successor, who proceeded to ruthlessly crush them and, in the process, razed the city-state of Thebes to instil fear among the unruly Greeks.

Known to history as Alexander the Great, the young Macedonian ruler embodied a conception of victory that is very different to that discussed so far. He strived for glory and fame and, above all, immortality for his name and legacy. Victory, for Alexander, was synonymous with the 'self' – his personal ambitions and great deeds. Once he consolidated his power over the Greek world, Alexander embarked upon 'world conquest' first by defeating and subjugating the Persian Empire and then by marching into Central Asia and India. Had his troops not tired and mutinied – forcing him back to his headquarters in Mesopotamia – he would probably have continued with his military exertions as long as there were new nations to be conquered and new lands over which he could project power. An exceptional military commander who in successive battles displayed superb skill and acumen, Alexander exuded personal charisma and leadership qualities until the last few months of his life when his judgement appears to have been affected by illness or excessive consumption. As he lay dying in Babylon, he was asked to whom he was leaving his empire. He is said to have replied, 'to the strongest'.[7]

If assessed solely through the lens of his 'world conquest', then Alexander's empire unravelled quickly in a swathe of bloodshed soon after his death. Conversely, if judged by his legacy – what Alexander himself strived for – then he was remarkably successful and his name lives on.[8] Modern scholars continue to debate Alexander's character and intentions: whether he genuinely believed he was the son of Zeus, the Olympian chief god, or whether this was a carefully constructed claim for the more gullible among his subjects; and whether he was a callous conqueror or a visionary whose ultimate goal was to unite the Graeco-Macedonians with the Persian nobility to create a unified people and end – once and for all – the cultural conflict between east and west. Perhaps the greatest legacy Alexander has bequeathed to posterity is that he has come to epitomise an ideal: many generals and would-be conquerors – from Julius Caesar and the Emperor Caracalla in antiquity to a host of figures from medieval and modern European history – have either styled themselves after the Macedonian king or have been flatteringly compared to him. This is because Alexander has become a symbol and the representation of victory incarnate: the man who was never defeated in a military campaign.

Alexander's death marked the start of an epoch in classical antiquity known as the Hellenistic Age. It was an era that saw the diffusion of Greek culture across the Near East, and in Egypt, and also the rise and fall of many 'wannabe' Alexanders. Following in the footsteps of the Macedonian conqueror, Hellenistic rulers regularly fought and competed with one another for land and wealth, as well as for glory and fame. Ptolemy I in Egypt, Antiochus III of the Seleucid-Syrian Empire, Pyrrhus of Epirus, Philip V and Perseus of Macedon and Mithridates VI of Pontus are among some of the best-known king-generals of the age. Each took to the field regularly to augment his position or to add territory and wealth to his kingdom. Their Hellenistic realms were essentially a cluster of states that shared the same heritage and worldviews and, broadly speaking, were evenly matched in terms of power and resources. Therefore, there emerged between them shifting alliances and an intricate balance-of-power system to regulate inter-state relations. Periodically a revisionist leader sought to upset the status quo but, as he became too powerful, other states would check his ambition by cutting him down to size. In a world where power was evenly distributed, victory was measured by small gains in wealth, territory and prestige and was borne out of opportunism, guile and treachery. This was the state of play in the east of the Mediterranean when a new colossus appeared. The warring Hellenistic kingdoms were to find in Rome a power unlike any other and with a new and terrifying conception of what it meant to win.

## The Roman conception of victory

The Roman conception of victory was deceptively simple but was predicated around intricate religious rites and maintained by complex socio-political structures. At its core was the collective will for geopolitical hegemony – whether by outright military conquest or by establishing, through a variety of other means, absolute supremacy over surrounding peoples. Symbolised by the sobriquet SPQR[9] – often inscribed on public buildings and carried as a standard by Rome's legions – it was the expression of a desire not to simply defeat but rather to break the will of adversaries and force them into complete submission. Confrontation with Rome could, therefore, never produce compromise, reconciliation or mutual peace because the Romans viewed all conflicts as a zero-sum game. No sacrifice was too great, no financial cost too exorbitant, no defeat too calamitous, and no enemy – or combination of foes – too powerful to be overcome. This singular and uncompromising approach made the Roman conception of winning wars significantly different from how it was viewed and utilised by their contemporaries.

This is not to suggest that there were no common elements. As with other civilisations in antiquity, projection of power, communal security, the acquisition of wealth and identity all played a role in the Roman conception. But in every instance these attributes assumed a uniquely Roman characteristic. Take, for example, identity: unlike the Greeks, the Roman notion did not revolve around ethnicity or birthplace but centred on allegiance and service to the state alongside devout observation of Roman rites and traditions. Roman identity was thus defined by culture, ideology and law. This meant that Roman citizenship was not closed off – as was the case with the Greeks – but could be extended to others. In a policy that was centuries ahead of its time, and certainly unprecedented in antiquity, newly conquered subjects could vie for Roman citizenship. Once the criteria were met, a non-Roman would be granted the full privileges of enfranchisement and, in some cases, citizenship was bestowed upon entire cities. Nor was it unusual for freed slaves to be granted the privilege. In one of the most extraordinary demonstrations of Roman social mobility, within a generation a certain Publius Helvius Pertinax, the son of a freed slave, first became a senator and then the Emperor of Rome.

Rome's munificent policy of extending its citizenship was integral to its ascendency. Aside from providing conquered peoples with a personal stake in the welfare of the Roman *imperium*, it provided the state with a stream of new blood and potential army recruits. The conventional route for a non-Roman to acquire citizenship was by enlisting and serving as a legionary for a contracted period of just over 20 years. The practical benefit of this policy was to provide Rome with a seemingly endless reserve of manpower. During the second and third centuries BC, the Republic could mobilise around 270,000 troops from its own citizen body alone. An additional 350,000 could be called to arms from 'allied' states – polities who for the time being were permitted to keep their independence in exchange for providing Rome with soldiers. In comparison, at the height of Macedonian power, Alexander the Great had led an army 30,000–40,000 strong.[10] As Pyrrhus and Hannibal or Gallic and Germanic invaders were to discover to their horror, a succession of Roman legions could be defeated in battle only for Rome to field yet another army of equal size and strength. By the 1st century AD, Augustus – the first emperor – commanded a standing army of 25 legions totalling 125,000 troops plus 150,000 auxiliaries and his personal Praetorian Guards.[11] A century later, these numbers had swelled to a combined force of 450,000 infantry and cavalry.

There is no doubt that Rome was remarkably good at winning wars and that, at its pinnacle, it presided over the best-trained, most disciplined and organised military to be found anywhere in classical antiquity. However, the key consideration should not be Rome's strength at its peak but how a small city-state became such a dominant power across the Mediterranean. After all, the ancient world was full of aggressive and warlike communities. It was an unforgiving environment where survival depended almost exclusively on strength and force of arms. During their formative years the Romans by no means had the best or the strongest of militaries. Yet, even by the standards of their contemporaries – including the warlike Greeks or the martial Latin city-states of southern Italy – the Romans were an exceptionally aggressive people. The slimmest of pretexts was sufficient for them to declare war, while no slight would be allowed to pass unavenged lest the perception of weakness might lead to an even greater threat. Faced with the choice of being victims or lording over others, the Romans opted for the latter. So victory followed victory and defeats were avenged with such persistence and savagery that they could eventually be declared incontrovertible triumphs.[12]

Rome's projection of strength was invariably accompanied by terror. This was not in itself unusual. As we have already seen, other ancient civilisations also relied on the mechanism of fear by practising wholesale enslavements and massacres. The Romans could be just as brutal and pitiless as the worse of their contemporaries but they could equally be generous and pragmatic – dangling the possibilities of mercy and other privileges in return for submission. What also helped set them apart was the gradual realisation by other nations that there was no escaping the Romans. Once the Romans set their eyes on conquering a people, they could not be bought off, negotiated with, or physically stopped. Rome would continue its wars of aggression until it prevailed. The knowledge of this produced a compliance, over time, from would-be victims to surrender and submit to Rome as early as possible to preclude unnecessary hardship and bloodshed. In his final will and testament, Attalus III – the Hellenistic King of Pergamon in Asia Minor – left his entire kingdom to the Roman Republic in anticipation of what he regarded as an inevitable outcome.

So what shaped the aggressive character of the Romans? The answer lies in part in the geopolitics of the early Roman Republic, when the floundering city-state was surrounded by enemies, and in part in the early myths and traditions that informed the Roman worldview. Tracing their origins to the survivors of the Trojan War, and to the god of war Mars, the Romans believed they were a 'chosen people' entrusted by the gods to carry out their divine will. This ideology put warfare at the core of Rome's cultural psyche as well as its traditions and religious practices. For almost all of its history the Roman Republic was annually at war. For a people who treated peace as an interlude, war was the norm and peace an aberration. This was symbolised by the Temple of Janus, whose gates were to be ceremoniously shut during peacetime. In the inscription that commemorated his deeds, the Emperor Augustus boasted that the Gates of Janus had only been closed twice since the foundation of the city in 753 BC, but that they were closed on three occasions during his reign (19 BC–AD 14).[13] That is to say only in five out of seven hundred years of history had Rome not gone to war. Far from perturbing the average citizen, this record would have been a source of pride for the Romans.

The Roman worldview engendered a relentless drive for imperialism and expansion as epitomised by the motto, *imperium sine fine*: 'empire without boundaries in time and space'. The intimation being, that even if some nations enjoyed freedom from Rome for now, one day their turn would come. This worldview was nurtured by Rome's religious practices, social hierarchy and political institutions. Roman society was deeply divided between the ruling elite and the masses – the quintessential 'haves' and 'have nots'. The Roman state, however, was imbued with mystical qualities that increasingly took the form of religion. Rome as an idea was a unifying force that created a strong bond between its citizen body regardless of their social class and wealth. The public devotion to the state was reinforced by numerous religious rites and festivities scattered over the annual calendar which always drew the enthusiastic participation of the masses. For its part, the nobility came to define itself by the public service it rendered to the state. From a young age Roman noblemen competed for political office and for military commands – with each requiring experience in the other arena so that the two went hand-in-hand. By the mid-Republic, leadership in politics and in war had come to be regarded as the highest form of public service and a discharge of one's religious duties.

With public office and military command came prestige, status and wealth. In Rome's fiercely competitive and tightly structured upper social strata, every member of the nobility was expected to safeguard the family name and its finances by either bettering the achievements of the previous generation or, at least, equalling its forbearers.[14] The by-product of this was that Roman noblemen sought military success as a means to acquire fame and riches to fund the next stage in their political ascent. This often necessitated new wars against adjacent 'barbarian' tribes along Rome's ever-expanding frontiers. During the Republic – though not the later Empire – the traffic in war was normally only one way. Roman noblemen always sought a pretext to justify their military excursions, and wars could be precipitated on the most spurious of grounds. By the late Republic arose a litany of ambitious and self-serving generals – Julius Caesar the most notable name among them – who embarked upon successive wars of aggression in order to cement their social and political status and to enrich themselves and their political allies. Yet regardless of its causes and motives, it was a uniquely Roman characteristic for every general to insist his actions were merely the rendering of public duty by 'upholding the majesty of the Roman state' and 'the dignity of the Roman people'.

More broadly, this insistence on the legality and legitimacy of its wars was a key tenet of the Roman conception of victory. It was derived from two components: *Iusta causa* (just war) and *parta victoriis pax* (peace through victory). The Greek historian, Polybius, observed that the Romans took special care not to give the impression of either beginning an unjust war or undertaking it from a desire for their neighbour's wealth. In fact, in every war they fought, the Romans maintained that they had been the wronged party, compelled to take up arms to defend themselves or their allies.[15] In *Iusta causa* we can find a principle that was a link between Roman religion, politics, the self-interest of the nobility and warfare. Performed as a religious ritual, the *Iusta causa* rite was a manifestation of the relationship the Romans believed existed between the gods and the Roman state and its people. It was based on the supposition that by fighting in self-defence, Rome was undertaking a just war sanctioned by the gods who would in turn give Rome victory. Regardless of the spuriousness of their cause, or the brutality they may have exercised in war, the ritual gave the Romans the veneer of righteous morality and provided their military actions with political as well as religious validation.

By the turn of the first millennium Rome's insistence on *Iusta causa* was wearing thin with much of the Mediterranean basin already under its hegemony. Claiming self-defence when their legions were dispatched as far afield as the banks of the Euphrates, or to the mountains of Armenia, or by the River Elbe in northern Germany, shows the extent to which the Romans sought to distort their acts of aggression by underwriting the narrative. They were indeed among the earliest and most sophisticated propagandists in terms of presenting themselves as victims even when they frequently initiated the conflict. Whether there was any objective truth to claims of self-defence or not, the Roman leadership often played on the fears and insecurities of the masses whilst simultaneously exploiting their desire for the potential rewards of war.[16] The only thing that changed between the early Republic and the later stages of the Empire was the party that was the main beneficiary of Rome's wars: the senatorial class in the early to mid-Republic, ambitious generals – such as Pompeius Magnus and Licinius Crassus – in the late Republic, and the emperor and his entourage during the imperial age.

A second justification that gave a veneer of legitimacy to Rome's expansionist wars throughout the ages was the claim that it was acting on behalf of the gods to bring lasting peace to humanity. In the *Aeneid*, the epic poem composed on the origins of the Roman people, Virgil had the gods proclaim: 'You have been entrusted O Roman with the task of ruling the world: to crowning peace with justice by sparing the vanquished and subduing the proud.'[17] Far from signifying the absence of war, the Roman understanding of peace was tantamount to pacification followed by possible Romanisation. Encapsulated by the expression *parta victoriis pax*, the Romans demanded unconditional capitulation and unquestioned recognition of Rome's suzerainty. Therefore, the 'Roman peace' was something that was imposed, rather than mutually agreed. Being an agent of the gods precluded any possibility of Roman compromise or negotiation, with victory being the only outcome guaranteed by *pax*. For the vanquished, the 'Roman peace' could entail terrible consequences: whether subjugated by force or surrendering by their own volition, any people on the receiving end of *pax* found themselves at the mercy of the Romans, who could do with them as they pleased.

## Turning defeat into victory

So far we have discussed how ideology, cultural traditions and national character shaped the Roman conception of victory. But without appreciating how the Romans dealt with loss, we cannot fully understand their interpretation of 'winning'. From extant records we know that just in the second century BC Rome suffered up to 43 military defeats.[18] Then there is the catalogue of losses which still carry their ancient stigma of notoriety: defeats to Pyrrhus at Heraclea and Asculum (280–279 BC); a hammering at the hands of Hannibal at Lake Trasimene (217 BC) followed by a crushing defeat at Cannae (216 BC); and a routing by the Parthians at Carrhae (53 BC). Nor was Rome immune from military humiliation during its imperial age. Under Augustus, three legions were lost to the Germanic Cherusci tribe at Teutoburg Forest (AD 9); the Emperor Valerian was defeated and captured by the Persians (AD 260); and Emperors Decius (AD 251) and Valens (AD 378) were killed with their entire armies by Gothic invaders.

Normally Rome responded to these defeats with resilience and an extraordinary demonstration of defiance. According to Polybius, 'the Romans were never more intractable or formidable than after suffering defeat'.[19] For instance, Pyrrhus's victories might have persuaded any other government to negotiate but the Romans refused and instead inflicted such heavy losses on his armies that he withdrew from Italy altogether. But few events can be as revealing of a nation's response to an existential crisis as that which befell Rome in the darkest days of the Second Punic War. Having lost up to 15,000 men to the Carthaginians at the battle of Trasimene Lake, Republican Rome suffered its most calamitous defeat a year later. The exact figures are disputed but Roman losses at Cannae are estimated at around 50,000–60,000 killed and 10,000–15,000 captured. The extent of the carnage becomes more stark if we consider that Rome's male population at the time was no more than 400,000. For comparison, the British losses on the first day of the battle of the Somme were 20,000 dead and around 40,000 wounded against a male population of over 21 million.

Livy relates that in the aftermath of Cannae there was 'neither any Roman camp, nor general, nor soldierly', so that allied Italian city-states 'now began to despair of Roman power' and waver in their loyalty, with some defecting to the enemy. 'Surely any other nation', he offered, 'would have been overwhelmed by such misfortune.'[20] Left almost defenceless, the Romans kept their nerve. In a show of defiance, the

Senate refused to pay the ransom for Roman prisoners of war and instead offered a public expression of thanks to the survivors of Cannae for 'not giving up on the Republic'. The people, too, rallied and, in spite of the prevailing gloom over personal loss and knowledge of an imminent siege, there was no public outcry for negotiation or demands for the state to sue for peace.[21] The Romans merely refrained from meeting Hannibal in open battle and bided their time. Within 14 years, the orphans of Cannae formed the spine of the Roman army that Scipio led to Carthage where he resoundingly defeated Hannibal at Zama.

'If Alexander had been worsted in one battle, the war would have been over', wrote Livy, 'what army could break the will of Rome when ... Cannae failed to do so?'[22] The aftermath of Cannae gives a good account of how the Romans responded to perilous defeat. It also highlights political defiance. The Senate's demeanour only makes sense once we realise that, for the Romans, defeat was not measured by military reversals or defined by events on the battlefield. It was a political decision. Only the Senate – and later the emperor – could designate an end to hostilities. In practice, every time the Romans suffered defeat, they integrated this defeat into a larger narrative of victory. Chronological timelines were manipulated – split or linked as necessary – with defeat becoming a mere event in a process that always ended with Roman victory. In this constructed narrative, Rome lost battles but never the war.

Roman victories were, therefore, not just the physical act of winning but also a matter of timing. The political leadership would determine when a conflict could be convincingly presented to the public as a victory. Once this was agreed, a triumph might be awarded to the successful general. This was a public celebration that officially marked the closure of war and was arranged as a commemorative festivity. Sometimes accompanied by a triumphal arch or a temple dedication, the triumph was an event that was both recognisable to the electorate and also invited public confidence that the war had been won.[23] Originally awarded when 5,000 or more enemy soldiers had been killed, the awarding of the triumph became highly politicised. If a general claimed a major victory, it was up to the Senate – or the emperor – to decide whether he merited the ceremony. As an alternative to the triumph, a general could receive a lesser prize known as an *ovation* and the conflict would be continued.

The politicisation of 'victory' had interesting ramifications on the battlefield. For example, during the war against the Spanish city of Numantia in 139 BC, the general, Quintus Pompeius Aulus, reached a negotiated settlement with the enemy but insisted that they should go through the sham of surrendering first to satisfy Roman tradition. When this became known to the Senate, it denounced Aulus, rejected the treaty, and continued with the war.[24] In the same conflict, another Roman commander negotiated terms with the Numantines to save the lives of his men. The Senate immediately repudiated the treaty and handed the general to the enemy bound and naked. Bizarrely the general himself, C. Hostilius Mancinus, spoke in the Senate in favour of the bill to turn him over because he had disgraced Roman honour.[25] Similar stories reveal the battlefield politics of Roman victory and defeat. In 171 BC the Macedonians dispatched an envoy to impose terms on the vanquished Romans. Refusing the terms, the Roman commander Publius Licinius imperiously advised that it was the Macedonians who in fact should be submitting their pleas to the Roman Senate.[26] Julius Caesar relates how his general, Quintus Cicero, responded to lenient peace terms offered by the Gallic force that had trapped his Roman army: It was not the habit of the Roman people to accept terms from an enemy but, if the Gauls would lay down their arms, Cicero would intercede with Caesar on their behalf.[27] What these episodes show is that, whatever the exigencies of the battlefield, the Roman state did not tolerate surrender and would politically seek to obfuscate military defeat.

None of this detracts from Rome's record of battlefield victories. By the 1st century AD, its legions were arguably the finest military force that existed until the early modern European era. The Romans were quick to learn from their mistakes and highly adaptable. Beaten by the formidable Macedonian phalanx at Cynoscephalae in 197 BC, they crushed the same formation decisively in 168 BC and conquered Macedon. But this military prowess was complemented by a pragmatic approach and a political astuteness that allowed Rome to remain the preeminent power in the Mediterranean for hundreds of years. Rome was skilful at making political and economic alliances which, in time, would turn allies first into client-states, then satellites and before long provinces of the Empire. Yet, unlike many other ancient civilisations, the Romans were also willing to openly incorporate the cultural traits, religious rites, and political and military features of their adversaries and subjects whilst extending their own citizenship to others. So despite its uncompromising hegemonic stance, Rome's conception of victory was also based on a progressive worldview and remarkable inclusiveness.

## The end of antiquity

Stretching from the shores of the Atlantic to the banks of the Euphrates River in modern-day Iraq, Rome reached its territorial zenith in AD 117. To contemporaries, it was the true embodiment of an 'empire unrestrained by time and space'. So why did Rome collapse? What happened to a society that could never accept defeat? Ever since Edward Gibbon's monumental 18th-century opus, *The Decline and Fall of the Roman Empire*, modern historians have been fascinated by these searing questions with which they continue to grapple.

Gibbon himself was of the view that Rome's prosperity, coupled with the rise of Christianity, planted the seeds of its decay and fall. Later historians identified a confluence of other factors, ranging from environmental degradation to an over-stretched military deployed along 7,500 kilometres of porous frontiers. The rise of the Neo-Persian Sassanid Empire in the east, and the continuous influx of 'barbarian' tribes across the European front, are central to arguments that external threats sapped Rome's resources and extinguished its hitherto irrepressible outward thrust.[28] Economic historians prefer to focus on internal causes: the inherent contradictions of the Roman economy – with its embedded wastefulness and predilection for slave labour – the steady devaluation of the currency, and the demise of the middle classes under the weight of heavy taxation. Cultural dilution and demographic change are among other popular theories. In the first three centuries of the new millennium, the senatorial nobility of the old Republic was slowly replaced by provincial elites. An even greater demographic change occurred in AD 212 when the *Constitutio Antoniniana* extended the franchise to all free residents of the Empire.

These factors undoubtedly had major socio-political and cultural ramifications. Yet there is no evidence that they fundamentally altered the Roman conception of victory. As absorbing as it might be to debate the reasons behind Roman decline, the answer to what became of Rome's indomitable ideology of *invictus* must be sought in its trinary transformation. The first of these occurred at the end of the 1st century BC, when the Republican government was institutionally transformed into an imperial regime. Many of the powers previously held by the Senate, the tribunes of the people, the magistrates and military commanders came to be invested in a single individual: the emperor. Thereafter the emperor became the personification of the state: the 'unconquerable' spirit of Rome symbolically resided within his person, signified by the title of *invictus*, which emperors now styled themselves and inscribed on their coins. Victory was now the exclusive perquisite of the imperial household, with the traditional triumphal celebrations recalibrated around the emperor. However, the emperor's personification as *invictus* was accompanied by unforeseen political and military consequences.

The emperor was acknowledged as 'invincible' only as long as he was victorious on the battlefield.[29] Beginning in the 1st century AD, and accentuated by events during the next three centuries, a long succession of ambitious generals plunged the Empire into civil war by seizing the imperial throne only to be challenged and overthrown by another general. Under an imperial system that had no mechanism for succession, Rome's thirst for victory, glory and plunder now turned inwards. Victory was just as likely to be sought against a competing Roman army as against the Persians or frontier barbarians. Rival claimants fielded Roman legion against Roman legion as internal conflict became the norm, interspersed with sporadic periods of peace. Additionally, with the emperor often forced to be on the march, his place of residence became the epicentre of state activity. As the bureaucracy grew mobile out of necessity, new centres of power emerged in Ravenna, Milan, Nicomedia and Constantinople. The metropolis of Rome gradually lost its pre-eminence so that, by the 4th century AD, victory was no longer identified with the Eternal City, or with the Italian peninsula, but become diffused as a parochial idea centred on local and provincial communities.

Rome's second transformation was its Christianisation. Once Christianity became the prevalent religion of the Empire, spiritualism progressively replaced antiquity's unabashed dalliance with the material world. Christianity's emphasis on otherworldliness and piety did not sit well with the Roman tradition of worshipping the state nor with victory in the name of Rome rather than the Christian God. When in AD 390 the Emperor Theodosius knelt before Ambrose, the Bishop of Milan, the supremacy of the Church was symbolically established over the State. Roman institutions and traditions were incrementally altered to serve the Christian church into which it was subsumed. Arguably the relentless expansionism of Christianity in late antiquity, and the hegemonic aspirations of the Holy See in the early medieval era, contained some of the old drive and spirit of the Roman *imperium*. In a sense, the Roman conception of victory did not altogether dissipate but was instead integrated into the institution of the Church and into a Papal State which for centuries was to dominate European Christendom.

The third transformation of Rome occurred when the Empire was physically divided in half. Ruling over a vast rustic empire became a near impossible task for any one individual so that, by the 4th century AD, appointing a co-emperor became an accepted practice. Although in principle the leadership of both empires identified themselves as Roman and recognised the other as one half of a united empire, in practice the Western and the Eastern empires grew apart as separate entities with their own preoccupations. Additionally, since the establishment of Constantinople in AD 324, the centre of gravity had shifted to the east with the ruling families, manpower and riches of the west moving to the Eastern Empire. So much so that the 5th-century historian, Vegetius, bemoaned the shortage of manpower in Iberia, Italy and Gaul and warned of an impending disaster awaiting the western provinces.[30]

In modern culture the 'fall' of the Western Roman Empire is associated with AD 476 and the deposition of the child emperor, Romulus Augustulus. But it was the Victorians who designated this date as the demarcation between antiquity and the middle ages. To 5th-century contemporaries, the events of 476 would not have signified Rome's 'collapse'. The inhabitants of Italy were accustomed to the removal of emperors and Roman presence continued to persist, albeit in small pockets, across the old Western Empire. Extant sources from the period suggest that, despite the breakdown of the imperial government, there was an expectation in the west that the Empire would reassert itself.[31] And in a way it did: within 60 years Justinian, the Eastern Roman Emperor, reconquered Italy and much of North Africa and Spain to reconstitute the Empire around the Mediterranean. Following his death, the Roman *imperium* in the west evaporated and the Empire was replaced by independent Germanic kingdoms, though the erosion of Roman identity occurred over many succeeding decades.

Eventually the ancient Roman Empire ceased to exist in corporeal form. Nevertheless, the essence of the Roman idea of victory was passed on to posterity.

Even after its demise, Rome remained a symbol of power and of victory. Charlemagne drew on this symbolism and Rome's cultural legacy when he was crowned the 'Emperor of the Romans' in AD 800 – to be followed for the next thousand years by a long line of Holy Roman Emperors. Nor should we forget that the Eastern Empire persevered for over a thousand years in the guise of Byzantium. Even though largely a Greek-speaking imperial project, its leadership and people continued to identify themselves as 'Romans' and exhibited throughout the medieval era – right up until the fall of Constantinople in 1453 – the Roman spirit of victorious defiance against great odds. The empire of the Russian Tsars had similar pretensions by claiming to be the 'Third Rome' and a direct successor state to the Eastern Roman Empire. From Napoleon to Hitler, and the British Empire to the American superpower, the idea of Roman invincibility has continued to draw comparisons, with the symbolism of Rome and its conception of victory still permeating through the narrative and visual paraphernalia of would-be conquerors and hegemonic powers.

## Conclusion

This chapter has argued that it is a combination of narrative, perspective and cultural and national attributes that determine the criteria for winning wars. For the ancient Greeks, victory was intertwined with their sense of political and cultural identity – maintaining the purity of their tribal bonds and political institutions in their individual city-states and that of the collective Hellenic identity against 'barbarian' influence and external threats. For Alexander the Great, victory came to be largely associated with fame and the aspiration to achieve immortality through great deeds. For his Hellenistic successors, it was about opportunistically scoring an advantage over rival kingdoms by seizing territory and fame but also establishing a reputation as a successful battlefield general. A behemoth that often crushed all before it, Rome's conception of victory centred around an indomitable collective will to expand and dominate. But the Romans were also master propagandists who constructed favourable narratives by focusing on success while obfuscating defeat within the larger story of victory. So, we must acknowledge that sometimes the lines between victory and defeat are so blurred as to be subjective. This subjectivity is captured by one of the most famous quotations from antiquity: 'Vanquished Greece took captive her savage conqueror' wrote the poet Horace.[32] Greece was conquered by Rome only for Greek culture to take possession of Rome.

### *Notes*

1. When discussing nearly a thousand years of history, a degree of generalisation and simplification is inevitable.
2 Philip de Souza, '*Parta victoriis pax*: Roman Emperors as Peacemakers', in Philip de Souza and John France (eds), *War and Peace in Ancient and Medieval History* (Cambridge: CUP, 2008), 77.
3 Herodotus, *The Persian Wars*.

4 See, for example, Andrew Lang, *The True Story Book* (London: Longmans, 1893); and Richard Jenkyns, *The Victorians and Ancient Greece* (Oxford: Blackwell, 1980).
5 Thucydides, *The History of the Peloponnesian War.*
6 Ibid.
7 This was in character with a man who once remarked, 'Heaven cannot brook two suns nor earth two masters'.
8 Alexander also founded a dozen cities, all named after himself, with the most famous of these still a thriving port in Egypt.
9 An abbreviation that stands for *Senatus populusque romanus*: 'in the name of the Senate and the People of Rome'.
10 Nathan Rosenstein, 'War and Peace, Fear and Reconciliation at Rome', in Kurt Raaflaub, (ed.), *War and Peace in the Ancient World* (London: Blackwell, 2007), 236.
11 Adrian Goldsworthy, *The Pax Romana: War, Peace and Conquest in the Roman World* (London: W&N, 2017), 309.
12 Rosenstein, 'War and Peace', 229; and Goldsworthy, *Pax Romana*, 55, 171.
13 Augustus, *Res Gestae Divi Augusti.*
14 Rosenstein, 'War and Peace', 233.
15 Polybius, *Fragment 99* [Buettner-Wobst]; and Livy, *Ad Urbe Condit* XLII:47.8.
16 De Souza, '*Para victoriis pax*', 77.
17 Virgil, *Aeneid* VI:851–55.
18 Jessica H. Clark, *Triumph in Defeat: Military Loss and the Roman Republic* (Oxford: OUP, 2014), 1.
19 Polybius, XXVII:8.8.
20 Livy, *Ad Urbe Condit* XXI:54–61.
21 Clark, *Triumph in Defeat*, 62–91.
22 Livy, *Ad Urbe Condit* IX:19.9–10.
23 Clark, *Triumph in Defeat*, 1–3, 96, 133–134, 208.
24 Appian, *Iberica*, 79–83; and Rosenstein, 'War and Peace', 227.
25 Rosenstein, 'War and Peace', 227.
26 Carlin A. Barton, 'The Price of Peace in Ancient Rome', in Raaflaub, *War and Peace*, 248.
27 Julius Caesar, *De bello Gallico* V:41.
28 Peter Heather, *The Fall of the Roman Empire: A New History* (London: Macmillan, 2005).
29 Michael Kulikowski, *Imperial Triumph: The Roman World from Hadrian To Constantine* (London: Profile Books, 2016).
30 Vegetius, *Epitoma rei militaris.*
31 Ian Hughes, *Patricians and Emperors: The Last Rulers of the Western Roman Empire* (Barnsley: Pen & Sword Military, 2015).
32 Horace, *Epistles* II.

CHAPTER 2

# The European Concept of 'Winning' in the Middle Ages

John France

People in developed countries nowadays draw a sharp distinction between peace, which we regard as normal, and war, which we see as appalling and abnormal. We envisage war as a series of highly dramatic battles which end with winners and losers, and the restoration of the peaceful norm. But for much of the world this is meaningless, as is our distinction between criminal violence and war. For so many people war is not an occasion but a condition to be lived with, and this was certainly the case in what we call the Middle Ages. In this period of European history, between the 5th and the 16th centuries, winning was rarely sharp and clearly defined. Most wars were about possession of land, about adjustments, a means of persuasion rather than making existential threats. Battle, with all its chances, was high risk and, in any case, often indecisive. Armies were not standing forces and were very fragile in their structures. Winners of wars were those who adapted to these conditions which, because they involved long periods of struggle, truce and peace, demanded patience and persistence rather than military brilliance.

Given the primacy of battle in our minds, it comes as a shock to discover that the writer on war most highly respected in the Middle Ages, Vegetius, advised that, 'Good generals never engage in a general engagement except when opportunity offers, or under great necessity.'[1] One reason for this, as Vegetius says, is because in battle 'fortune tends to have more influence than bravery' and this was as true in the Middle Ages as at any time. Battle is an inherently chancy business. In 1097, the First Crusade was besieging Antioch when the garrison made a sally, but was pinned back against the gate. A horse escaped from the scrum, and when it was pursued by a small group of knights, the crusader infantry thought they were being deserted by the cavalry, panicked, fled and were massacred.[2] Even the greatest of medieval soldiers were cautious about exposing themselves to the risks of battle. Richard the Lionheart (1189–99) is acknowledged to have been a supreme soldier, yet he fought only three battles, and of those two were in the Holy Land and only one – St Mégrin near Angoulême in 1176 – was in Europe.[3] His father, Henry II (1154–89), accounted to be a great warrior, was never present at a battle. This may seem strange, but quite apart from the vagaries of chance, medieval generals had to be especially conscious of the fragile nature of their armies.

Medieval European armies originated in the various 'tribes', Ostrogoths, Visigoths, Franks, Lombards etc, which crossed into the Roman Empire in the 5th and 6th centuries. But they were not racial groups. As the Roman defences in the western part of the Empire collapsed, largely as a result of internal conflict, groups of opportunists gathered around the most successful leaders. Amongst the notable followers of the Frankish King Clovis (481–511) was Wiomad, who seems to have been a Hun. The great men who became barbarian kings needed followers who in turn needed alliance with others and leadership to give them the strength to overcome the remnants of the Roman army and to overawe local populations who often despised the barbarians who they feared.[4] These invaders did not come to destroy, but to enjoy the benefits of Roman civilisation.

Sometimes the original invaders were destroyed by others: in Italy the Lombards expunged the Goths in the 6th century. In Britain, violent resistance by the British elite devastated the infrastructure of Roman government and left a network of small statelets, some Native, some Germanic. Even in Gaul, where the newcomers were less destructive and able to come to terms with the existing elites, there was severe administrative disruption. The complex Roman machinery of government had taxed the people and paid the legions, but this virtually collapsed. The newcomers were armies, and they replaced the old Roman army. But they were paid by a much simpler method – they were given land. The great men around the king got vast estates, while their followers received much smaller allotments. In time these soldiers, great and small, married into Native populations of both great aristocrats and petty landowners. This militarised the Native notables, so that soldiering was no longer an ethnic function but one arising from landholding. In fact, this process created a vital link between landholding, freedom, immunity from taxes (which soldiers had never paid), military function and social status. Those who held land were expected to rally to the ruler in time of war, but their first concern was the threat to their own lands. So they felt at liberty to judge when that was in their interests and very generally to use force in pursuit of what we see as their private ends. Monarchs, therefore, had to persuade them to make common cause.

This evolution had political and military consequences. Kings had to listen to their followers whose landed wealth gave them control of local government. The *placitum*, an assembly of the notables of a kingdom, was a common institution across Europe. Kings could gather armies from beyond their personal territories only by agreement, though obviously they were the richest of the elite and so, depending on personality, some were more dominant than others. Dukes, counts and lords, like kings, were landowners, and they employed soldiers well-trained and equipped with armour and expensive weapons of iron and steel. They preferred to travel and fight on horseback, though they were quite prepared to dismount. Crowds of lesser people recruited as foot soldiers supported them. All this emphasised that the kings of Europe needed to consult and discuss important matters with their leading men, for they, too, were landowners and the power they claimed as sovereigns was paralleled by that of their great men. So any ruler had to negotiate on affairs of state. In these circumstances a major army was a composite of many smaller armies, and, therefore, lacked the cohesion of a standing force. Medieval people were very aware of linguistic and other differences between peoples, as Regino of Prüm remarked in the 10th century: 'The various nations differ in descent, customs, language and law.'[5]

But for the elite family relationships and alliances concerned with land ownership were much more important than ethnic loyalties. It was very difficult to control armies whose leaders were aristocrats with an immensely powerful sense of aristocratic privilege. As late as the 15th century, when English armies were threatening Paris, its citizens complained that aristocratic factions were weakening the kingdom, but the Duke of Berry responded: 'We fight each other when we please and we make peace when we see fit.'[6]

Only very slowly, in the later Middle Ages, did such people come to identify their interests with the state and nationality. In these circumstances, the structure of an army rather resembled an onion. At its core kings and great nobles had small military households for military administration and leadership, and they commanded men from their own lands on whose loyalty they could count together with mercenaries whose adherence depended upon their being paid. Then there were the contingents of their great men, who might or might not be enthusiastic for a particular conflict. Beyond them were allies who, of course, had their own agendas.[7]

These were some of the general conditions under which medieval commanders fought. It is rare to find a medieval commander amongst the lists of 'Great Captains', though Richard I of England (1189–99) has achieved some recognition. This is because of these and other restrictions within which medieval commanders operated. Europe is a close and seamed land, which in this period was studded with fortifications which dominate its poor communications and roads. Technology was limited and pretty much unchanging across the period – there were no great leaps forward. These physical conditions and the nature of their armies discouraged the kind of bold movement which was possible, say, for Napoleon. This is not to say that commanders were stupid; some were because military ability was not necessarily commensurate with inherited social standing. But war was so expensive and consumed so much of their available resources that kings and other leaders thought carefully about it. The word strategy is a modern one, but medieval generals were perfectly capable of thinking in strategic terms. In 1214 King John of England constructed a wide coalition of allies from Flanders and Germany to attack France from the north while his own forces advanced from the south.[8] Careful thought was given to the tactics to be adopted, though we barely recognise the fact because the pattern of warfare was so different form our own experience, above all in the relative rarity of battle.

The contingent nature of loyalty in medieval armies naturally increased the hazards of battle. Moreover, medieval society depended on agriculture, so great men and their close dependants had many preoccupations as landholders. Because wealth was limited, only a very few people could be retained as truly professional warriors. Armies, therefore, came together only briefly; they were temporary assemblies of noble retinues with other less well-armed supporters, and so rarely had time to train together. As wealth increased so did the number of professional soldiers, sometimes by about 1300 organised in 'companies' whose members were professional soldiers and were recruited as a coherent group. But regular standing armies did not emerge in Europe until the very end of the Middle Ages, and even then they were composed of distinct groups welded together (or not) by a strong leader and the hope of loot.

The conquest of England in 1066 illustrates the importance of having a cohesive army. William the Conqueror's invasion fleet was stuck at Dives for lack of a good wind. His army was raised from Normans, French, Flemings and Bretons, many of them being mercenaries, but he was a masterful general and he used the opportunity to train them together. This partly explains why his army at Hastings in 1066 could sustain a very unusual battle. It lasted a full day, while most such confrontations were settled quickly when one side took to its heels. Furthermore, Duke William was able to rally his army after sustaining an early repulse with heavy casualties.[9] It was decisive because there was no obvious successor to Harold and the massive casualties inflicted upon the Anglo-Saxon elite came on top of losses in earlier battles against the Danes.[10] By contrast, in 1119 Louis VI of France (1108–37) became impatient at Brémule in 1119 and led the cavalry of his army in a wild disordered charge which was defeated in a very short affair by the Anglo-Norman forces most of whom were dismounted.[11]

When battle did occur, victory usually went to the better organised army. At the battle of Bouvines in 1214 Philip II King of France (1180–1223) defeated a coalition of forces from England, Flanders and Germany, and an eyewitness remarked:

> The King had his echelons put in formation and they rode forward. You could see among them many noblemen, much rich armour and many noble banners. The same was true for the opposite side, but I must tell you that they did not ride as well and in as orderly a manner as the French, and they became aware of it.[12]

An important reason why major leaders were cautious about committing to battle was that they were expected to fight in person and they were aware that random events in the confusion of battle could cost them their life – Harold chose battle at Hastings in 1066, and died on the day by the strike of a chance arrow. Richard the Lionheart made a policy of leading from the front; his luck ran out at the siege of Châlus when he was mortally wounded by a crossbow bolt. Peter II of Aragon was killed fighting on the front rank at the battle of Muret in 1213. But leadership depended upon commitment, so that at the battle of Crécy in 1346 Edward III refused to rescue his son, the Black Prince, from savage fighting, declaring 'let the boy win his spurs'.[13]

The commander was the very reason why an army came together. His death usually signalled the dissolution of his army. And great leaders had to beware of treachery which all sides commonly promoted amongst their enemies. William the Conqueror, in alliance with King Henry of France, defeated the Norman rebels in a close-run battle at Val-ès-Dunes in 1047, largely because Ralph Tesson defected to him.[14] At Bosworth Field on 22 August 1485 Richard III of England lost, partly because Lord Stanley held aloof from the battle, waiting to join whoever was victorious.[15] In 1153 Henry of Anjou (later Henry II of England) invaded to claim the throne of England, but though his army twice confronted King Stephen's, there was no engagement.[16] Both leaders knew that loyalties were uncertain, because, after all, if one king won, the barons who had risked their lives in battle would be subject to him.

Battle offered glory but did not always result in winning a war. A key reason for this was the large number of castles in Europe. In the late autumn of 1184, the lands of Count Baldwin V of Hainaut were invaded by the allied armies of Count Philip of Flanders, Archbishop Philip of Cologne and Duke Godfrey of Louvain. Baldwin did not try to fight them in the open field. Instead he cleared out the countryside of food and retreated into his stronghold of Mons, instructing his major vassals to do likewise and rallying them by saying: 'Take comfort and be strong, because our enemies will withdraw at some time, and leave our lands to us, because they cannot take the lands with them.'[17]

In the event the three allies had made very little logistical provision for what they assumed would be a short operation not far from their home bases. As a result, their armies went hungry and were forced to retreat.

When battle did happen, it often had only limited results. The Emperor Frederick II (1220–50) was determined to conquer Milan and the cities of the Lombard League. In 1237 he began a highly successful campaign in which a number of cities of eastern Lombardy capitulated. But when he approached Brescia the army of the Lombard League placed itself between his army and the city preventing any siege. It was November, when most armies went into winter quarters. Frederick pretended to disband his forces, and when the League army themselves moved home he ambushed them and won a stunning victory at Cortenova.[18] The League then offered to submit to his rule on terms, but Frederick wanted total submission. Faced with, as they saw it, an existential threat to their liberties, the League fought on, and in 1238 Brescia defeated Frederick's siege. The League had been mauled at Cortenova, but not fatally. Frederick had overplayed his hand, stimulating resistance. The English

King Edward III (1327–77) won staggering victories at Crécy in 1346 and Poitiers in 1356 (where the French king was captured), and his armies devastated much of France. But he was obliged to renounce his claim to the throne of France and to conclude a compromise peace at Brétigny in 1360.[19] Leaders had to accept the limitations of their situation if they were to win wars. It might be argued that the 'Hundred Years War' lasted as long as it did because the English won enough battles to keep their hopes alive, but never enough to win the war. The fragility of armies imposed upon commanders the need for careful calculation, persistence and cunning.

In the 12th century prosperity increased, as did the scale of warfare and its frequency, with a resulting demand for practised professional soldiers. Henry II of England was a great employer of mercenaries, especially as foot soldiers. They were so often drawn from the Low Countries that the term 'Brabanter' became synonymous with mercenary.[20] Henry and other kings paid their cavalry, but they were drawn from the ranks of the landed gentry and nobles who had a monopoly of high command, and were kept together only for short periods. By the end of the 13th century, warfare in the prosperous cities of Italy was largely the business of full-time soldiers organised in coherent fighting 'companies'. In the 14th and 15th centuries professionalism become dominant in European armies, and victories like Crécy depended greatly on the high skills and disciplined solidarity of the English. But these armies of professionals were not standing armies training and working together as unities. Moreover, large sections of them were raised and controlled by aristocrats, the traditional leaders in war, with all the limitations of control that implied. In Italy the emergence of professional forces produced the *condottiere*, professional contractors like the Englishman John Hawkwood, who gathered 'companies' – sometimes very major forces, but used them in their own interests rather than simply those of their employers.[21]

By contrast, in Egypt the army officers seized power and established the Mamluk Sultanate, which was essentially elective amongst themselves, a kind of military republic. By ruthlessly exploiting the land they created a standing army which defeated the Mongol invasion of the Middle East in 1260 and destroyed the last crusade state in 1191.[22] In the late 14th century the Ottoman Turks created a huge standing army which enabled them to threaten Christian Europe down to the end of the 17th century.[23]

It was only in the 15th century that the French monarchy began to recruit soldiers organised in standing companies, and these greatly assisted their victory in the 'Hundred Years War'. This institutionalisation continued so that eventually a permanent army of 20–25,000 emerged and by the end of the century France had become the greatest military power on the continent. Thereafter, most of its rival kingdoms followed suit.[24] All these standing armies of Europe and the Middle East continued to be supplemented in time of major war by casual recruitment, so armies continued overall to be somewhat loose, but their cores became increasingly strongly bound to the monarchies.

Qualitative superiority, having the better disciplined and organised army, was essential, but of course numbers always mattered. It is very difficult to discover the size of medieval armies, but simply gathering together many people increased the problems of generalship. Crusader armies could be enormous. But on the Second Crusade both the French and German armies suffered terrible losses because of failures of command and control and these were by no means rare in Europe.[25] And logistics were a major limit on the size of armies. Storing and transporting food for men and horses was difficult. In Western Europe with its rich agriculture it was usually possible for armies to feed themselves by purchase or plunder, as long as they were moving. It was down to individuals within their units to forage, plunder or purchase, and we can trace this rather ramshackle process on the Second Crusade.[26] If an army settled down to a siege, however, they quickly exhausted local supplies. The First Crusade reached Antioch in October 1098 and was starving by December.[27] By contrast St Louis of France prepared carefully for his crusade, 1248–54, creating great stockpiles of food in advance on Cyprus for his troops.[28] Things tended to be better organised in the later Middle Ages, but the problems of transport and preservation of food were difficult, and without food there could be no victory.

Before the development of gunpowder, there was only very limited technological innovation and it was mainly applied in siege warfare. When we think of siege what we have in mind is the besieger surrounding a fortification, and either starving it out or mounting an assault. In fact, both were very difficult. If the attacker chose blockade, perhaps because the place was important and its garrison strong and determined, he needed a lot of troops to surround the place and cut its communications. In loosely organised armies, sieges tended to be porous. King Stephen of England trapped his rival for the throne, Matilda, in Oxford Castle and by December 1142 its capitulation seemed likely. But the guards he placed were negligent and she escaped through a small postern gate and fled to safety.[29] In blockade, logistics assumed enormous importance. The besiegers lacked accommodation, while the besieged had shelter. Normally a castle would have stocks of food, while the besieger would have to organise his own supplies. In fact, siege, like battle, demanded a high degree of organisation in armies which normally lacked it.

If the besieger chose to storm the city or castle the simple fact was that, faced with strong walls, an attacker can go over, under or through, and all pose difficulties. In an age when close-quarter fighting was the norm, the balance of advantage lay with the defender. Mounting the walls on siege ladders was hazardous in the extreme, and while showers of arrows could keep the defenders' heads down, they too had archers. Mining demanded skill, and ground conditions were not always favourable. Wise commanders used combinations of these methods. But bribery and corruption were very frequent, and even more common was the resort to compromise by which the defending chief transferred his allegiance but was left in possession. This was especially common in borderlands, like the Vexin which lay between Normandy and the kingdom of France, where any castle holder could make a legal case for holding from one side or the other at need. Attackers could use technical devices, but they were very limited and the defender could deploy them as well. The *Perrier* or traction trebuchet threw small stones and essentially supplemented archers in making defenders keep their heads down. The counterweight trebuchet was capable of firing much bigger shaped balls, but had a slower rate of fire, and even the biggest had only a limited wall-smashing capacity.[30]

As in battle, the key to winning in a siege was organisation and control. Château Gaillard, which Philip II of France attacked in 1203, was one of the greatest castles of the age. Philip only resorted to siege after its supporting castles had been taken – notably Verneuil which was betrayed to him. He brought a huge army, established good lines of communication for supply, and built a strong siege camp. This fortification enabled his troops to fight off a major relief force. In the winter the garrison of Château Gaillard expelled all civilians to conserve food, but Philip refused to allow them to pass and hundreds died of exposure and starvation in the ditch. Philip persisted in the siege, and the castle surrendered in March 1204. This was a huge victory for persistence, organisation and a ruthless sense of purpose.[31]

The most important military innovation of the period was gunpowder, and too often it is thought of as revolutionising war. The formula for gunpowder was known in Europe by the mid-13th century, and the first certain evidence of a gun is a manuscript picture of 1326.[32] But integrating the new weapon into tactics was difficult and guns were so heavy and clumsy that they were difficult to deploy in field warfare. Their primary value lay in sieges, and they were very important in capturing fortified places in the 'Hundred Years War'. But for a long time guns were not really able to smash walls and simply battered cities and their populations, so sieges continued to be long drawn-out affairs. However, major improvements in metallurgy and the manufacture of gunpowder produced guns with longer barrels and iron shot capable of breaking through the strongest fortifications. To keep them at a range from which they could do little harm defenders built boulevards, wood and earth fortified strongpoints equipped with guns.[33] Handguns were essentially tiny cannon mounted on long wooden stocks, capable of making much noise but little else. However, in the late 15th century the arquebus, a weapon like a modern rifle with a barrel about 70 calibres, made its appearance and was clearly of enormous value in both siege and field warfare.[34] At the very end of the century something like the cannon with its narrow bore, long barrel and big wheels appeared. These weapons could be used in the field and in sieges, and greatly helped the French in their Italian wars 1494–1559.[35] Visibly the speed of technical innovation was accelerating, and commanders had to be prepared to innovate.

But the real challenge of gunpowder for commanders was integrating the new weapons into strategy and tactics. This was especially problematic because, like the crossbow, they were very slow to load. Simply having advanced weapons was not the key to winning. In the late 15th century Charles the Bold, Duke of Burgundy, had created the greatest and most modern artillery force in Europe. Yet at Grandson in March and Murten in June 1476 his forces and their magnificent artillery were crushed by the Swiss, whose army, although it had some gunpowder weapons, was primarily made up of pikemen. On 5 January 1477, Charles was defeated and killed by the Swiss at Nancy.[36] The Swiss were highly disciplined and tactically intelligent, moving quickly into battle before their enemy could deploy. But essentially they made war with the simplest and oldest of weapons, the spear.

The strength of fortifications and the risks of battle imposed a pattern of warfare which also accommodated the relative fragility of the political and military structures of medieval Europe. Across most of Europe aristocrats were the local (and sometimes only) government. Even in tightly ruled England the royal writ did not run in the Welsh marches, while the north was a world of its own where the prince-bishop of Durham was a great power. Kingdoms met not at closely defined frontiers but in ill-defined zones known as 'Marches', where the loyalty of self-seeking lords could be won by trickery, by main force or by bribery, or, most commonly by a combination of all three. The ruling elites of Europe were divided by factional and personal quarrels so that when great men, like the contending parties of 1184–85, fought, it simply intensified disputes which simmered anyway. Threats were rarely existential, so the obvious way of making war was destruction – to pillage the economic base of your enemy and so to impoverish him and to challenge his legitimacy.

William the Conqueror's biographer praised his subject:

> This was his chosen way of attack: to strike fear into the settlement by frequent, lengthy expeditions in that territory, to lay waste the vines, fields and domains, to capture fortified places and put garrisons in them wherever it was desirable; finally to attack the region relentlessly with a great multitude of troubles.[37]

His was war on a large scale, but others did the same in a less spectacular way. In 1089, during the weak rule of Duke Robert Curthose, a feud erupted in Normandy between the families of Breteuil and Goël which became a struggle for control of the castle of Ivry. At first the Goël won, but William of Breteuil built up his resources 'with the help of ransoms of captives and plunder taken from the country people' and so he bought the support of Duke Robert and the King of France. This enabled him to acquire Ivry by negotiation, but one of his allies, Robert of Bellême felt he had not been rewarded for his contribution, and so the war continued in another form.[38] This tendency for one thing to lead to another was inherent in the war of destruction. After the retreat of the allies in 1184 the Count of Hainaut took revenge upon a vassal who had defected to the enemy during the invasion:

> With much looting and setting of fires, he burned 72 towns, both his own and from his fief, and he seized Condé which was totally burned.[39]

In this way war became endemic – something which merged with banditry and had to be lived with, a relentless force which could swallow whole communities.

The stronger states of the later Middle Age were capable of waging war on a much larger scale because they enjoyed greater riches. However, this did not mean an end to wars of destruction, nor did it bring conflicts to an end any quicker. The 'Hundred Years War' is remembered by the great battles, Crécy, Poitiers, Agincourt and others. But the reality was a war of *chevauchées*, highly destructive raids designed to pay for the war, to weaken the allegiance of French lords to their monarchy and so to undermine the legitimacy of the French crown.[40] It was so expensive to maintain an army that the English crown was happy to pay off companies of soldiers and leave them to earn their own pay by seizing small towns and raiding the countryside. A contemporary recorded:

> In the year of Our Lord 1358 the English came to Chantecoq … and burned almost all the town. Then they brought the whole region around under their control, ordering every village great or small to ransom itself and buy back their bodies, good and stores of every inhabitant or see them burned, as they had been in so many other places … Those who stood in their way the English killed, or locked away in dark cells, threatening them daily with death, beating and maiming them, and leaving them hungry and destitute.[41]

This was not, to use a modern term, 'collateral damage', but the essence of medieval warfare. Henry V may or may not have remarked that, 'War without fire has no value; no more than sausages without mustard', but if he did not, he certainly should have. Such methods took time to bite, but persistence offered a real hope of winning without undue risk.

At the end of the Middle Ages England dissolved into civil war, the 'Wars of the Roses', but their genesis was caught up in the local rivalries of the great noble houses of northern England.[42] So even when there was no war in the sense in which we understand it, there was violence. The estates of great men were not clearly defined blocks of territory, but aggregations of lands gained by inheritance, marriage, agreements and purchase. Rights like monopolies which forced peasants to bake in the lord's oven or buy from his mill, were often separated from real estate. In a world where government was weak and had no monopoly of violence, disputes amongst this elite could easily degenerate into violence, and formal hostilities merged with mere criminality and the settling of local scores. Symptomatic of the attitudes arising was Ralph of Cravent whose first act, on being dubbed a knight in the early 12th century, was to rob a monk.[43] All too often winning was a matter of endurance, of inflicting and accepting terrible sufferings in the dogged pursuit of some desirable end. Of course, those with the power of decision were not the people who bore most of the sufferings.

War was controlled and directed by the European elite whose 'chivalry' valued the personal qualities of bravery which were so important in the skirmishes which arose from the war of destruction. However, this tended to increase the value attached to individual qualities rather than discipline and organisation which were essential in large-scale operations. Chivalry was not a moral code, but essentially an elaborate and self-regarding assertion by the social elite of their right to practise violence at their own discretion and in their own interests.[44] But the European elite could not be immune from the influence of the church whose pervasive influence moulded European culture. In so far as chivalry had positive values – the sparing of poor and women from massacre – it derived from this Christian influence. But the church had the most powerful impact upon the conduct of the European elite with the call to crusade which produced war with a special ethos. When the First Crusade (1095–99) was launched by Pope Urban II in 1095 there is little evidence that Europeans had a special hatred of Islam.[45] In fact the army entered into agreements and treaties with many of the Muslim powers of the Middle East, most notably Egypt. The appeal of the crusade to the European elite was that it offered them a novel path to salvation. By fighting for the liberation of Jerusalem, the place where Christ lived and died, a soldier could have all his sins forgiven, and, if killed, enter into paradise as a martyr.

The First Crusade was a remarkable military campaign. Five main armies, each grouped around a great prince, set out numbering between 40,000 and 60,000.[46] Each army was, of course, made up of the close followings of its leader with others more loosely attached, so the army, like most European forces, was a loose gathering. The problems of winning with such a force were evident on arrival at its first engagement, the siege of Nicaea (May–June 1097), for as they arrived many 'were inspired by every sort of heroism and warlike feeling, and they rushed towards the city and attacked … those who died were the ones who had heedlessly dared to attempt battles next to the walls with a blind rush and a sudden attack'.[47]

After the city's fall, as they marched east they were ambushed by the Turks and only rallied by the great efforts of their leaders.[48] As the army shrank though the ravages of war, disease and starvation, it became more coherent and disciplined. This was vital because their enemies, the Turks, were horse-archers who aimed to cause confusion by surrounding their enemies and firing clouds of arrows before closing in for the kill. The crusade succeeded because the leaders imposed enough discipline and profited from the growing experience of the men. Working together over a period of time welded the diverse forces into a true army.

But once established in the east the westerners were never accepted because Jerusalem was holy to Muslims, and their Turkish rulers felt its loss as a blow to their power and prestige. The result was the growth of real religious hatred and fanaticism on both sides and battle became relatively common. In the 1170s one of the settlers, William, Bishop of Tyre, would comment:

> For in conflicts of this nature, resentment inspired by sacrilege and scorn of laws always acts as an incentive to bitter hatred and enmity. War is waged differently and less vigorously between men who hold the same law and faith. For even if no other cause for hatred exists, the fact that the combatants do not share the same articles of faith is sufficient reason for constant quarrelling and enmity.[49]

This new ideological warfare was more savage and more ruthless than anything experienced by Europeans in earlier ages, and the church whipped up hatred through careful propaganda as a means of convincing soldiers to undertake this appalling experience. It tested the military capacity of western leaders, and what is surprising is not that crusading failed, but that westerners persisted for so long. This was war fought under the most difficult circumstances. Sheer distance created problems of logistics which could only be solved at enormous expense. They were fighting wealthy, well-organised and -equipped enemies who were also inspired by religious fanaticism. Despite many victories, the crusades to the east failed.

But this fanaticism could be turned to other targets, for the army of God was a papal army which could be turned against any perceived as an enemy of the faith. As the First Crusade departed, participants massacred Jews, the enemy within. Heresy in southern France was destroyed by the Albigensian Crusade (1209–29) with horrific massacres.[50] At the start of the 13th century the Baltic area was still dominated by pagan tribes. This was an affront to the church, which also feared they might be converted by the Russians who were Greek Orthodox. German merchants wanted to dominate the rich northern trade in furs. This produced a fanatical northern crusade which harnessed the superior technology of crossbows, castles and shipping to conquer and convert the whole area and to bring it within the cultural orbit of Western Europe. The spearhead of this process was the Teutonic Order of fighting monks whose crusading fervour enlisted help from all Europe. The determined resistance of local peoples, the distance from European centres and the appalling winters meant that this was a long process, not complete till the late 14th century. But the Teutonic Order was a corporate entity able to pursue its ends consistently over generations, and in the process they established a state, the *Ordenstaat*, which later became East Prussia. Fanaticism on its own is not necessarily a winning force, but its power to motivate and discipline men, to weld them into coherent fighting forces, was formidable.[51]

The crusades represented warfare on a new and vastly greater scale than Europe had known before 1095. They failed to retain Jerusalem, where the last Christian outpost collapsed in 1291, but they provided a new dimension to western warfare. It was not simply experience of managing and maintaining large numbers of troops over vast distances. It was the fanaticism which they inspired that made European armies so difficult to defeat. This drove conquest across the Mediterranean where the Byzantine Empire for a time was destroyed and much of Greece brought into a western orbit, along with Crete and Cyprus. The Ottoman Empire only finally destroyed these western outposts in the 16th century. In northern Europe, fanaticism won with the establishment of states on a western model all along its eastern littoral. But spectacular as this was, most of Central and Eastern Europe came into being through a barely chronicled process of piecemeal seizure of land by force, negotiation and conversion. The glamour of the great crusading armies should not blind us to the fact that war was usually on a much smaller and more local scale.

Winning wars in the Middle Ages demanded all the usual skills of the general, intelligence, strategic vision and tactical awareness. But the greatest challenge was the construction of a coherent and disciplined army. Religious enthusiasm was immensely powerful but could only be applied in particular circumstances. Charismatic leadership could impose itself on masses of men to inspire them – Charlemagne and Richard Lionheart were notable examples, though, of course charisma had to be accompanied by wealth. Charisma could be the more easily transmitted through the medium of a common set of values. Richard Lionheart embodied contemporary ideas of chivalry by leading from the front and cultivating a reputation of generosity to brave men. The late medieval fashion for founding royal chivalric orders, of which Edward III of England's 'Knights of the Garter' was one, was conscious propaganda with precisely this purpose.

But there were other groups for whom chivalry was not applicable and for whom there were different kinds of solidarity. Tribalism outside the Baltic area was rarely a factor in European warfare except in a few areas. At the battle of the Standard in 1138 the Galwegians attacked the English line in fury. But the Scottish pillaging of northern England had aroused the solidarity of its peoples who slaughtered them in the battle.[52] In the cities of Italy ordinary citizens fought with their relatives and neighbours to protect the liberties of their communities, acquiring a good reputation as steady soldiers. Milan was divided into areas, the 'Gates', and numbers were mobilised according to need. At Legnano in 1176 the Milanese defeated the army of the Emperor Frederick Barbarossa, and they continued to be a powerful factor in the warfare of the cities until the 14th century when war between the cities of Italy became so persistent that professional 'companies' replaced them. For the same reason the greater cities tended to absorb others and to deploy their wealth in the creation of larger mercenary forces.[53] The cities of Flanders developed a similar solidarity, and in 1302 at the battle of Courtrai triumphed over the flower of French chivalry at the battle of the Golden Spurs.[54] The threat of conquest by neighbouring states brought the Swiss Cantons together and their solidarity produced brilliant soldiers in flexible armies. In England the men of the 'companies' who fought so successfully in the 'Hundred Years War' were paid soldiers whose association produced integrated tactics with archers and men-at-arms, on horseback or on foot, working together.[55]

For most commanders the problem was that all these solidarities were ephemeral or sectional, needing to be integrated with others. Even the overarching power of religious enthusiasm tended to wane and to give way to other factors, like divisions over possession of Antioch which split the First Crusade, or the endless bickering over loot which characterised the Fifth.[56] Winning a war depended on a great variety of factors, not least the strengths and weaknesses of the enemy, but overcoming the problems posed by the structure of armies was fundamental to victory.

In general, the mark of winning was the possession of disputed real estate, or, at least, the recognition of a right to possess it. Even the crusades to the east boiled down to who should hold Jerusalem. To win, good commanders had to work within the limitations of the armies at their command. Negotiation rather than dictatorship was the rule, and practical problems of logistics demanded great care. As Europe became more prosperous, professional soldiers emerged, sometimes gathered in companies which were coherent fighting forces. But they were so expensive that they were only elements in composite and disparate armies. Such conditions placed a heavy emphasis on siege and its accompaniment, the war of destruction, rather than battle. Winning in war demanded patience and a ruthless use of every means that came to hand to achieve the end desired. Major conflicts, like the 'Hundred Years War', were usually protracted. Germany's frontier with the Slavs crept eastward over three centuries, while the absorption of the Baltics into western culture was not complete until the late 14th century. The conquest of Wales by powerful and nearby England took two centuries, while Scotland was never subjugated. And these wars embodied many other, often very local conflicts. Winning was primarily a matter of adapting to these complex conditions. There was rarely a tidy end, rarely a VE Day – this was a world of war. And perhaps the examples of Northern Ireland, Iraq, Afghanistan and Palestine/Israel represent something more like the experience of Europe in the Middle Ages – worlds in which war is a norm and winning hard to define.

## *Notes*

1 Vegetius, *De Re Militari* (*Epitome of Military Science.*) *Vegetius: Epitome of Military Science*, translated by N. P. Milner (Liverpool: University of Liverpool Press, 1993), 108–10.

2 Raymond d'Aguilers, *Historia Francorum qui ceperunt Iherusalem*, translated by John H. and Laurita L. Hill (Philadelphia: American Philosophical Society, 1968), 33.

3 Roger of Hovenden, *Annals*, translated by Henry T. Riley, 2 vols (London: Bohn, 1853), AD 1176, 1: 412; Ralph V. Turner and Richard R. Heiser, *The Reign of Richard the Lionheart, Ruler of the Angevin Empire 1189–99* (London: Longman, 2000), 63.

4 Peter Heather, *The Fall of the Roman Empire. A New History of Rome and the Barbarians* (Oxford: Macmillan, 2005).

5 Simon Maclean (ed), *History and Politics in late Carolingian and Ottonian Europe: The Chronicle of Regino of Prüm and Adalbert of Magdeburg* (Manchester: Manchester University Press, 2009), 47.

6 Jonathan Sumption, *The Hundred Years War, Volume IV: Cursed Kings* (London: Faber and Faber, 2015), 657, n. 41.

7 J. O. Prestwich, 'War and Finance in the Anglo-Norman State', *Transactions of the Royal Historical Society* 4 (1954): 19–43; John France, *Western Warfare in the Age of the Crusades 1000–1300* (London: UCL Press, 1999), 128–38.

8 Anonymous of Béthune, *Chroniques des rois de France,* tr. Georges Duby, *The Legend of Bouvines: War, Religion, and Culture in the Middle Ages*, tr. Catherine Tihanyi from a French original of 1985 (Berkeley: University of California Press, 1992), 195.
9 Stephen Morillo, 'Hastings: An Unusual Battle', in Jim Bradbury, *The Battle of Hastings* (Woodbridge: Boydell, 1998), 219–27.
10 John France, *Western Warfare*, 136.
11 Jim Bradbury, 'Battles in England and Normandy', *Anglo-Norman Studies*, Vol. 6 (1983): 1–12.
12 Georges Duby, *The Legend of Bouvines: War, Religion, and Culture in the Middle Ages*, 195.
13 John Gillingham, 'The Unromantic Death of Richard I', *Speculum*, Vol. 54 (1979): 18–41; France, *Western Warfare*, 167–69; Robin Neillands, *The Hundred Years War* (London: Routledge, 2001), 102.
14 David C. Douglas, *William the Conqueror. The Norman Impact upon England* (Berkeley: University of California, 1964), 50.
15 Charles Ross, *The Wars of the Roses* (London: Thames and Hudson, 1986), 128–35.
16 Jim Bradbury, *Stephen and Mathilda the Civil War of 1139–53* (Stroud: Sutton, 1996), 158–59.
17 Gilbert of Mons, *Chronicle of Hainaut* translated by Laura Napran (Woodbridge: Boydell, 2005), §114, 95.
18 France, *Western Warfare,* 153–55.
19 Neillands, *Hundred Years War*, 151–62.
20 John France, 'People against mercenaries. The Capuchins in Southern Gaul', *Journal of Medieval Military History*, Vol. 8 (2010): 1–22.
21 William Caferro, *John Hawkwood. An English Mercenary in Fourteenth-Century Italy* (Baltimore: Johns Hopkins, 2006).
22 Reuven Amitai-Preiss, *Mongols and Mamluks. The Mamluk-Ilkhanid War 1260–81* (Cambridge: Cambridge University Press, 1995).
23 Rhoads Murphey, *Ottoman Warfare 1500–1700* (London: UCL Press, 1999).
24 Philippe Contamine, *War in the Middle Ages*, translated by M. Jones (Oxford: Blackwell, 1984), 165–72.
25 Jonathan Phillips, *The Second Crusade* (London: Yale University Press, 2007), 168–84.
26 John France, 'Logistics and the Second Crusade', in J. H. Pryor (ed), *Logistics of Warfare in the Age of the Crusades* (Aldershot: Ashgate, 2006), 77–94.
27 John France, *Victory in the East. A Military History of the First Crusade* (Cambridge: Cambridge University Press, 1994), 236–37.
28 Joinville, *Life of St Louis* translated by M. R. B. Shaw (Harmondsworth: Penguin, 1963), 197.
29 Jim Bradbury, *Stephen and Matilda: The Civil War 1139–53* (Stroud: Sutton, 1996), 123.
30 Michael Fulton, *Artillery in the Era of the Crusades: Siege Warfare and the Development of Trebuchet Technology* (Leiden: Brill, 2018).
31 H.-Francois Delaborde (ed), *Oeuvres de Rigord et de Guillaume le Breton*, 2 vols (Paris: Renouard, 1885), 2: 200–260.
32 For a modern survey of the early history of gunpowder see Kay Douglas Smith and Robert Douglas Smith, *Rewriting the History of Gunpowder* (Nykøbing Falster, Denmark: Middelaldercentret, 2010).
33 Clifford J. Rogers, 'The Military Revolutions of the Hundred Years' War', *The Journal of Military History,* Vol. 57 (1993): 241–278, reprinted with revisions in C. J. Rogers (ed), *The Military Revolution Debate* (Boulder: Westview Press, 1995).
34 Bert S. Hall, *Weapons and Warfare in Renaissance Europe* (Baltimore: Johns Hopkins, 1997), 95–99.
35 Ibid., 157–200.
36 Robert D. Smith and Kelly Devries, *The Artillery of the Dukes of Burgundy 1363–1477* (Woodbridge: Boydell, 2005), 188–201.
37 William of Poitiers, *Gesta Guillelmi* translated and edited by R. H. C. Davies and Marjorie Chibnall (Oxford: Clarendon Press, 1998), §38, 61.

38 France, *Victory in the East*, 41–42.
39 Gilbert of Mons, *Chronicle of Hainaut*, §116, 97.
40 Mollie M. Madden, *The Black Prince and the Grande Chevauchée of 1355* (Woodbridge: Boydell, 2018).
41 Jonathan Sumption, *Trial by Fire: The Hundred Years War, Volume II* (London: Faber & Faber, 1999), 379–80.
42 Ralph A. Griffiths, 'Local Rivalries and National Politics: The Percies, the Nevilles, and the Duke of Exeter, 1452–55', *Speculum*, Vol. 43 (1968): 589–632.
43 Ordericus Vitalis, *Historia Aecclesiastica*, edited by M. Chibnall 6 vols (Oxford: Clarendon, 1969–79), 3: 242–3.
44 Richard W. Kaeuper, *Medieval Chivalry* (Cambridge: Cambridge University Press, 2016).
45 Nicholas Morton, *Encountering Islam on the First Crusade* (Cambridge: Cambridge University Press, 2016).
46 France, *Victory*, 122–42; Jonathon Riley-Smith, 'Casualties and the Number of Knights on the First Crusade', *Crusades*, Vol. 1 (2002): 13–18.
47 Albert of Aachen, *Historia Ierosolimitana*, Susan B. Edgington (ed) (Oxford: Clarendon, 2007), 95–95.
48 France, *Victory*, 173–84.
49 William of Tyre, *A History of Deeds Done Beyond the Sea by William of Tyre*, translated by E. A. Babcock and A. C. Krey, 2 vols (New York: Columbia University Press,1943) 2: 25.
50 Sean McGlynn, *Kill Them All: Cathars and Carnage on the Albigensian Crusade* (Cheltenham: History Press, 2015).
51 The horrors of the northern crusade were chronicled by a contemporary, Henry of Livonia, *Chronicle* translated by James Brundage (New York: Columbia, 1961).
52 Henry of Huntingdon, *Historia Anglorum* edited by D. Greenway (Oxford: Clarendon, 1996), 718–19 and see John Gillingham, 'Conquering Barbarians: war and chivalry in twelfth-century Britain', *The Haskins Society Journal*, Vol. 4 (1992): 67–84.
53 Michael Mallet, *Mercenaries and their Masters: Warfare in Renaissance Italy* (London: Bodley Head, 1974).
54 Jan F. Verbruggen, *The Battle of the Golden Spurs*, translated by David R. Ferguson from a 1952 Dutch original (Woodbridge: Boydell, 2002). The battle is so called because the victors hung the spurs of the French knights in their church as tokens of victory.
55 Kelly Devries, *Infantry Warfare in the Fourteenth Century: Discipline, Tactics and Technology* (Woodbridge: Boydell and Brewer, 2006).
56 John France, 'The Crisis of the First Crusade: From the defeat of Kerbogah to the departure from Arqa', *Byzantion*, Vol. 40 (1970): 276–308; James M. Powell, *Anatomy of a Crusade 1213–21* (Philadelphia: University of Pennsylvania, 1986).

CHAPTER 3

# The Early Modern Period in Europe, 1500–1715

David Parrott

The waging of war was the principal occupation of European rulers and their states throughout this 200-year period. The numbers of troops mobilised during the 16th century might appear distinctly pre-modern, with military establishments of 50–60,000 troops maintained by the greater powers. Yet by the end of the 17th century, military establishments had become behemoths. In the mid-1690s, France, with a population of around 20 million, was sustaining a wartime army of around 340,000 troops. This military expansion came at a high cost: the Thirty Years' War, which raged across the German states of the Holy Roman Empire between 1618 and 1648, cost the lives of around 15–20% of the population and resulted in decades of subsequent economic dislocation.[1] Given the increasing human and material costs of war, what did belligerent powers aim to achieve? In what ways did they understand winning through waging war? In an early example of sceptical pacifism, Desiderius Erasmus asked his contemporary rulers to: 'Sit down before you draw the sword, weigh every article, omit none, and compute the expense of blood as well as treasure which war requires, and the evils which it of necessity brings with it; and then see at the bottom of the account whether, after the greatest success, there is likely to be a balance in your favour.'[2]

There was little sign that early modern rulers were prepared to heed Erasmus' advice, so what were the inducements leading them so frequently into warfare? Liberal thought in the modern west draws a sharp distinction between the states of peace and war. The former is overwhelmingly desirable on humanitarian, economic and social grounds, and the latter must be avoided unless the state faces a readily perceived threat to security, territorial integrity or its core values. This does not prevent modern wars happening, but it frames the cultural context in which war is discussed. In contrast, war in early modern Europe was not seen as an unnatural event, forced upon an international order whose normal condition was peaceful co-existence. Rulers did not regard avoiding war as inherently desirable; indeed they regarded the pursuit of peace at the cost of their international reputation and personal prestige as more damaging than resorting to war – even in cases where the chances of military success appeared slight. The Spanish monarchy was well aware of the operational difficulties and expense of campaigning against its rebellious subjects in the Low Countries. Restarting the lengthy war with the Dutch Republic in 1621 at the expiry of the Twelve Years' Truce was more likely to involve further rounds of crippling expense and indecisive siege warfare than any outright victory. However, the truce had been deemed dishonourable and damaging to the reputation of the Spanish monarchy, so the decision to resume the war had been a foregone conclusion for a half-decade before 1621.[3]

There was some recognition that the state of war required containment within a theoretical framework. Christian theology and ethical teaching, while by no means espousing pacifism, did seek to define the circumstances under which war could be justly undertaken (*jus ad bello*) and the ways and means by which it could be waged once begun (*jus in bello*). Through the notion of laws of war, theologians sought to define and secure the rights of non-belligerents and to establish some principles of proportionality in the use of violence and destruction to achieve military objectives. But in practice rulers and statesmen acted with considerably greater latitude in defining what they considered just cause for war, and assessing the benefits that they hoped to obtain through waging it.

Wars in this period were not fought on behalf of an idea of the 'state' or the 'nation' as some collectively understood concept. War was neither represented nor waged as a struggle for national survival. The claim that the security of the state could demand an ultimate, absolute commitment or sacrifice from its subjects, was born in the 1790s. It was at very most invoked on a local scale in the early modern period, a call to last-ditch measures in a besieged town or city, or in defence of a community defined by its common religious practice or culture. Nor should we be misled by so-called 'reason of state' thinking and policy justification in the later 16th and 17th centuries, often presented as a path to modern thinking about the state, war and peace. In largest part the contemporary writings about reason of state were casuistical responses to the notorious challenge thrown down by Niccolò Machiavelli (1469–1527) about moral norms versus practical imperatives in political conduct.[4] Such discussion concerned not the abstract ideal of the state or nation; it sought to facilitate the interests of rulers and their dynastic ambitions by justifying actions and policies which would not bear rigorous moral or theological scrutiny. A Catholic kingdom making alliances with Protestant rulers to fight another Catholic power – the situation of France in the 1630s in her war with Spain, for example – provides a good example of the application of 'reason of state' justifications. Such thinking, which traditionally went in hand with an insistence that these matters should anyway be concealed from ordinary subjects as 'secrets of state' (*arcana imperii*), served more to distance most subjects from policy and war-making, rather than drawing them into any sense of collective identity with these policies.

The state as a distinct source of identity, national sentiment and loyalty did not exist for most of this period, and for clear reasons. For the strongest force determining the collective allegiance of groups of subjects was the identity of their ruler. The most frequent motive for warfare, and the usual determinant of what constituted winning, was dynasticism. Throughout this period a majority of European states were ruled as hereditary principalities or kingdoms, where the principle of family right and inheritance was the basis of sovereignty. Legitimacy came from the blood line, while territory and its populations belonged to the ruler as a dynastic inheritance and as family property.[5] Externally as well, dynasticism was the currency in which international relations were conducted: territorial rights were established and maintained through familial or marital inheritance. In much of Eastern and Northern Europe, and in the case of the papacy and the emperors of the Holy Roman Empire, monarchy was elective. Yet once elected, most such rulers sought to pursue family interests as the primary motivation for foreign policy. Even where republican governments prospered, their ruling elites were frequently constituted by a handful of powerful, intermarried families. To this picture of policy-making in the hands of rulers who saw themselves as dynastic proprietors of their peoples, should be added the widespread development of 'composite' monarchies.[6] Few early modern rulers simply ruled over a single territory defined by a national language, territorial contiguity and common cultural identity. The majority, through inheritance or marriage, had come to rule over multiple territories, frequently detached from one another, or linguistically and culturally diverse, whether the kingdoms of Denmark and Norway, or the British composite monarchy from 1603 of England, Ireland and Scotland. Composite monarchy reached its apotheosis in the 'state' of Charles V of Habsburg (1500–58), which brought together the inheritances of his Austrian Habsburg, Spanish and Burgundian ancestors and relatives. His inheritance created a territorial conglomerate that stretched from Vienna to Seville, and Palermo to Brussels, a rich and hugely diverse quilt of peoples, governments, languages and cultures.[7]

The upholding or assertion of dynastic right was the most widely accepted justification for warfare, and the measure of the success or failure of belligerence. Notions, for example, that kingdoms had 'natural frontiers' or that wars might be fought for overtly geostrategic concerns, belong more typically to the 19th, not the 17th, century. Yet a cynical view of early modern dynasticism would argue that it was a malleable rhetoric, deployed to justify whatever opportunistic or aggrandising projects rulers chose to adopt. The Tudor monarchy's claims to the crown of France might seem little more than a pretext, exploiting France's military preoccupations elsewhere to try to expand England's grip on the territory around its one remaining ancestral possession across the Channel, the Calais Pale.[8] In turn the French crown's invasion and occupation of Piedmont from 1536 to 1559, an Italian territory over which the crown had the most tenuous family claims, required a spate of *post-hoc* dynastic invention and justification from writers and diplomats. A century later as France became involved in the Thirty Years' War, dynastic claims were sharpened up to justify the occupation of the independent duchy of Lorraine (1632–59).[9]

Yet to see dynasticism as no more than a cynical cloak for *Realpolitik* is to underestimate its embedded character in the assumptions of the international community. France might have won the war militarily in both Piedmont and Lorraine, but on each occasion she was constrained to return these territories to their rightful dynastic owners at subsequent peace settlements. In practice dynasticism brought as many challenges and unwanted obligations as it generated opportunities. It was a betrayal of family obligations and honour to renounce dynastic rights, or even to fail to pursue them with enough vigour. A mass of putative claims on territories brought by consanguinity or marriage were stored up in the minds – and often the chanceries – of rulers and their advisors. To renege on any one of these without some appropriate trade-off would be an indication of weakness, however impractical these rights were to assert or maintain. Losses and defeat in the pursuit or defence of dynastic objectives were universally perceived as grave, not primarily on a cool strategic assessment of their significance but as a humiliating blow to family status and reputation. It might be considered that the loss of Calais by the English crown in 1557, a vulnerable outpost surrounded by French territory, would be a blessing in disguise: the opportunity for an under-resourced monarchy to shed a territory which was constantly vulnerable to its much more powerful cross-Channel neighbour. But the loss of Calais cankered the remains of Mary Tudor's already unpopular reign, and haunted the policymaking of Elizabeth I and the early Stuarts.[10] Ironically it was the regicidal Cromwellian Protectorate of the 1650s which finally (if temporarily) made good this dynastic setback by demanding the port-city of Dunkirk as the price for military intervention on behalf of the French.[11] No less certain to draw rulers into unwanted conflict was the likelihood that several ruling dynasties would have acquired claims through marriage or lineage to the same territory. Even when one claimant might be perceived to have an objectively stronger claim, other parties might choose to demand lesser territorial or financial compensation, or perhaps a trade-off in form of recognition of claims elsewhere. This was a defining feature of European wars, seen in the numerous conflicts explicitly titled 'wars of succession' from Mantua in 1613–17 and 1628–31, through those fought far into the 18th century.

Dynastic claims could prove burdensome obligations rather than strategic opportunities. They also however permitted a sliding scale of possibilities of what it meant to win in war. The French crown's claims in Italy focused above all on the kingdom of Naples and Sicily, claimed through their Angevin predecessors. These claims clashed directly with those of the Spanish House of Aragon, whose collateral branch sat on the Neapolitan throne. France invaded Italy in 1494 marching south to take possession of Naples. Thus began the dynastic struggle known as the Italian Wars, lasting until 1559 and pitting the family claims of France against those of Spain, and Spain's composite ruler from 1519, Charles V of Habsburg. The notion of southern Italy ruled from Paris may appear historically counter-intuitive, but was no more so than the Netherlands ruled from Madrid. Yet faced with the resources and determination of Charles V, the French crown's hopes of winning and ruling the kingdom of Naples were slight. By 1530 it might seem that the conditions for Habsburg hegemony over much of Italy had been established on the basis of a series of crushing defeats of French armies. Yet France continued to expend military resources and political and diplomatic credit for another three decades down to 1559 trying to overturn the verdict. The French crown may have retained some hope that a battlefield victory could be followed up by successful campaigning that would prise one or more Italian territories out of the hands of the Habsburgs. But the reality was that the monarchy was primarily concerned to keep its claims alive – claims which now also extended to the duchy of Milan – and to preserve dynastic prestige and status in the eyes of independent Italian and European rulers. Winning in war was not just about emerging with a major territorial gain but ensuring that the Habsburg tightening of practical military and political hegemony in Italy did not also ensure that their dynastic rights were simply accepted without opposition.[12]

Prestige and (family) honour were thus as much at stake in this type of warfare as substantive territorial gains. Honour, often at a highly personalised level, was a driving force in much belligerence. At the Peace of the Pyrenees in 1659 which finally settled a 25-year conflict between France and Spain, the Spanish negotiators were prepared to jeopardise the settlement and concede one or two extra conquered places on the Flanders frontier to achieve a different success: to ensure that the French crown was forced to pardon and re-establish the rebel prince, Louis de Bourbon-Condé, who had entered the service and placed himself under the protection of the Spanish crown in 1652. The honour of the king of Spain was at stake, but as Condé was the cousin of Louis XIV it was also about gaining prestige over the rival king of France.[13]

Dynasticism was the chief determinant of war and its aims in this period, but what of ideological reasons for conflict? Political ideologies in general assumed a minor role. There were no wars in the period explicitly to impose 'absolute rule', nor to defend 'republican liberties'. Where historians have sought to impose such ideological positions, the struggles tend to reveal either underlying religious motivation as the cause of division. Or they show an essentially non-ideological struggle between elite factions who supported or opposed central policies with little sense that those policies represented anything other than different ways of dividing the cake of political benefit. The British Civil Wars (1642–49) provide a good example of strong underlying religious motivations behind the conflicts in all three realms: it was confessional differences, aggravated by distrust of King Charles I, that proved fundamentally intractable and led to conflict.[14] A few years later, the *Fronde* in France

(1648–53) was a good example of an internal power struggle fought overtly over the political and patronage opportunities that would be provided by dominating the council of the young King Louis XIV.[15] Winning in war, insofar as it had any implications in terms of political ideology, was far more likely to be presented by the winning side as restoring the *status quo ante*, disrupted by an opponent's actions.

For cases where ideology, broadly defined, drives early modern war and its aims, we should look instead at the impact of the Protestant and Catholic Reformations and their permanently divisive impact across Europe, and perhaps also at the continuing struggle throughout this period between Europe and the Islamic Ottoman Empire.

In striking contrast to dynastic conflict, the 'wars of religion' of the post-reformation period come closer to the notion of a total war: a war fought without constraints and applying whatever force was required to annihilate an opponent or compel them to accept your will. The Catholic majority in France through the 1560s to 1580s sought an unconditional victory over the Protestant (Huguenot/Calvinist) minority in which all Protestants should be killed or driven out of France. In turn, Protestants were prepared to advocate doctrines of tyrannicide and separatism, fighting to create a Huguenot 'republic' in the south and west of France where Calvinism would become the majority and intolerant religion. On both sides in the French Wars of Religion, ideas of religious purification and God's providential support for their causes could blend into millenarian or apocalyptic visions in which righteousness had to triumph at all costs. Prevarication, calls for toleration or the prioritisation of political unity, were acts of enemies, scarcely less culpable than heresy itself.[16] Similar extreme views of the aims and waging of warfare can be seen in the struggle which broke out in the Netherlands after 1567, following Spain's heavy-handed treatment of political and religious dissent. Calvinist minorities, faced with a struggle for religious and personal survival, became the not-always welcome backbone of a resistance movement which was simultaneously trying to draw in moderate support to negotiate a more favourable settlement with Spain. On the other side however, the strongly Catholic Philip II of Spain declared his outright refusal to rule over heretics, and offered Protestant rebels no possibility of compromise. The resulting conflict generated atrocities by both sides, mobilised occasional last-ditch resistance, and eventually split the Netherlands between a Catholic, royalist south and the northern Dutch Republic, dominated by a Calvinist ruling elite.[17]

These examples from France and the Netherlands suggest a very different, inflexible interpretation of winning in war, where anything other than total victory was failure. Yet some circumspection is required: it was by no means inevitably the case that religious difference would drive an intensification of war aims and violence. The Thirty Years' War provides a surprising demonstration of this. It is widely assumed that the most long-lasting and – until World War II – destructive war in European history finally exhausted and extinguished extreme confessional antagonism after confessional strife had led to the slaughter and immiseration of millions. In fact, beyond partly motivating its outbreak in the Bohemian revolt of 1618, religion played a modest role in what was more obviously a protracted dynastic struggle between the Habsburg monarchy, the greater territories in the Holy Roman Empire, and various external powers, most notably France, Spain and Sweden. The unanticipated scale of the military successes of the Emperor during the 1620s tempted him into what is typically regarded as the high-point of the 'religious' Thirty Years' War, the 1629 Edict of Restitution, which sought to reclaim all Catholic lands and institutions secularised by Protestant powers since 1552. The political and military consequences of imposing the edict, which was as much princely land-grab as confessional policy, were disastrous, and by 1635 the emperor had effectively abandoned – 'suspended' – the edict to achieve peace with the German princes.

It was not just the relative lack of religious motivation in the war aims; equally notable was the absence of extreme confessional violence. The armies of the Thirty Years' War were notorious for their brutality, extortion and destruction, but very little of that reflected religious motivation. Violence was functional, concerned with securing the soldiers' subsistence and profit; it was rarely deployed to enforce religious conformity or to punish heterodoxy. The armies themselves were, with a handful of exceptions, melting pots of soldiers recruited from different confessions with little inclination to act as enforcers of religious conformity. Generalising beyond the Thirty Years' War it is notable how unusual extreme religious violence was in fighting wars across early modern Europe: the circumstances of France, the Netherlands or Ireland need special explanation for the infiltration of intense religious hatred into the aims and methods of the belligerents. Even when wars were fought to defend or impose some form of religious settlement, as was the case with the civil wars fought in England, winning the conflict did not mean extermination or gunpoint-enforced conformity. The defeated Anglican royalists were marginalised not persecuted, while the victorious Parliamentarians squabbled, but did not fight, over competing Presbyterian and more radical 'Independent' stances on church reform.[18]

On the borders of Christendom, it might be assumed that the struggle between Europe and Ottoman imperialism would polarise and intensify war, given the high stakes involved. The abrogation of the laws of war in fighting the infidel indicates the ideological shift, while general fear of an Islamic 'Other' could briefly unite Christian confessions – seen for example in spontaneous popular Protestant celebrations at news of the 1571 Catholic victory over the Ottoman fleet at Lepanto.[19] Yet beneath the rhetoric, and some intense moments of combat in central Europe and in the Mediterranean, relations and war-aims between European powers and the Ottomans were a lot more normal than the religious/ideological gulf would seemingly dictate. Neither side fought an unconstrained or unlimited 'war of civilizations'. The European states were of course fragmented both politically and religiously. From the perspective of France, the Ottoman Empire was a distant threat but a useful ally against the French kings' dynastic rivals, the Habsburgs. From Francis I to

Louis XIV, French foreign policy sought to turn Ottoman military force to political advantage: winning wars against the Habsburgs in alliance with the mortal enemy of Christendom did not unduly concern the 'Most Christian King of France'. This was equally the case when it came to policymaking amongst the Protestant powers, especially the Dutch and English, who sought both trading and military alliances with the Ottomans.[20] Even the Republic of Venice, one of the key players in the Holy League at Lepanto, saw more advantage following the victory in surrendering Cyprus to the Ottomans in return for maintaining Levantine trading privileges. Lepanto led to *détente* in the Mediterranean rather than any ratcheting up of war aims and objectives.

It is easy to see some of the slow-burning Ottoman–European conflicts as early examples of 'intervention for effect'. Launching a full-scale crusade to recover the Holy Land had long since moved from the agenda of practical military/political objectives to a virtue-signalling aspiration. A few idealists still contemplated crusading expeditions, if not to reconquer the Holy Land, at least to push back Ottoman conquests elsewhere. Most famously in 1578 King Sebastian of Portugal was defeated and killed leading his nobles into battle in Morocco in pursuit of his crusading ambition to drive the Moors, allies of the Ottomans, out of North Africa. Sebastian's death without a direct heir enabled Philip II of Spain to assert his claim, through his mother, to the inheritance of the Portuguese state and empire, which became part of Philip's already immense composite monarchy in 1580. More usually, neither Ottomans nor the European powers sought much more than to burnish the credentials of involvement in 'God's Wars'. The 'Long Turkish War' of 1593–1606, fought in the Hungarian borderlands, was remarkable for the absence of real will to bring about military transformation or to radically change the geo-political situation of the two powers. The Habsburgs, who bore the brunt of the defence, drew on large numbers of volunteers and military contributions from allied powers: Hungary became a 'school of war' for international Catholic (and sometimes Protestant) nobles, who could demonstrate active piety while simultaneously gaining military experience.[21] The same could be said for much Ottoman/European warfare in the Mediterranean, where again a crusading/jihadist ethos could be set alongside the day-to-day realities of permanent and small-scale corsair warfare, waged as a profitable exercise in prize-taking at sea and as coastal expeditions seeking plunder and slaves.[22] Winning in war combined self-financing military operations with visible engagement in an ideological struggle.

In general, however, concern to balance the costs and benefits of war had not been a feature of the waging of larger-scale conflicts. Dynastic warfare, presented in terms of family honour and reputation, was self-consciously dismissive of any such financial calculations. By the end of the War of the Spanish Succession in 1713 Louis XIV had managed to keep his grandson on the throne of a (diminished) Spanish empire but at the cost of accumulated debts of 2000 million *livres*, at least ten times the crown's annual revenues.[23] The reigns of Louis XV and XVI unfolded in the shadow of this fiscal incontinence.

There was some awareness, at least among ministers and councillors, that winning – let alone losing – wars at such cost to state and economy was unsustainable. One potential response was to wage war for economic objectives, ways in which the state's prosperity – and therefore tax-paying capacity – would be enhanced. The argument for economic warfare was fuelled by prevailing mercantilist doctrines which represented trade and commerce as a zero-sum activity in which states competed to obtain control of a fixed quantity of global trade. Moreover, domestic prosperity was achieved by favourable balances of trade, which allowed the build-up of wealth within the state, above all in the form of bullion. A favourable balance could be sought by appropriating the commerce of other powers, by domestic programmes of import substitution, and by aggressive tariff wars. All of these were adopted by Louis XIV's minister of finance, Jean-Baptiste Colbert, in an effort to strengthen France's economy in the 1660s and 1670s.

Behind the evident falsity of most mercantilist assumptions, there lay a few grains of truth. The Dutch Republic had achieved immense prosperity through cost-efficient expansion and exploitation of commercial activity, built on supplying a wide and diverse range of European markets and when circumstances permitted establishing and aggressively maintaining trade monopolies. By the mid-17th century the Dutch were far ahead of any other European power in commercial capacity.[24] The first two Anglo-Dutch wars (1652–55, 1665–67) concerned the relative repositioning of Dutch and English naval and mercantile power. They resulted from the steady accumulation of friction as rival trading activity led to clashes in the Mediterranean, the English Channel and North Sea and, ultimately of greatest significance, in the world beyond Europe, both the Caribbean and the Far East. English victory, and in 1667 defeat, was only initially about gaining and maintaining local naval hegemony; the main outcome of naval success was to facilitate the wider commercial activity of the victorious state at the expense of the defeated.[25] In a commercial environment where trade and privateering frequently overlapped, most merchant ships were armed for both defence and attack; the taking of prizes was a profitable activity, a form of low-level conflict, and, when required, a *casus belli.* Mediterranean warfare, outside a few massive naval encounters and heroic set-pieces, was a centuries-long exercise in corsair activity, waged by largely private forces. Legitimate targets loosely reflected the enmity or alliances between different powers, so that over lengthy periods Venetian galleys carrying out commerce-raiding would scrupulously respect Ottoman merchant shipping to avoid destabilising trading and diplomatic agreements with Istanbul.[26] But attempts to subordinate profit-motivated privateering to larger economic and political objectives were never more than partially successful, and the issue raises larger questions about the practicalities of winning in early-modern warfare.

This discussion has so far examined the motives for waging war, and the extent that these motivations determined conceptions of winning, of desired outcomes: it has sought to show how dynastic imperatives, ideology – especially religious conviction – and even some economic ambitions, were causal factors. But it is no less true that the manner in which early modern wars were waged shaped the meaning of winning. How far did the practical problems of mobilising and then deploying military resources constrain what could be achieved by war as a tool of policy?

This has particular salience to the notion of decisiveness, achieving military outcomes that could allow a victor to impose an extremely unfavourable, even unconditional, set of peace terms on the defeated. For the most notable feature of warfare from 1500–1700 is its lack of such decisiveness. A handful of conflicts had indisputably great and lasting consequences: the 1526 Ottoman campaign and victory at Mohacs against Louis II, king of Hungary and Bohemia, would certainly qualify: the crushing defeat of his army and the death of Louis extinguished the Jagellonian line, led to the partitioning of Hungary between Ottomans and Habsburgs, and established the Habsburgs as elected, later hereditary, rulers of Bohemia.[27] But such transformative victories were rare. A year earlier the harsh terms imposed after the decisive defeat and capture of the king of France at Pavia by the army of Emperor Charles V did not outlast the release of Francis I and his formal repudiation of Charles' demands once back in his kingdom. Similarly, Charles V might have crushed the Lutheran German princes at the battle of Mühlberg in 1547, but by 1552 his victory had been completely overturned thanks to military cooperation between some of those princes and the French crown.[28] Winning an apparently outright battlefield victory or aggregating successes in campaigning regularly failed to achieve the outcome of a territorially or strategically advantageous settlement.

Key to this failure was the capacity of early modern states to organise and mobilise resources for war. These were not modern states, characterised by centralised administrations with extensive 'reach' into provincial and institutional power-structures. In 1500 most policymaking, resource mobilisation, and military administration was carried out under the overall authority of the ruler, but through the informal involvement of courtiers, entrenched institutions, and the influence and cooperation of provincial elites and their networks of clients and adherents. The price of this dependence involved respecting the political privileges and financial immunities of the elites in these societies. While the populations of Europe were expanding and it proved easier to fill the ranks of armies, rulers lacked the capacity to tap the wealth that was becoming ever more concentrated in the hands of landed and mercantile elites. Rulers could raise troops, usually through drawing on their greater subjects' local influence, and sometimes their resources. But rulers still had to face the bulk of the financial and organisational challenges of supporting warfare: meeting the costs of supplying and paying troops, keeping them in munitions and equipment, meeting all the extraordinary costs from transport to siege works. For this they were financially extremely ill-equipped. To misquote Charles Tilly's well-known aphorism, making war in these conditions was a lot more likely to break the state than to make it.[29]

These problems were compounded on the threshold of the 16th century by some radical changes in the nature of warfare. The proliferation of artillery may not have transformed warfare as greatly as its enthusiastic exponents have suggested, but it certainly added a heavy new expense to military budgets.[30] Even more costly was the move from traditional armoured cavalry, preserve of a self-trained and often self-equipped noble caste, towards professional infantry – the massed pike square. Relying on the weight and impact of ordinary foot-soldiers, the cohesion and discipline to deploy such formations in battle was easy to recognise, but exceptionally hard to achieve in practice. Pioneered by the Swiss cantons, the only serious competition came from the German *Landsknechte*. Monopolising military skills considered essential to winning wars, the providers of these military specialists sold their services extremely expensively. But rulers of both major and secondary states eagerly sought their services regardless of cost: by the 1550s the proportion of foreign troops in French armies regularly approached 80% of their overall strength.[31] Recruitment contracts needed to be paid in advance of the levy, and additional, regular payments would be the price of keeping these hired soldiers in service.

Rulers and states tackled the problem of the inflationary costs of warfare and the massive strains that this placed on their financial resources and revenues by an expedient which was probably less a conscious decision than a simple bowing to necessity. In the period down to 1550, large-scale 'classic' warfare was waged intensely but for very limited duration. The Italian Wars were especially notable here; warfare rarely lasted more than 2–3 consecutive campaigns, and in cases only for a single campaign. The belligerent powers acknowledged that reckless financial over-extension based on heavy borrowing against anticipated tax revenues could not be sustained for any longer without the risk of financial – therefore likely military – collapse. As a result, wars were fought around the imperatives of short (and few) campaigns: winning would be determined by what were in effect just one or two throws of the military dice. All belligerents recognised that waging war for extreme, open-ended objectives was simply not realistic given the constraints on their ability to mobilise resources. On three occasions, in 1495, 1501 and 1528, French armies fought their way down into the kingdom of Naples, but each attempt to establish a permanent foothold failed. Rejecting this verdict and fighting all-out to wrest Naples and Sicily from Spanish grasp would have required a protracted, year-on-year series of military campaigns, placing unsustainable burdens on France's fiscal and organisational capacity.

This caution about being dragged into lengthy war served rulers who thought primarily in dynastic rather than geostrategic terms, and could shape their expectations of outcome accordingly. A single campaign might, for example, be spent using a large army to conquer a couple of fortified frontier cities which could then be bargained and returned against a recognition of dynastic right or status elsewhere. This reluctance to turn warfare into a mutually destructive process of attrition seems to have been general in the first half of the 16th century; large-scale war was short and sharp, and belligerents did not double down to try to achieve more ambitious aims.

The breakdown of these tacit conventions came during the 1550s as a result of the last Italian war, which began in 1552 and lasted, disastrously, until 1559. Eight years of conflict resulted in exactly what might have been predicted. In 1556–57 both main belligerents, Henri II of France and the Emperor Charles V, were forced into a massive rescheduling of accumulated debt – effectively state bankruptcy – which shattered their financial markets, temporarily derailed military activity and left both rulers floundering for ways to continue funding their military commitments. Limping through to the Peace of Cateau-Cambrésis in 1559, and despite an imposing Spanish battlefield victory at Saint-Quentin in north-east France in 1557, both sides were anxious above all else for a settlement: winning was a lot less important than gaining peace before a further financial breakdown.[32]

The mid-16th century proved to be a watershed in the relationship between states, armies and warfare. It was marked above all by a shift from the direct state-hiring of prohibitively expensive mercenaries towards the state's outsourcing of warfare into the hands of military entrepreneurs and contractors. The possibility that the prestige of military command could be combined with a remunerative financial investment was an attractive combination for wealthy and socially ambitious subjects. In return for taking on the financial responsibility for raising and sustaining their troops, unit commanders would receive overall administrative, financial and judicial autonomy in running their regiment, company or naval squadron. The benefit to the ruler was evident: the huge expenses and administrative challenges of recruiting and equipping an army would be met through the cash, credit and organisational efforts of its officers. Moreover, these financial commitments – the officers' 'investment' in the army – would be reimbursed later by what became a parallel system of outsourcing, this time of the tax base: this entailed the collection of much heavier taxes – so-called 'contributions' – either directly by the soldiers themselves, or by local administrators backed by the threat of military enforcement. These taxes would then pass directly to the officers to meet their costs and reimburse them for their expenses.[33] Ideally such contributions might be collected from enemy or neutral territory. But wartime necessity could justify the imposition of such systems on the ruler's own subjects, sweeping away privileges and restrictive systems of fiscal immunities that had hampered previous revenue-raising.[34]

One additional factor shaped this upsurge in military enterprise. The 1550s had ended the epoch in which wars were short and followed by immediate demobilisation: for the next century wars were to be waged over unprecedented timespans. Individual periods of 'official' conflict both in France and the Netherlands in the second half of the 16th century grew increasingly interconnected, and there was little willingness to demobilise forces between these upsurges in warfare. The 14 years of the 'Long Turkish War' from 1593–1606 gave way to the Thirty Years' War from 1618–48, the longest period of uninterrupted large-scale warfare in European history. The relationship between lengthier wars and the flourishing of military entrepreneurship was symbiotic. Without the ability to outsource the costs of raising and maintaining troops, none of the belligerent states could have afforded these lengthy periods of war. And for the proprietor colonels and senior officers looking for a return on their 'investment', lengthier periods of war were attractive in offering greater opportunity to reimburse themselves and their creditors through war-taxes, looting, ransoms or other extortion. What became possible, and for the military proprietors positively desirable, became the norm: for rulers, the ability and willingness to wage large-scale warfare for many years, even for decades, in pursuit of their objectives became vital to success. Winning in war meant a willingness to match opponents in their commitment to drawn-out attritional struggles that could only be sustained by outsourcing huge parts of the military machine to private capital and organisation.[35]

The consequences of this ability to extend the duration of war were considerable, as the Thirty Years' War demonstrates. A belligerent power might start with limited war aims: Gustavus Adolphus of Sweden landed in North Germany in 1630 with an army, some Dutch and French subsidies, and a few inchoate plans to push Imperial troops back from the Baltic coast and to free German princes from the emperor's assertive imposition of the 1629 Edict of Restitution. As Gustavus' military success burgeoned so did his war aims, which by 1632 had extended to a comprehensive reshaping of the Imperial constitution and a permanent Swedish territorial presence in the Empire. After Gustavus' death at the battle of Lützen in November 1632, the Swedish council acting for his young daughter, Queen Christina, continued to wage a war that imposed modest demands on the Swedish treasury, and could be sustained as long as required through an outsourced military system. Under the overall control of Sweden, this system drew heavily on the services of German contractor-officers, sustaining their troops by occupying and taxing German territory. Fortified by this military outsourcing Swedish war aims from the late 1630s sought the annexation of a great swathe of territory in North Germany, basing this demand not on dynastic claims, but merely as 'satisfaction' for the war-effort that they had waged and could continue to wage for as long as necessary to obtain their objectives.[36] Similarly Spain's response to what were judged unacceptably harsh French peace terms was the threat to continue warfare indefinitely. This threat was implemented in early 1648 when Spain simply withdrew from negotiations with France, and the Franco-Spanish war continued for another eleven years until 1659.[37]

This more extensive notion of waging and winning wars was possible because rulers were less concerned about the impact on their government and their fiscal resources of allowing warfare to run for years or even decades. But the price of this was paid in war aims that needed to recognise the rulers' obligations to their military contractors. In 1635 the colonel-proprietors of the (mostly German) regiments in Swedish royal service seized the Swedish chancellor, Axel Oxenstierna, and forced him to agree that in any peace negotiations undertaken by Sweden the financial interests of the colonel proprietors would be fully taken into account.[38] Down to the final settlement in 1648 the notion of a Swedish indemnity to be imposed on the German states to pay off the officers of the Swedish army was a non-negotiable peace demand. Though Sweden achieved some of her territorial ambitions in North Germany, these had to be negotiated in conjunction – and potentially in competition – with the priority of obtaining this huge financial indemnity to pay off its military contractors. As one Swedish minister sardonically suggested, the only way to avoid meeting the demands of their own officers was to lose the war. Long wars stoked up the aims and ambitions of the belligerent powers; but their officers were effectively shareholders in the military policies of governments and expected their wishes to be taken into account in waging war and determining its objectives.

The end of the Thirty Years' War allowed rulers and their governments to step back from this overly close relationship with military outsourcing. It had allowed them to wage warfare for decades but at a heavy cost to control of resources and policy: the state's 'monopoly of violence' had never seemed more flimsy. But rulers had learnt an important lesson about financing war: the wartime 'contributions' which had been imposed by their contractors showed just how much greater was the taxable capacity of their states and populations. In addition, the Thirty Years' War had generated a searing collective image of death and devastation attributable to war waged by unconstrained military contractors. The two perceptions could come together in a different form of bargain between rulers and subjects. Territorial rulers after 1648 would claim that the only way to provide military security and to prevent a recurrence of the horrors of the Thirty Years' War was to raise substantial forces recruited by and under the direct control of the state. Paying for this military protection would require much higher regular levels of direct taxation, now justified as the lesser evil compared to the 'contributions' and other arbitrary exactions by military contractors.[39] But these higher, permanent fiscal burdens, even where they were accepted by subjects, were not sufficient alone to support the costs of warfare. Most rulers shifted from reliance on explicit military contracting to subtler mechanisms for drawing their elites into supporting military activity without fully outsourcing military control. This could involve open sale of military office, seen for example in France, or regimental/company proprietorship, where initial costs borne by the officer came with the potential to make a financial return from close management of the regiment or company. Either way, administrative, judicial and, to some extent, operational control by the state was strengthened; the role of officers as 'shareholders' in the waging of war and its objectives was reduced; central war offices and their provincial and field administrators played a correspondingly larger role in recruitment, management and support of armies.

In combination, these developments permitted the extraordinary military expansion of European powers in the later 17th century. Powered by an outstandingly successful formula for exploiting domestic resources and manpower, Louis XIV's France led the way in massively expanding its armies, forcing successive coalitions of enemies to match the scale of French mobilisation.[40] Thus states by the late 17th century were raising armies under their direct control that were three or four times larger than those operating in the Thirty Years' War. They could sustain at least part of these forces as unprecedented, permanent armies, and could wage wars of a duration that would have destroyed the fiscal capacities of their 16th-century predecessors: the last two wars of Louis XIV's reign ran, with only four years' intermission, from 1688 to 1713, while the Great Northern War was waged from 1700 to the Peace of Nystad in 1721.

Yet this military expansion led to no radical change in notions of what constituted winning in war. The state may have resumed direct control of the waging of war, but in most cases it was a dynastic state, whose military policies were even more strongly shaped by the wishes and priorities of the ruler and his family. Thus a transformation in the scale and organisation of warfare underpinned war aims that prioritised dynastic interests and the pursuit of dynastic goals as strongly as any 16th-century ruler would have done. The state's war aims remained personalised, and goals such as economic expansion or geostrategic security remained of secondary importance. Louis XIV agreed to war against the Dutch in 1672 in part-response to the argument of his finance minister, Colbert, that defeating the Republic would undermine its mercantile supremacy in world trade, and give France a significant economic advantage. But the war quickly took on a more conventional character as the king redirected his armies to assert his dynastic claims over the Spanish Netherlands and other territories across the French frontiers.[41] The main effect of recruiting and maintaining larger armies and navies was simply to ensure that the same dynastic goals were pursued more tenaciously: prising away some of Louis XIV's territorial acquisitions from the 1680s and forcing him to recognise the right of William of Orange and Mary Stuart to the British throne took nine years of war and mobilised well over half a million combatants in total.[42]

War might be fought under more direct state control and in greater accordance with the operational and strategic priorities of rulers. But lengthy and resource-consuming campaigning remained the rule. Few rulers believed that their war aims could be achieved in any rapid and decisive campaign; winning in war was largely envisaged as a slow, attritional process. Perhaps the best-known battles of this period, the imposing victories won by the allied generals Marlborough and Prince Eugene of Savoy in the 1700s at Blenheim, Ramillies, and Oudenarde, averted allied defeat, but failed to deliver the decisive blow that could end the War of the Spanish Succession with

France.[43] Winning in wars fought for essentially dynastic motives did at least offer the time-honoured option that the belligerents' objectives could be scaled down from extensive territorial ambitions to exercises in preserving family prestige and status.

Substantial change over 200 years in the scale, mechanisms and effectiveness of resource mobilisation was paradoxically accompanied by considerable continuity in the perceptions of what war was fought for, and what constituted winning. While the ubiquity of war in its relationship with diplomacy was never seriously questioned, there remained a wide-ranging recognition that it was a blunt instrument for the outright achievement of political goals. On those occasions when conflicts were fought for ideological – religious – objectives, even the most fanatical proponents of unconditional war against religious foes were forced to confront the impossibility of achieving the elimination of their opponents through military means. Changing patterns in the way that armies were raised and supported might have an impact on the length and scale of wars, but 'winning' remained largely defined by the achievement of traditional dynastic goals which were flexible and could be realistically adapted to the difficulties of achieving decisive outcomes on the battlefield or on campaign. The 'total' wars that would be fought without compromise between the mobilised citizens of nation-states remained a grim prospect for a later epoch.

## *Notes*

1 P. Wilson, *Europe's Tragedy: A History of the Thirty Years War* (London: Harvard University Press, 2009), 781–812.
2 Erasmus, *The Complaint of Peace (1521)* (Chicago and London: The Open Court Publishing co., 1917).
3 J. Israel, *The Dutch Republic and the Hispanic World, 1606–61* (Oxford: Clarendon Press, 1982), 28–85; J. Elliott, *The Count-Duke of Olivares: The Statesman in an Age of Decline* (New Haven and London: Yale University Press, 1999), 51–65.
4 R. Bireley, *The Counter-Reformation Prince: anti-Machiavellianism or Catholic Statecraft in early modern Europe* (Chapel Hill: University of North Carolina Press, 1990); W. Church, *Richelieu and Reason of State* (Princeton: Princeton University Press, 1972).
5 H. Rowen, *The King's State: Proprietary Dynasticism in Early Modern France* (New Brunswick, NJ: Rutgers University Press, 1980); P. Monod, *The Power of Kings: monarchy and religion in Europe, 1589–1715* (New Haven: Yale University Press, 1999); P. S. Fichtner, *Protestantism and Primogeniture in early modern Germany* (New Haven and London: Yale University Press, 1989).
6 J. Elliott, 'A Europe of Composite Monarchies', *Past and Present*, Vol. 137, No. 1 (1992): 48–71.
7 G. Parker, *Emperor. A New Life of Charles V* (New Haven and London: Yale University Press, 2019).
8 S. Gunn, *The English People at War in the Age of Henry VIII* (Oxford: OUP, 2018), 1–16; D. Potter, *Henry VIII and Francis I: The Final Conflict, 1540–47* (Leiden: Brill, 2011).
9 M. Mallett, C. Shaw, *The Italian Wars, 1494–1559: war, state and society in early modern Europe* (Harlow: Pearson, 2012); R. Babel, *Zwischen Habsburg und Bourbon: Außenpolitik und europäische Stellung Herzog Karls IV. von Lothringen und Bar… (1624–34)* (Sigmaringen: Jan Thorbecke Verlag, 1989).
10 C. S. L. Davies, 'England and the French War, 1557–9', in J. Loach, R. Tittler (eds), *The Mid-Tudor Polity, c. 1540–1560* (London: Macmillan, 1980), 159–85.

11 C. Korr, *Cromwell and the New Model Foreign Policy: England's Policy towards France, 1649–58* (Berkeley and London: University of California Press, 1975).

12 Mallett and Shaw, *Italian Wars.*

13 J. Inglis-Jones, 'The Grand Condé in Exile: Power Politics in France, Spain and the Netherlands', Oxford University DPhil Thesis, 1994; K. Béguin, *Les princes de Condé. Rebelles, courtisans et mécènes dans la France du grand siècle* (Paris: Champ Vallon, 1999).

14 J. Morrill, *The Nature of the English Revolution: Studies in Religion and Politics, 1603–42* (London: Routledge, 1993).

15 D. Parrott, *1652: The Cardinal, the Prince and the Crisis of the Fronde* (Oxford: OUP, 2020).

16 M. Holt, *The French Wars of Religion, 1562–1629* (2nd ed., Cambridge: CUP, 2005); M. Greengrass 'The Psychology of Religious Violence', *French History,* Vol. 5 (1991): 467–74; G. Murdock, P. Roberts, A. Spicer (eds), *Ritual and Violence: Natalie Zemon Davis and Early Modern France* (Oxford: OUP, 2012).

17 G. Parker, *The Dutch Revolt* (2nd ed; London: Penguin, 2002); M. van Gelderen (ed), *The Dutch Revolt* (Cambridge: CUP, 1993).

18 C. Durston and J. Maltby (eds), *Religion in Revolutionary England* (Manchester: Manchester University Press, 2006).

19 N. Capponi, *Victory of the West: the story of the battle of Lepanto* (London: Macmillan, 2006).

20 N. Malcolm, *Useful Enemies. Islam and the Ottoman Empire in Western Political Thought 1450–1750* (Oxford: OUP, 2019).

21 J. P. Niederkorn, *Die europäische Mächte und der 'Lange Türkenkrieg' Kaiser Rudolfs II (1593–1606)* (Vienna: Verlag der Österreichischen Akademie der Wissenschaften, 1993).

22 P. Williams, *Empire and Holy War in the Mediterranean: the galley and maritime conflict between the Habsburg and Ottoman empires* (London: Bloomsbury, 2014).

23 G. Rowlands, *The Financial Decline of a Great Power: war, influence and money in Louis XIV's France* (Oxford: OUP, 2012).

24 J. Israel, *Dutch Primacy in World Trade, 1584–1740* (Oxford: Clarendon Press, 1989); J. de Vries and A. van der Woude, *The First Modern Economy: success, failure, and perseverance of the Dutch economy, 1500–1815* (Cambridge: CUP, 1997).

25 J. R. Jones, *The Anglo-Dutch Wars of the Seventeenth Century* (London: Longman, 1996).

26 A. Tenenti, *Piracy and the Decline of Venice, 1580–1615* (Berkeley, CA: University of California Press, 1967).

27 G. Pálffy, *The Kingdom of Hungary and the Habsburg Monarchy in the Sixteenth Century* (New York: Colombia University Press, 2010).

28 Parker, *Charles V.*

29 C. Tilly, *Coercion, Capital and European States AD 990–1990* (Oxford: Blackwells, 1990).

30 For scepticism, see for example: S. Pepper, 'Castles and cannon in the Naples campaign of 1494–95', in D. Abulafia (ed), *The French Descent into Renaissance Italy, 1494–95* (Aldershot: Routledge, 1995).

31 R. Baumann, *Landsknechte. Ihre Geschichte und Kultur vom späten Mittelalter bis zum Dreißigjährigen Krieg* (Munich: C. H. Beck, 1994); D. Potter, *Renaissance France at War: Armies, Culture and Society, c. 1480–1560* (Woodbridge: Boydell and Brewer, 2008); J. B. Wood, *The King's Army: Warfare, soldiers and society during the Wars of Religion in France, 1562–1576* (Cambridge: CUP, 1996).

32 H. Hauser, 'The European Financial Crisis of 1559', *Journal of Economic and Business History,* Vol. 2 (1929–30): 241–55. M. Rodriguez-Salgado, *The Changing Face of Empire. Charles V, Philip II and Habsburg Authority, 1551-59* (Cambridge: CUP, 1988).

33 F. Redlich, *The German Military Enterpriser and his Work Force, 14th to 18th Centuries* (2 vols) (Wiesbaden: F. Steiner, 1964); J. Fynn-Paul (ed), *War, Entrepreneurs, and the State in Europe and the Mediterranean, 1300–1800* (Leiden: Brill, 2014).

34 F. Redlich, 'Contributions in the Thirty Years War', *Economic History Review*, 2nd ser., Vol. 12 (1959–60): 247–54.

35 D. Parrott, *The Business of War: Military Enterprise the Military Revolution in Early Modern Europe* (Cambridge: CUP, 2012).

36 Wilson, *Europe's Tragedy*; M. Roberts, *Gustavus Adolphus: A History of Sweden, 1611–32* (2 vols) (London: Longmans, Green, 1958); K-L. Böhme, 'Geld für die schwedischen Armeen nach 1640', *Scandia*, Vol. 33 (1967): 54–95.

37 A. Malcolm, *Royal Favouritism and the Governing Elite of the Spanish Monarchy, 1640–1665* (Oxford: OUP, 2017); P. Sonnino, *Mazarin's Quest. The Congress of Wesphalia and the Coming of the Fronde* (Cambridge, MA: Harvard UP, 2008).

38 The 'Powder Barrel Convention' of 21 August 1635: Wilson, *Europe's Tragedy*, 576.

39 P. Wilson, 'The Power to Defend or the Defense of Power? The Conflict between Duke and Estates over Defence Provision, Württemberg, 1677–1793', *Parliaments Estates and Representation*, Vol. 12 (1992): 25–45; P. Wilson, *German Armies, War and German Politics, 1648–1806* (London: UCL Press, 1998).

40 G. Rowlands, *The Dynastic State and the Army under Louis XIV: Royal Service and Private Interest, 1661–1701* (Cambridge: CUP, 2002); H. Drévillon, *L'impôt du sang. Le métier des armées sous Louis XIV* (Paris: Tallandier, 2005).

41 P. Sonnino, *Louis XIV and the Origins of the Dutch War* (Cambridge: CUP, 1988).

42 J. A. Lynn, *The Wars of Louis XIV, 1667–1714* (London: Longman, 1999).

43 J. Ostwald, 'The "Decisive" Battle of Ramillies, 1706: Prerequisites for Decisiveness in Early Modern Warfare', *Journal of Military History*, Vol. 64, No. 3 (2000): 649–78.

CHAPTER 4

# From the Age of Reason to the European Nation State, 1750–1850

Jonathon Riley

This chapter examines warfare during the 18th and early 19th centuries, the age of horse and musket, before the full effects of the industrial revolution had been felt. Technology advanced a little, but tactics and the operational art evolved so considerably that the effects of winning and losing on the battlefield changed. Strategically it was an age which began and ended in Europe with centralised, authoritarian, monarchical systems of government, but one in which revolution, liberalism and the ideas of the Enlightenment came to prominence. It is therefore no surprise that the effects of winning and losing encompassed not only territory and resources, but also ideas. In this chapter we will examine what winning and losing meant in the Seven Years' War and, in brief, its immediate predecessors: the Wars of the French Revolution and the Napoleonic Wars.

The Seven Years' War or, as it was known in North America, the French and Indian War was something different from what had preceded it in the history of warfare. The Peace of Westphalia in 1648, which ended the Eighty Years' War between Spain and the Dutch Republic, and the Austrian–Swedish phase of the war in Germany, had many consequences – religious, territorial and dynastic. Among them were the strengthening of France and the emergence of the power of individual German states and their rulers, freed from the central authority of the Habsburg Empire. In doing this, it sowed the seeds of rivalry between the Houses of Bourbon and Habsburg.

The wars that followed and which were in some way directly related to that rivalry – the Nine Years' War in the Low Countries (1688–97),[1] also known as King William's War in North America; the War of the Spanish Succession (1701–14), or Queen Anne's War;[2] and the War of the Austrian Succession (1740–48),[3] or King George's War – all ended in compromise. Territories changed hands, rival claimants to various thrones (including the Stuarts) came and went; the power of France, Austria, Sweden, Spain and Britain waxed and waned. As in 1648, Europe was devastated by the effects: industry and agriculture in ruins, towns and cities burned and looted. But in the end, the Westphalian system continued and the old order did not change.

The character of continental European war from the late 16th century was dominated by major innovation in the science of fortification. The Italians were the first to experience the new technology of gunpowder weapons and it was in Italy, after about 1530, that the development of fortification began to take account of it, both in mounting and resisting cannon. The defensive works that resulted were built on a thick but low profile of rammed earth and rubble faced with stone and the trace, or ground plan, was designed on mathematical lines, hence the name these works were given: the *trace italienne*. Looked at from above, the ground plan resembled a star which allowed interlocking and mutually supporting fields of fire for guns that were fully casemated and thus protected from counter-battery fire; there was no cover for an attacker and the besieger's artillery could be kept at arm's length until extensive systems of entrenchments were constructed to bring fire within a range sufficient to breach the defences.

It was the French engineer Sébastien le Prestre de Vauban who combined all these developments into a single code of practice, building or remodelling some 60 fortresses during his career.[4] Vauban was Louis XIV's chief engineer; as well as developing fortification and siege techniques, he also founded the corps of engineers in the French army. Warfare thus became as much science as art and as it did so, the thinking of men like Vauban became more scientific. In the late 17th century, we see an increasing emphasis on observation, reconnaissance, intelligence gathering, analysis and evaluation – on rationality and calculation.[5] Vauban also developed a comprehensive scheme for the defeat of fortresses by siege warfare, for only formal sieges could break such fortifications and they needed heavy artillery to create a breach, supported by heavy mortars to lob shells inside fortifications to kill defenders, cause fires, blow up ammunition and reduce the will to fight. Meanwhile, the attackers would dig lines of circumvallation, saps and parallel lines of trenches until they were within storming distance.

Wherever war seemed likely, defences would be built – from Lombardy to the Low Countries – and the huge increase in fortification stultified warfare. Even a relatively small fortified town would be strong enough to resist a siege, even with heavy artillery, for weeks or months provided it was well stocked with food and munitions. The heavy armoured cavalry, once the unrivalled king of the battlefield, was eclipsed by the infantry, the gunners and the sappers who were needed to form garrisons or invest forts. In 1639, for example, no less than 208 fortified areas required garrisons in the Spanish Netherlands involving over 33,000 men, from Dunkirk with its 1,000-man garrison, to a mere ten men at La Grande Misère near Ghent.[6] As an example of the national investment required, each of Amsterdam's 22 bastions cost half a million *florins*, while Vauban took six years and five million *livres* to fortify Ath. On the other hand, the siege of La Rochelle cost 40 million *livres*.[7] This was the nub of the concept: although the initial investment was huge, fortification was cheaper than paying a large standing army. A defended, fortified town with a garrison of 1,000 men needed at least ten times that number to reduce it. Thus a well-designed system of fortifications could delay and obstruct an attacker at little cost. They might not be able to *win* wars, but they could prolong them and increase the attrition on powerful opponents. For a field commander, it became very difficult to reach a decision by battle. Moreover, for a government, attrition caused war weariness and increased the willingness to reach a compromise peace.[8]

The result was that warfare in Europe became positional, linear and highly formal. The emphasis was on the formal reduction of fortresses and battles were rarely truly decisive, fought for a tactical position rather than in pursuit of an operational or strategic objective. Examples of this can be found among the Duke of Marlborough's victories in the early 18th century: Ramillies, Malplaquet, the breaching of the *Ne Plus Ultra* lines. However, within the context of the political order and its rivalries, military strategy remained, as Professor John Childs points out, a fundamental component in achieving the objectives of national policy and in reaching a decision to use war to do so. Soldiers like Turenne, Condé and Vauban informed the political and decision-making of Louis XIV of France alongside his civilian ministers. James II and William III of England, like Lord Protector Oliver Cromwell before them, combined the functions of head of state and supreme warlord for exactly this reason.[9]

The emphasis on fortress warfare was still very much present during the Seven Years' War (1756–63) – some 15 major sieges were conducted during the conflict, from Torgau to Pondicherry, but the emphasis in European warfare was slowly changing in favour of manoeuvre as can be seen from battles like Detingen, Warburg, Hochkirch and especially Rossbach, among many others. However, 18th-century generals continued, as before, to distinguish rigidly between manoeuvre and giving battle: the operational art – or strategy as it was then called – frequently demanded an evasion of general engagements. The notable exception to this was Frederick the Great, from whom Napoleon later drew so much inspiration. To Frederick, battle was 'the bloody decision'.[10]

Apart from this move away from positional warfare – and indeed the crippling costs for governments – there were other factors that distinguished the Seven Years' War from what had gone before. It was different first in terms of scale. On one side was a coalition led by Great Britain which included Portugal, Prussia, and some of the German states; and on the other an alliance consisting of France, the Austrian Empire, Spain, Sweden and, until 1762, Russia. Secondly it was different because of the span of conflict. Previous wars had not been confined to Europe by any means, but the Seven Years' War encompassed Europe, West Africa, the West Indies, North America, India and the Philippine Islands. In the outlying theatres of war, the coalitions expanded to include Native rulers and tribes and it is thus, rightly, dubbed the first world war. Thirdly, it was different because, unlike these preceding conflicts, one belligerent was decisively defeated: there were clear winners on one side and clear losers on the other.[11]

What therefore did winning and losing look like on the battlefield of the mid-18th century? The climax of battlefield technology of the time was the gunpowder weapon – musket and cannon. This had evolved slowly from the mid-17th century, with the flintlock replacing the matchlock, some refinements in gunpowder, higher strengths and better techniques in barrel casting, iron ramrods instead of wood. But fundamentally the weapons were the same: muzzle-loading with a slow rate of fire, low rate of accuracy, limited range and high obscuration from smoke. Winning usually left the victor in possession of the field and of at least part of the enemy's army and its supplies – but rarely, if ever, was it decisive. Battles are decisive in two ways: first, they tip a conflict inevitably towards the victory of one party (such as Gettysburg); or they bring a long period of warfare to a definite conclusion (such as Waterloo).[12] The only battle of the Seven Years' War which might fall into the first category was Valencia de Alcantara in 1762. War was decided therefore not by a single great battle, but by the attrition of the military and financial resources of the parties engaged.

Battle was more likely to be decisive outside Europe, where fortifications were less well developed. It is possible to show this from examples elsewhere in this first world war – such as in India and North America. Robert Clive, leading the army of the British East India Company against the combined armies of Rai Durlabh and Murshidabad owing allegiance to Siraj-us-Daula, along with his French allies, facing what appeared to be overwhelming odds, emerged victorious. The magnitude of the victory was not realised at the time because Clive's set-piece action achieved success with few casualties and an elegant simplicity that gave the appearance of ease. It resulted in the consolidation of the East India Company's dominant position in Bengal and its expansion throughout India over the next century – not to mention the eclipse of French ambitions.[13] On the other side of the world in Quebec, the battle of the Plains of Abraham in 1759 was the culmination of a three-month siege by the British, commanded by General James Wolfe. This was the final attempt of four since 1685 by the British and New England forces, to remove French presence from their borders.[14] The British scaled the heights of Abraham and then successfully resisted a sortie from the fortress by a column of French troops and Canadian militia under General Louis-Joseph, Marquis de Montcalm. Both generals were mortally wounded during the battle. In the wake of the battle, the French evacuated the city and removed their remaining military force in Canada, resulting in the rest of north-eastern North America coming under British rule.[15]

So much for tactical victories. But what did victory – and defeat – look like for nations and empires at the end of this first global conflict? In Europe, Frederick I, King of Prussia, whom history knows as 'the Great', regained much of the prestige that Prussia had lost in 1748 along with all her former territories at that date. Prussia emerged as a new European great power, one that could never again be ignored. Austria failed in its original goal of recovering the territory of Silesia from Prussia, but the power and position of Austria, which had been part of the Hapsburg superpower since the 16th century, cannot be said to have been diminished. Portugal, Spain and Sweden did not benefit greatly in return for their involvement and all continued to decline from their former status as great powers. This was particularly marked in the case of Spain, now separated from her former territories in Italy and Austria.[16]

The most marked evidence of victory and defeat however lay outside Europe. The war was successful for Great Britain, which destroyed French power in North America and consolidated its position on the continent. Britain also took, but did not in every case keep, part of the Spanish territories in the Floridas and the Caribbean, along with Senegal in West Africa which was the western terminus of the slave trade. Spain on the other hand gained Louisiana from the French and held on to its valuable colonies in Cuba and the Philippine Islands, which had been captured by the British during the war but were returned at the Peace of Paris in 1763. The war also saddled France with a heavy burden of debt which its financial system – much less well-developed than that of the Netherlands and of Great Britain – could not cope with. In India, the victory of Plassey opened the way for British domination of the sub-continent and its resources, marking the final eclipse of the Dutch by the British as the pre-eminent trading nation of the world after a century of rivalry.[17]

In North America, the Native peoples were among the losers, as they continued to decline in numbers and influence in north-eastern North America, a process that had begun almost as soon as Europeans began to arrive on the continent – the product of new diseases, loss of traditional hunting and gathering habitats and warfare. They remained a factor in warfare, however, until the end of the Napoleonic period.[18] As for the British colonies – now 21 in number[19] and of a territorial size that was extremely difficult to manage – the threat from New France, and, to a lesser degree, Spanish and Swedish settlements further south, had been removed. The colonists saw little need for British military presence and resented being made to pay for it, while not being represented in parliament at home. This sense of grievance reignited an older one. When the colonies were first established, they were free-standing, self-governing colonies each with a charter which made them largely independent. They had co-operated in the various wars, but on their own terms. After the Restoration in 1660, these original charters were steadily replaced by new ones, making them subject to the British Crown and effectively removing that independence – the very thing which the original colonists had sought in leaving the Old World. Outside threats had submerged the sense of grievance but now it was back and it can be of little surprise therefore that the American Revolutionary War broke out only 13 years after the end of the Seven Years' War. The surprise is that it took so long: with good cause, the American historian Stephen Saunders Webb has called 1676, not 1776, 'the end of American independence.'[20] Connected to this movement was the desire for revenge on the part of France in particular but to a lesser extent Spain and the Netherlands, resulting from the Seven Years' War and hence their involvement with the American cause after 1776. Without the Seven Years' War, therefore, the American Revolutionary War would have been long delayed and the French Revolution, inspired in part – a supreme irony – by American notions of liberty, might not have occurred when it did, and in the way it did. Nor would British power and influence have been turned away from North America and towards India.

Winning from the mid-17th century to the mid-18th century was therefore about territorial and financial gain. The gain of territories in Europe and with them greater security, a larger population and a more extensive agricultural and industrial base. Outside Europe, the result of winning was empire. Empire meant cheap raw materials such as cotton, timber, bulk foodstuffs; high-value commodities such as gold, silks, sugar tobacco and spices; and slaves. The foundation of the British, Portuguese and Dutch outposts in particular, in West Africa, gave access to the huge resources of manpower provided by Arab and Black African rulers in return for trade goods. This labour in turn led growth in the valuable colonies of the West Indies and southern North America where Black slaves quickly replaced White indentured labour. Increased revenues and the growth of processing industries at home in turn produced greater wealth and through taxation, the maintenance of larger military forces to defend and, if necessary, expand the empire further.[21] Losing meant the converse of all that, except in the case of indigenous people. Although colonialism sat lightly at first in India, for the Native peoples of North America losing meant the loss of territory and with it, the livelihood from hunting and gathering; exposure to diseases for which they had no immunity; and battle against a huge technological overmatch.

If the Seven Years' War was different in outcome, if not so much in character, from what had gone before, the French Revolutionary War was different again. The revolution broke out in July 1789 and rapidly led to the overthrow and destruction of the *ancien régime*. King Louis XVI's failed escape attempt and after it, the declaration of war by Prussia and Austria on his behalf, led to the execution of the king on 21 January 1793: 'The kings of Europe advance against us,' thundered Georges Danton, first president of the Committee of Public Safety, in the National Constituent Assembly, 'we throw at their feet, as a gauge of battle, the head of a king.' The war that followed, from February 1793 until the Peace of Amiens in March 1802, was as much a war of ideology as of territory. War as the export of revolutionary ideology as well as for expansion was inevitable for any French regime. The reckless optimism of 1792 resulting from the ideas of liberty, equality and fraternity was bound to cause upheavals and struggles against the old order throughout Europe, and the agency of this upheaval was the zeal and energy of the French people. Clausewitz remarked on this by saying that:

> Clearly the tremendous effects of the French Revolution abroad were caused not so much by new military methods and concepts as by radical changes in policies and administration, by the character of government, altered conditions of the French people, and the like.[22]

The Committee of Public Safety and the Directory sought first to expand and extend the ideas of revolution throughout Europe, both in order to undermine the authority of their enemies but also from purely ideological motives; then to build up barriers of territory beyond the so-called natural frontiers of France, in order to ensure the security of the homeland and the revolution. A strategic appreciation prepared by the young Napoleon Bonaparte for the Committee of Public Safety described Revolutionary France's military situation. France was at war with England, Austria, Piedmont and Spain – and thus effectively surrounded by land and sea. French armies were drawn up along the frontiers in a posture of strategic defence. Maximilien Robespierre was anxious to go onto the offensive, however, and his report to the Committee of Public Safety on 25 December 1793 was explicit about exporting the principles of revolutionary government, which he described as a necessary and provisional form of war against the enemies of liberty.[23]

Napoleon's appreciation recommended remaining on the defensive in the Pyrenees and in the Low Countries; combining the armies of Italy and the Alps to subjugate Piedmont; then advancing through Lombardy and into the Austrian Tyrol, in cooperation with the Army of the Rhine. This was a strategy embraced, after an interval, and which led to the expulsion of Austrian power from Italy. It provided Napoleon with his first senior field command in 1796 at the age of 26. Revolutionary war was a young man's game: in Italy the average age of Napoleon and his subordinate generals, most of them from lower- or middle class backgrounds, was 37; that of his professional and often aristocratic Austrian opponents was 61.[24] Only the conditions of revolution could have allowed the likes of Napoleon to rise so rapidly from obscurity; and only the conditions of revolution could have prevented his execution after the disastrous campaign in Egypt and Syria.

Alan Forrest remarks that as early as 1797, the whole French economy was geared to war, and domestic ambitions were predicated on military success.[25] The means of military success were the French Army and Navy: but both were severely disorganised by the effects of revolution and the terror that had followed it under the Committee of Public Safety. This meant in particular the loss of professional officers from the army. In the pre-Revolutionary French Army there had been around 13,000 officers but relatively few of these remained, and very few indeed in more senior ranks: figures are hard to come by but in the last four months of 1791, for example, more than 2,000 former royal officers fled France, never mind those imprisoned and executed;[26] between 1793 and 1799 more than 500 general officers were cashiered as being unreliable or too old. Only one of Napoleon's marshals, Christophe Kellermann, had been a general in the pre-Revolutionary Army.

What of the army itself? Faced with the combined power of every European nation, reforms and expansion had to be introduced and improvised rapidly. This task fell to Lazare Carnot, known to French history as 'the architect of victory'. As a military engineer, Carnot initially favoured fortress warfare and the defence. However, with the threat of invasion from several directions and the need to export the ideas of revolution, Carnot switched to the doctrine of the offence. From him came the organisation that turned the tide of war from 1793 to 1797. Carnot's method was to develop a huge force divided into several armies able to move more rapidly than the enemy and attack from the flanks rather than head on. These tactics of rapidity and the indirect approach proved highly successful against the more traditional tactics and organisations of France's opponents.[27] They were different from the methods used by these opponents, but also very different from the earliest essays of the Revolutionary Army which were based on frontal attacks by massed bodies of semi-trained troops aimed at overwhelming an enemy force before its firepower could be brought to bear.[28] They were also a complete contrast to the earlier models of fortress warfare. Finally, building on previous experimental organisations employed in particular by Marshal Maurice de Saxe, the French army instituted an all-arms divisional organisation in 1796 which laid the foundations of the later Napoleonic model.

To raise this sort of force, Carnot turned to conscription as the means of generating the strategic mass that would carry out the work of dominating Europe.[29] Conscription began as a one-off act of necessity in the *levée en masse* of August 1793 under the direction of Carnot in the face of imminent danger to the young republic.[30] From its inception as an annual process in 1799 until April 1815 there were 32 levies on native-born Frenchmen. The law of 10 Fructidor year VI (5 September 1798) replaced the levies of the Revolution by a regular method of conscription which, with a few modifications, remained in force until 1815. Of the 5,692,164 men belonging to the 18 classes affected by this law, 2,716,567 were called up and 2,022,201 were actually incorporated in the army. Troops levied in the 12 years 1800–11, of whom slightly more than 75 per cent came from areas which were part of France in 1792, accounted for no more than 50 per cent of those mobilised between 1798 and 1815. The remainder came from conquered territories, client states and later, the Empire.[31]

No other policy intruded the Revolutionary, and later Napoleonic, state so forcefully into the lives of people or the fabric of nations, for it was imposed throughout France, the empire and the client states during the entire period of French rule as Michael Broers points out.[32] This use of foreign soldiers was not an innovation, since the French had employed up to 50,000 foreign troops since 1654 during the wars of Louis XIV and XV.[33] They were needed: for even though France was the most populous nation in Europe with a population of around 30 million, maintaining armies of this size was a severe strain on that part of the male population between 16 and 50; and a massive competition with the demands of industry and agriculture. And certainly, too, no other policy engendered so much hatred. It was without doubt a central issue of the French strategic system, and as such, a continuous problem for mayors and heads of departments who were obliged to enforce it, not to mention a devastating imposition for the hundreds of thousands – if not millions – of families touched by it. Not surprisingly, the scale of avoidance – either legally through buying substitutes or illegally through desertion – was huge, especially in the annexed territories, and the policy was only maintained through heavy-handed enforcement by the *Gendarmerie*, whose officers and men were carefully selected for their energy and commitment to the regime.[34]

It was the armies produced by this unprecedented influx of manpower that enabled France to fight offensive campaigns across Europe. Sieges dwindled, to be replaced by pitched battles – only four major sieges occurred during the war, those of Mantua, Genoa, Toulon and Acre. Two coalitions were formed, from European powers and Great Britain, to counter the spread of French ideas and territory and to match them in terms of manpower – but these coalitions met with no success on the continent of Europe. Success was gained at sea and through the conquest of French and other colonies elsewhere. In the end, a peace settlement was negotiated which resulted in the treaties of Lunéville and Amiens. Greeted with joy everywhere in Europe, these treaties legitimised the French Republic and its new client republics and allowed many French *emigrés* to return. Prisoners and hostages were exchanged. Britain returned the colonies of the Cape and Dutch Guiana to the Netherlands – now the Batavian Republic and a French satellite – but retained Ceylon. Britain also gained Trinidad from Spain and retained Gibraltar, although, surprisingly, it returned Minorca to Spain. Britain also withdrew from Egypt, balanced by a French withdrawal from the Papal States and the Kingdom of Naples in Italy. Elsewhere in the Mediterranean, the Maltese Islands were returned to the Knights Hospitaller and the Ionian Islands recognised as an independent republic.[35]

The end of hostilities in 1802 was, without doubt, a victory for the French Revolution and the French nation: winning made France far larger in terms of its metropolitan territory, its allies and clients – effectively vassals – and its remaining colonies in the West Indies, like Martinique and Guadeloupe. However, it also fixed the ideas of the revolution and of republican government, as already established in America. On the other hand there were unintended consequences, as at the end of the Seven Years' War and the American Revolutionary War: these emerged quickly as the new French government, under First Consul Napoleon Bonaparte and from 1804, the Emperor Napoleon, revealed itself as a less democratic and more autocratic system with its sights set on external conquests greater even than those of the revolution.

Once Napoleon had secured his power and authority as emperor of the French on 18 May 1804, his design for the domination of Europe by France became a realistic prospect. This policy dictated a strategy which was almost entirely based on military superiority rather than any other instruments of power, since maritime power projection and economic leverage were denied by the British. The chief instrument of Napoleonic strategy, therefore, was his *Grande Armée*. In 1804, building the all-arms divisional organisation of 1796, Napoleon created ten army corps, each of up to five divisions. These included the Imperial Guard, the reserve, and the cavalry. By 1812, there were 17 corps including those formed by, or contributed to by, his clients and allies.

In terms of technology, however, there was no great innovation during the Napoleonic period. The evolution of artillery especially the Gribeauval system of standardising artillery, its ammunition and its wagons; and the small arms of the infantry and cavalry all saw no real changes. With no innovation in technology, there was no impetus for any innovation in the basic tactical formations and movements of troops on the battlefield. Where Napoleon can be said to have innovated was in the introduction of army corps and their operational level manoeuvres, in the use of massed firepower to break through enemy formation,[36] and in the return of massed heavy cavalry as a weapon of shock action and pursuit.

It was the corps system that lay at the heart of Napoleon's design of the Imperial Army. Each corps was in effect a miniature army organised for a given task and the assessed abilities of its commander, normally a marshal. A corps would contain varying quantities of infantry, cavalry, artillery, engineers and bridging, with integral supply and medical units. The size of the corps could vary enormously, sometimes amounting to between one third and one half of the army, but whatever its size, it could operate independently for a period, allowing dispersal on the march and concentration on the battlefield.[37]

To achieve the concert of such an army, to direct and control it, to frame and transmit orders, to gather intelligence and provide supplies required an apparatus of command and control more advanced than anything previously attempted. Here was a classic demonstration of the correct relationship between doctrine and organisation: Napoleon had decided what kind of wars he was going to fight; his organisation would reflect that style of fighting. Napoleon was, as both emperor and commander in chief, responsible to no-one, called no councils of war, heeded no 'red cards' or caveats. He was able, on the basis of a general plan of operations, to direct campaigns as well as issue orders for specific engagements. His headquarters was, therefore, geared entirely to Napoleon personally – not to the commander, regardless of who that might be.

An early demonstration of the effectiveness of the corps system was made in the campaign of 1806 against Prussia. Napoleon had determined his strategic objective – the subjugation of Prussia; he then framed a campaign plan that would bring his huge army on several dispersed routes to a point where he could achieve the destruction of the Prussian Army in a decisive battle; then, as a result, bring about the surrender of his enemy.[38] During the war with Prussia in 1806, Napoleon pre-empted his enemies; took risks with a bold advance through rough country; and exercised effective command through the staff and corps system which allowed his army to advance prepared for action on any axis using what he termed the *battalion carrée*, or square of battle.[39] Berthezène quotes him as saying: 'I see only one thing, namely the enemy's main body. I try to crush it, confident that secondary matters will then settle themselves.'[40] In this he was clearly following the precepts of one of his heroes, Frederick the Great, in that: 'It is battles that decide the fate of a state.'[41] In his *Maxims,* he writes of the careful balancing of means and results,[42] a tailoring of all available military and political power to the requirements of his objective, with no distraction from the business of destroying the enemies' means to resist. Warfare for Napoleon had therefore to be a strategic offensive, in order to gain and maintain the initiative:

> To make war with success, the following principles should never be departed from: To be superior to your enemy in numbers, as well as in *morale*; to fight battles in order to spread terror in the country; to divide your army into as many corps as may be effected without risk, in order to undertake several objects at the same time; to treat WELL those who yield, to ILL treat those who resist; to secure your rear, and occupy and strengthen yourself at the outset in some post which shall serve as a central point for the support of your future movements; to guard against desertion; to make yourself master of the great rivers and principal passes, and to establish your line of communication by getting possession of the fortresses, by laying siege to them, and of the open country, by giving battle; for it is vain to expect that conquests are to be achieved without combats; although when a victory is won, they will be best maintained by uniting mildness with valor.[43]

Napoleon therefore fused the ideas of winning on the battlefield and winning campaigns and wars: destroy the enemy's armies and his territory would fall into Napoleon's hands as a direct consequence. Thus the concept of winning through the strategic, or decisive, battle was brought to its apogee: Marengo, Austerlitz, Jena, Friedland, Ulm, Borodino are all examples of those that succeeded; but there were also failures, like Aspern-Essling, Bautzen and Dresden. Although Napoleon wrote extensively of them in his *Maxims*, sieges all but disappeared. He himself conducted none, other than the investment of the Lines of Torres Vedras in Portugal; the British conducted only three in Spain (Badajoz, Ciudad Rodrigo and San Sebastian); and the Spanish only one, at Pamplona. Fortresses became largely supply bases and the means of guarding important defiles, or of subduing the population of an occupied territory.

By these means, the wars which began to safeguard the revolution and which went on to be the means of exporting it became the means of expanding and consolidating the Napoleonic Empire and the legitimacy of its ruler. Revolutionary zeal was, during this process, almost inevitably replaced by more recognisable forms of diplomacy and power projection. Competition for trade and colonies overseas emphasised this process. Winning in war was followed by treaties, or dynastic marriage, or the imposition of client rulers; but Napoleon was never able to convert former enemies into allies except in the case of small, wholly dependent states like Denmark and Saxony. Treaties of alliance produced subjects, like the Confederation of the Rhine or Holland, or restive satellites like Prussia. All were tolerated only as a source of men, money and materiel for further wars prompted by hatred of the only enemy never to submit to French power, Britain.

By 1811, Napoleon was aiming not just at a stable limit to his empire in Europe through peace with England following victory, but total domination of the world: his empire would have no limits. The hostility of the old monarchies, and especially of England, forced him to keep expanding until no opponents were left. In 1811 he remarked that 'in five years, I shall be master of the world: there only remains Russia, but I shall crush her.'[44] Napoleon and his system existed only through greater and greater success, as the means to a favourable and lasting settlement: that is, one that saw Napoleon and his empire in control of the international system. It was a path that led only in one direction – hence the formation of another four coalitions against Napoleon, financed by the economic power of Britain. Napoleon himself can be argued to have contributed to the momentum of opposition to his design by his three great strategic miscalculations: he may have won on the battlefield; he may have been a master of campaigning; but he is also proof that no matter how good a nation is at such things, if strategy is flawed, failure will surely follow. These miscalculations were first, the Continental System which almost cracked the economic power of Britain but which broke down because it was never a blockade of the British – this was impossible – but a blockade of Europe *by itself*. The consequences were not tolerable for most Europeans. Secondly, his occupation of Spain and the imposition of his brother as king, which ignited the 'Spanish Ulcer'. Third, the invasion of Russia in 1812 which created the conditions for his defeat in 1813.[45]

The matter of the effect of the Continental System on the British economy bears some further examination because of its influence on the business of winning the war against Napoleon. Britain, because of its world-wide empire, economic base, and strategic reach – through naval power – was arguably the only major power able to conduct strategy through means other than military power; however because of the blockade that was the Continental System, the British national debt became enormous: in 1812 it had reached £609 million and by 1814, £725 million – multiply by about 1,000 for today's rates – at a time when GDP was about £330 million per year. It did not approach the same level again until World War I. Only the export trade could service this kind of debt and thus maintain Britain's ability to sustain war. Following the collapse of the system brought about by Tsar Alexander, by the spring of 1813, British exports were at their highest level for years: total exports were valued at £118 million for the year 1812–13, while taxation yielded £68 million, or five times the pre-war figure. It was this vast financial resource that allowed Britain, with a population of only 18 million, to finance the armies of the 6th Coalition that brought Napoleon down. In 1813 alone, England was able to subsidise her allies in Europe to the tune of almost £7.5 million: £2.486 million to Portugal, £1.335 million to Sweden, £1 million to Russia, £700,000 to Austria, £666,000 to Prussia, £877,000 to Spain and £440,000 to Sicily.[46] It was the vast financial resources of the British Empire, as much as anything, which underpinned the allied will to continue the struggle against Napoleon and win. This, combined with its naval power, more than compensated for the relatively small size of Britain's army during the period.

Warfare in the Napoleonic period also spilled over into North America: the War of 1812, the result of a number of factors. These included a series of miscalculations at sea by the British regarding American shipping and trade, in consequence of the Continental System; miscalculation over how easy the conquest of British North America would be, by the Americans; and in part the expansion of the United States through Napoleon's sale of the re-acquired Louisiana territory, doubling the size of the United States and playing to the developing notion of its manifest destiny to control the entire continent. The sale also, incidentally, contributed to the outrage against Napoleon in Spain as it was a deliberate flouting of the terms of the Treaty of San Ildefonso. After two years of warfare that ended with the major public buildings of Washington DC in ruins, swathes of US territory in British hands, the US exchequer bankrupt and its fleet and merchant marine confined to port, the war was concluded at the Congress of Ghent. It might be expected that the British would take the fruits of success through winning the war; however the government was far more interested in the Congress of Vienna and, having won the war, it lost the peace. Matters returned to the *status quo ante bellum.* On the credit side, however, the war was the last episode of armed conflict between Britain and the United States and established the world's longest demilitarised frontier. The real losers were the Native peoples, who ceased to exercise power as a balancing force in European rivalries and, on the US side at least, were set on a path to disaster and impoverishment.[47]

It should be remembered, when considering the matter of winners and losers in the Napoleonic wars, that there were two big losers: Napoleon himself and France. It took many years and much bloodshed for his opponents to master his system of battle and campaign and then beat him: Vitoria effectively decided the war in Spain; the five-day battle at Leipzig in 1813, rightly called 'the battle of the nations', was the greatest land battle in Europe before the Great War and truly decisive in shaping the inevitable course of the war towards its end in 1814; Waterloo, although smaller, finally cemented the victory of the united powers of Europe against Napoleon and his system.

The French Revolutionary Wars had begun as the export of the ideas of liberty and liberalism. Not that Napoleon was in any way in favour of such notions, being an example of how revolutions tend to engender governments more repressive and authoritarian than those they supplant: Napoleon had no wish to go the same way as Louis XVI and his rule made sure that order was rigidly kept. The results of his defeat and of the victory of the 6th Coalition can be seen at the Congress of Vienna. France was deprived of all territory conquered by Napoleon. The French monarchy was restored under the rule of Louis XVIII. Austria was given back most of the territory it had lost and also received lands in Germany, as well as Lombardy and Venice in Italy. Russia gained Finland and control over Poland. Prussia gained much of the territory of Napoleon's ally Saxony and important parts of Westphalia and the Rhine Provinces. The British expanded their colonial territories and they maintained a clear dominance at seas. The Dutch Republic was united with the Austrian Netherlands to form a single kingdom of the Netherlands under the House of Orange. Norway and Sweden were joined under a single ruler, rewarding Sweden for her part in the 6th Coalition. Switzerland was declared neutral; Spain was restored to independence and the rule of Ferdinand VII.[48]

However, as important as gains of territory, as the reward of winning, was the restoration of aristocratic and monarchical government everywhere except in Britain. The Austrian chancellor, Metternich, and Tsar Alexander I of Russia were determined to excise all notions of democratic governance and in this they largely succeeded. There were revolutions in Europe in 1830 and 1848, in Russia in 1905, civil wars in Spain, and the struggle by peoples in the Balkans to free themselves from Ottoman rule; but in general, Metternich succeeded in establishing a European order through a system of alliances that maintained peace for forty years and a style of government that cannot be said to have ended until 1918.

In only one way did Napoleon win – but it was an important legacy in the history of warfare. When one looks at the curriculum for military history at the British Army Staff College in 1913 – and indeed its equivalent in the US, the Command and General Staff College – one is struck by the emphasis on the Napoleonic Wars. The dominance of Napoleon was marked, and in France, probably amounted more to worship than mere dominance. Every general clearly wanted to be him: to crush his enemy's army, march into his capital, and thus attain the goal of decisive victory.[49] This elusive ideal has persisted right down to the present.

For much of his career Napoleon had operated against poorly coordinated coalitions, inefficient armies, and elderly, second-rate generals. Faced with this sort of opposition, Napoleon did not have to be faultless, he just had to be better than the other side. Given this, and given the edge that superior French organisation and a unified command brought, it is not surprising that the legend grew to the size it did.

Because of this legend, the evolution of the nature of modern warfare over the next century and more became obscured. European armies after Napoleon were large organisations raised through conscription; and the full impact of the industrial revolution equipped them with weapons far closer to those of today's battlefield than of Leipzig or Waterloo. But there is, in warfare, a relationship between the introduction of new technologies, and the employment and deployment of troops. This relationship is not constant and needs careful and frequent revision. When it is not attended to, trouble follows. Thus, by the American Civil War, although the armies were equipped with powerful, rifled muskets and heavy artillery, and could be moved by rail, the tactics were still those of Waterloo. The results, for generals seeking the Napoleonic grail of the decisive battle, were the casualty rates of battles like Antietam, Fredericksburg, Gettysburg; and the spread of trench warfare – a return, almost, to the era of siege warfare with which this chapter began. One can argue that the same process continued through the Franco-Prussian war, the early stages of the South African War, and the first four months of World War I, which cost the French Army 800,000 casualties. Even then, the spirit of Napoleon still lived.

## *Notes*

1 See John Childs, *The British Army of William III* (Manchester: Manchester University Press, 1987), 240–258 and *The Nine Years' War and the British Army 1688–97* (Manchester: Manchester University Press, 1997), 5–97.
2 The most recent definitive work is Richard Holmes, *Marlborough: England's Fragile Genius* (London: Harper Press, 2008), 196–458; a good summary of British involvement in Spain is Nicholas Dorrell, *Marlborough's Other Army: The British Army and the Campaigns of the First Peninsular War, 1702–1712* (Solihull: Helion, 2015).
3 See the study by M.S. Anderson, *The War of the Austrian Succession 1740–1748* (Harlow: Longman, 1995), 5–20, 210–223.
4 Christopher Duffy, *Fire and Stone: The Science of Fortress Warfare 1660–1860* (New Jersey: Greenhill Books, 1975), 11.
5 Jurgen Brauer and Hubert Van Tuyll, *Castles, Battles & Bombs: How Economics Explains Military History* (London: University of Chicago Press, 2008), 129.
6 Geoffrey Parker, *The Army of Flanders and the Spanish Road 1576–1659* (Cambridge: CUP, 1972), 11–12.
7 John Childs, *Warfare in the Seventeenth Century* (London: Cassell, 2003), 113.
8 Jonathon Riley, *The Last Ironsides: The English Expedition to Portugal, 1661–1668* (Solihull: Helion, 2014).
9 Childs, *Nine Years' War*, 6–7.
10 David Fraser, *Frederick the Great* (London: Fromm International, 2000), 208–209, 211.
11 Fred Anderson, *Crucible of War: The Seven Years' War and the Fate of the Empire in British North America, 1754–1766* (New York: Vintage Books, 2000), xvi–xvii.
12 For a full discussion see Jonathon Riley, *Decisive Battles* (London: A & C Black, 2010), ch. 1.
13 Robert Harvey, *Clive: The Life and Death of a British Emperor* (London: St. Martin's Publishing Group, 1998), 210–212, 221–223, 384–385.
14 Wienand Drenth and Jonathon Riley, *The First Colonial Soldiers: A survey of British overseas territories and their garrisons, 1650–1714. Volume 2: the Americas and the Caribbean; part 1: New England and the Middle Colonies* (Eindhoven: Drenth Publishing, 2015), 73–74, 85–86, 96–98.
15 Anderson, 344–373.
16 Alice Clare Carter, *The Dutch Republic in Europe in the Seven Years' War* (London: University of Miami Press, 1971); Christopher Duffy, *Instrument of War: The Austrian Army in the Seven Years

*War* (London: Emperor's Press, 2000); *By Force of Arms: The Austrian Army in the Seven Years' War, Vol II* (London: Partizan Press, 2008); Franz A. Szabo, *The Seven Years' War in Europe 1756–1763.* (London: Routledge, 2007).

17 Anderson, 297–376.

18 Christian Ayne Crouch, *Nobility Lost: French and Canadian Martial Cultures, Indians, and the End of New France* (Ithaca, NY: Cornell University Press, 2014).

19 Virginia, Massachusetts, Pennsylvania, Delaware, North Carolina, South Carolina, New York, Georgia, Connecticut, Rhode Island, New York, New Jersey, Maryland, Bermuda, New Brunswick, Newfoundland, Nova Scotia, Upper Canada, Lower Canada, Prince Edward Island, Hudson's Bay.

20 Stephen Saunders Webb, *1676: The End of American Independence* (Syracuse: Syracuse University Press, 1995), 156–163.

21 Drenth and Riley, xxi–xxvi.

22 Carl von Clausewitz, *On War* (ed and tr by Michael Howard and Peter Paret) (Princeton: Princeton University Press, 1976), 604.

23 *Report on the Principles of Revolutionary Government, made in the name of the Committee of Public Safety by Maximilien Robespierre on 25 December, 1793*, www.tees.ac.uk, accessed 29 June 2019.

24 Compiled from various data in David Chandler, *The Campaigns of Napoleon* (London: Macmillan, 1966) and Alan Forrest, *Napoleon's Men* (London: Continuum International Publishing, 2005).

25 Alan Forrest, 5.

26 William Scupham, '"Friends, Fellows, Citizens. and Soldiers": The Evolution of the French Revolutionary Army, 1792–1799', *Primary Source,* Vol. V, Issue I (Fall 2014): 24.

27 S. J. Watson, *Carnot* (London: Bodley Head, 1954); Huntley Dupre, *Lazare Carnot, Republican Patriot* (Oxford, Ohio: Mississippi Valley Press, 1892), 5–20; Dino De Paoli, 'Lazare Carnot's Grand Strategy for Political Victory', *Executive Intelligence Review* (1996).

28 John A. Lynn, *The Bayonets of the Republic: Motivation and Tactics in the Army of Revolutionary France, 1791–94* (Abingdon: Taylor and Francis, 1984), 4.

29 John Terraine, 'Big Battalions. The Napoleonic Legacy' in John Terraine, *The Western Front 1914–1918* (Barnsley: Pen and Sword, 2003), 208–209; B. H. Liddell Hart, *The Ghost of Napoleon* (London: Yale University Press, 1935), chapter III.

30 Hew Strachan, *European Armies and the Conduct of War* (London: Routledge, 2005), 3.

31 Jonathon Riley, *Napoleon and the World War of 1813: Lessons in Coalition Warfighting* (London: Routledge and Keegan Paul, 2000), 22–30; Jonathon Riley, *1813: Empire at Bay. The Sixth Coalition and the Downfall of* Napoleon (Barnsley: Praetorian, 2013), 5–7; David Chandler, *The Campaigns of Napoleon* (London: Wiedenfeld and Nicholson, 1967), 333–34; Owen Connelly, *Napoleon's Satellite Kingdoms* (New York: Free Press, 1965), 51, 341; Michael Broers, *Europe under Napoleon* (London: Arnold, 1996), 75–76, 92; R. B. Mowatt, *The Diplomacy of Napoleon* (London: Arnold, 1924), 159.

32 Michael Broers, *Europe Under Napoleon* (London: Arnold, 1996), 35–36, 70–77.

33 Guy Rowlands, 'Foreign Service in the Age of Absolute Monarchy: Louis XIV and His Forces Étrangères', http://journals.sagepub.com, accessed 18 November 2017; see also Jonathon Riley, *1813: Empire at Bay.*

34 Michael Broers, 'The Napoleonic Gendarmerie. The State on the Periphery made real', *Crime, History and Society,* Vol. 20, No. 1 (2010).

35 David Johnson, 'Amiens 1802: the phoney peace', *History Today,* Vol. 52, No. 9 (2002): 20–26.

36 Andrew Roberts, *Napoleon the Great* (London: Penguin, 2014), 133.

37 David Chandler, *The Campaigns of Napoleon*, chapters 1, 3, 4; Russell F. Weigley, *The Age of Battles: The Quest for Decisive Warfare from Breitenfeld to Waterloo* (Bloomington, IN: Indiana University Press, 2004), 260–270.

38 Van Crefeld, *Command in War*, 78–85; Chandler, 452–479. Napoleon dictated 102 separate letters and documents detailing the plan for the war and the assembly of the army which are contained in the *Correspondence de l'Empereur Napoleon*. See also Von Clausewitz, 617 *et seq*, which is a treatise developed as a result of the author's experience against Napoleon.
39 Terraine, 208 *et seq*. See also Jonathon Riley, *Napoleon as a General* (London: Bloomsbury Academic, 2007), 178.
40 E. F. Berthezène, *Souvenirs Militaires* (Paris: Librairie Militaire, 1855), 243.
41 Fraser, 208.
42 Napoleon Bonaparte, *The Officer's Manual: Napoleon's Maxims of War* (English edition, Charleston, 1862), Maxim VII.
43 Bonaparte, notes to Maxim VII.
44 A conversation with Abbé Dominique de Pradt, Ambassador in Warsaw, in W. H. Ireland (ed), *The Napoleon Anecdotes. Illustrating the mental energies of the late Emperor of the French from the chronicles and actions of his contemporary statesmen and warriors* Volume III, (Boston, 1830) 8.
45 Riley, *Napoleon as a General,* Chapter II.
46 Ibid.
47 For detail of the War of 1812 and its relationship to the wider Napoleonic Wars see Jonathon Riley, *Napoleon and the World War of 1813: Lessons in Coalition Warfighting* (London: Routledge, 2000); see also Mark Zuehlke, *For Honour's Sake: The War of 1812 and the brokering of an uneasy peace* (Toronto: Vintage Books Canada, 2007).
48 See the accounts by Harold Nicholson, *The Congress of Vienna* (London: Phoenix, 1946) and by Henry Kissinger, *A World Restored: Metternich, Castlereagh and the Problems of Peace 1812–1822* (Boston: Houghton Mifflin, 1957).
49 John Terraine, 'Big Battalions', 208–209; Liddell Hart, *The Ghost of Napoleon*, chapter III.

CHAPTER 5

# 'Winning' in World War I, 1914–1919

Lothar Höbelt

The 19th century still cherished a romantic notion of warfare. It introduced conscription and mass warfare, yet still clung to elitist and chivalrous notions. But in general, a 19th-century public can be excused for believing that wars in general could very well provide a solution to long-standing political problems, at affordable prices, in term of both human and financial costs. Maybe the Austrian defeat in 1866 offers the best illustration of this: in the 18th century, Maria Theresa and Frederick the Great, *la belle et la bête*, had fought an inconclusive Seven Years' War (and several others); in 1866 the German question was to all intents and purposes solved within less than seven weeks, with cholera still claiming far more victims than even the famous needle gun. The brief campaigns of Italian unification, even the Balkan Wars of 1912–13, all pointed to the same conclusion.

Civil wars were a different matter, of course: the Carlist Wars in Spain dragged on endlessly, the American Civil War turned out to be a fairly bloody affair, to say nothing of the Taiping rebellion in China. Asymmetrical colonial conflicts followed a different rhythm: the heavy costs involved often were not offset by any possible returns. Overseas logistics usually drove up the financial cost of operations, tropical diseases the human cost of campaigns. The permanent administration of territories occupied as a result of fighting might not be worthwhile from an economic point of view. South African gold of the Rand was responsible for any number of theories of imperialism – because it was an exception to the rule. Even in the classic 'age of imperialism', many overseas campaigns assumed the character of punitive expeditions, maybe resulting in regime change, rather than wars of conquest. The Afghan wars or the British campaign to Magdala in 1868 spring to mind.

## Victors and losers: the US, Britain and Russia

The 20th century was to present quite a different picture, with World War I resulting in casualties of up to ten million while solving none of the great questions that had prompted the war in the first place. Let's approach the problem by asking the simple question: who actually won World War I? One obvious answer is: the United

States. Entry of the United States ensured the victory of the Entente over the Central Powers even though Russia dropped out of the war in 1917. American forces in Europe at the end of the war ranked only fourth among the Entente powers in terms of numbers. But in economic terms American participation was decisive: US loans saved the Entente from bankruptcy and helped overcome the challenge of the U-boats. Even if US production did not yet outstrip either France or Great Britain in terms of war material, tanks or artillery, it provided the raw materials of success. Whereas Germany produced roughly as much steel as the rest of Europe combined, the US produced approximately as much steel as the rest of the world combined.

The United States had become the potential number one world power long before 1914, a fact perhaps obscured by the fact that it had not been fighting any of the major European powers since the War of 1812, partly because they had not really been fighting each other. After all, the century that witnessed the rise of nationalism as a political movement turned out to be a surprisingly peaceful one. World War I, however, seemed to show that the United States could not escape being drawn into any major war in Europe because blockade and submarine warfare would always interfere with her commerce. Thus a system of collective security was intended to do away with that danger for good. President Wilson had outlined a programme for a new world order, based on the establishment of the League of Nations. But Wilson was disowned by the US senate that rejected his plan (partly because he had not seen fit to conduct his foreign policy in a bi-partisan manner). The standard phrase that the US 'withdrew' into isolationism, is perhaps a little bit overdrawn. The US continued to show a lively interest in the Far East, but rejected any suggestion that it should adopt any new commitments in the Old World, e.g. by guaranteeing France against Germany (or by protecting the Hejjas against the Saudis, as a horrified Senator Henry Cabot Lodge put it).

The United States had not acquired any extra territory as a result of the war. It had turned down the poisoned chalice of taking over a mandate in the Middle East, in Armenia, the ravaged mountain land that was to be squeezed between Turks and Soviets soon after. The US had won parity with Britain in naval terms. The Washington Treaties of 1921 granted each of them 15 dreadnoughts (with nine for Japan and five each for France and Italy). For all the symbolic rivalry, this kind of 'parity' represented no particular breakthrough: Britain had long since relinquished any thought of ever fighting the US again.[1] For the US, Canada as a hostage was a more important pledge of security than the parity in battleships. In fact, the more important result of US influence was that the special relationship between Britain and Japan was discontinued after 1921, to the chagrin of the CIGS, Sir Henry Wilson. Britain was supposed to favour a balance of powers in Europe. The US favoured a balance of powers in the Far East. 'It was not that the Japanese had greatly offended but rather that the Japanese alliance did not fit in with Britain's need for American goodwill and understanding.'[2]

Of course, in financial terms the US seemed to occupy an almost hegemonic position after World War I.[3] Without American help the Entente would no longer have been able to pay its bills by mid-1917. The war had turned America from being the biggest debtor nation to being the biggest creditor nation. The first part of that development had taken place while the US was still enjoying the fruits of its selective neutrality: by the end of 1916 the debt piled up during a century of American growth had been wiped out. Contrary to popular conspiracy theories, the US did not enter World War I to save its debtors from defeat. It only started lending to the Entente massively after it entered the war. Whatever the doubts of US policy makers about the merits of intervention in Europe after 1918, they certainly wanted their money back – and they failed to get it. Britain, theoretically still a net creditor, suggested a general waiver. The trouble was most British and French claims had been obliterated by the Russian revolution and the Bolshevik refusal to honour the debts of the Tsarist regime.[4]

As a substitute, Germany was expected to pay reparations, in theory to help rebuild France, in practice to enable France to service its debt to its partners. John Maynard Keynes was maybe going too far in his sweeping denunciation of the peace treaties, but he was certainly right when he advised his friends to be patient with the US: 'Give them time to discover that they are at our mercy as we are at France's, France at Germany's. It is the debtor who has the last word in these cases.'[5] Germany's unwillingness to pay ended in the Ruhr crisis of 1923. It was not the State Department but Wall Street that imposed a compromise solution with the Dawes Plan 1924: Germany was granted big loans to pay France, so that France could repay its debt to the US.[6] During the Great Depression this whole merry-go-round was overtaken by a moratorium, then by World War II, then the Marshal Plan.

The British Empire won. In 1935, the report of the Defence Requirements Committee still proudly asserted: 'The British Empire, as perhaps the extreme illustration of the "haves" of life, is perhaps not unnaturally exposed to "the envy of less happier lands"'[7] But at the same time the historian G. M. Trevelyan wrote: 'Great Britain was relatively more important in the world in 1713 than in 1815 or 1919.'[8] Britain had seen off the German challenge. No High Seas Fleet and no Lettow-Vorbeck was going to cause trouble for her any longer, it seemed (only U-boats and Rommel). With the addition of mandates in Africa and the Middle East, the boundaries of the British Empire were spread wider than ever before. However, it was an open question whether these extra responsibilities actually served to make Britain mightier yet. From the very beginning, Churchill and the War Office both had doubts whether Iraq was worth the trouble. The port of Basra would surely have been sufficient to seal off the Persian Gulf and keep an eye on the oil of the Anglo-Persian Oil Company (British Petrol). Wilson as CIGS would gladly have given large parts of Mesopotamia to Turkey.[9]

Britain had also got rid of Russia, to some extent. If Russia had counted as an essential element in the continental balance of powers, its expansionist drive had been less than welcome in Asia. In fact, the nuisance value of Russia in the east had been as important in persuading Britain to opt for the Entente of 1907 as the German challenge in Western Europe.[10] Now both challenges had for the time being lost much of their power. In 1915, Britain had been compelled to sign away Constantinople and the Straits. Lenin's coup d'etat and his separate peace with Germany saved Britain from having to keep its promise. To all intents and purposes, the Great Game had been decided in favour of Britain. For a time, Lord Curzon – a former viceroy of India, now running the Foreign Office – even had high hopes of keeping Baku out of Soviet hands.[11]

The Bolsheviks were regarded as obnoxious but for the time being they represented less of a danger than the ambitions of the Tsarist regime. World War I had certainly acted as a catalyst as far as the emancipation of the old-established dominions was concerned. From Ireland to India, independence movements followed suit and caused headaches for Imperial authorities, but the Komintern had fairly little to do with them. It was only in China that the communists were in the ascendant, until the Chinese nationalists of the Kuo Min Tang turned on them in 1927. Until shortly before, Britain had actually been more anxious about Chiang-Kai-shek's advance than the Japanese. But then, as Leo Amery, the Colonial Secretary, put it: 'In China there was no absolute good, but only a choice of evils.'[12]

Thus, the Anglo-Saxon sea powers had not really won hugely in material terms, but at least they had done fairly well compared with most of their competitors. The reverse side of the coin was that Russia, that most continental of great powers, had certainly lost, by almost every possible definition. So had Austria-Hungary. The Habsburg monarchy simply vanished from the map of Europe. Russia lost all the western borderlands that it had gained as a result of its victories over Charles XII and Napoleon in the two preceding centuries. For a number of years, Russia was subject to a bloody civil war. The Red Army managed to re-conquer the Ukraine in late 1919 but failed to overrun Poland in 1920. The world revolution that Lenin had confidently expected to follow in the footsteps of Red October, failed to materialise (except for a few isolated outbreaks in Hungary and Saxony). Stalin drew his own conclusions from that failure when he re-directed Soviet strategy to a defensive posture of defending the 'Socialist sixth of the world.'[13]

## Small, but big winners

Austria-Hungary had disintegrated in the autumn of 1918. For a time, its place in Central and Southeast Europe was taken by a *troika* of minor winners, minor in the eyes of the great powers but not at all in the size of spoils that they got out of the Great War. Newly established Czechoslovakia represented the most powerful of the successor states, with almost 15 million people. As Churchill put it, tongue-in-cheek, in 1938, 'No doubt, a small country – with an army only two or three times as large as our own, and a munitions supply three times as great as that of Italy.'[14] Romania had re-entered the war on 10 November – a week after the Austrian armistice – and acquired territory at the expense of both the Central Powers and Russia. Serbia had turned itself into the 'Kingdom of the Serbs, Croats and Slovenes', three times the size of pre-war Serbia. If you were looking for winners, winners they were, in line with Wilson's aim of empowering the smaller nations of Europe.

Those three smaller but big winners were loosely allied in the so-called Little Entente from 1920/21 onwards. With 40 million against 10, the Little Entente served as a perfect deterrent against Hungary's revisionist aspirations (or against the threat of a Habsburg restoration that might undermine their internal stability). But the Little Entente could do very little to safeguard Czechoslovakia against a resurgent Germany, or Romania against Russia or Yugoslavia against Italy that dearly wanted to turn the Adriatic into an Italian lake. Moreover, the Achilles' heel of all three was over-digestion, the millions of unruly minorities that provided an opening for every enemy eager to destabilise them. In Yugoslavia, communists gained a surprisingly large share of the vote – not because of proletarian solidarity permeating rural society but because Macedonians and Montenegrins chose that ticket as a way to vent their frustration. Croats who had been co-opted in 1918, soon started to learn the German phrase 'Los von Belgrad'.[15] In Czechoslovakia, the second biggest ethnic group after the Czechs, were not the Slovaks but the Sudeten Germans (three million of them). As a British diplomat once put it: The country resembled 'a patchwork quilt sewn together by an impatient hausfrau.' In the language of finance, the windfall profits of the Little Entente were soon threatened by a bear market.[16]

The same held true for Poland. A brief look at the map of Europe might convince the casual observer that World War I should be christened a War of Polish independence. In fact, the independence of Poland had figured prominently in the war aims of both the Central Powers and the Entente after 1916/17. If Poland had not quite managed to regain its old borders of 1772, as a country of almost 30 million inhabitants it could almost lay claim to being a great power in its own right. Yet, it was sandwiched between the two defeated giants, Germany and Russia. As early as 1920, when the Red Army stood at the gates of Warsaw, Trotsky had suggested a fourth partitioning of Poland to the Germans (and Churchill had written a memorandum advocating a policy of 'Kill the Bolshie, kiss the Hun' to guard against that danger).[17] The French insisted that Poles and Czechoslovaks should learn to cooperate. Instead, both gave the impression that they blithely assumed they could deflect German aggression onto the other.[18] France was (perhaps) willing to guarantee Czechoslovakia against Germany, but certainly not Poland against Russia. When the French military attaché paid his compliments to Pilsudski in 1932, the marshal bluntly told him: 'I know, you are going to betray us.'[19]

## Great expectations unfulfilled: '*Vittoria mutilata*': Italy and Japan

Italy and Japan were among the powers that had not been dragged into the conflict by their alliance commitments and by the July Crisis. They entered the war deliberately for reasons of great power politics. All the others pretended to act in self-defence. Neither Rome nor Tokyo could plausibly put forward any such claims.

Italy's disenchantment with the results of World War I has been encapsulated in Gabriele d'Annunzio's controversial phrase about the mutilated victory ('*vittoria mutilata*'). True, Italy's fond hopes that her intervention would turn the tables on the Central Powers, had come to naught. Even worse, some of her allies might scoff that in fact Italy been saved from defeat by her allies after the battle of Caporetto in late 1917. Contrary to popular folklore, Italy had not entered the war to liberate oppressed Italians from Austrian tyranny but to consolidate Italy's status as a great power. Her statesmen started from the assumption that a power that did not actively intervene in the Great War would not be taken seriously by the victors. Italian Foreign Secretary Sidney Sonnino had been a life-long supporter of the Triple Alliance. But geography dictated that Italy could not risk antagonising the Royal Navy by supporting Germany. Thus, if she wanted to exercise any influence on events she had no option but to join the Entente.[20]

The 1915 Treaty of London had promised Italy not just the Brenner frontier in the north, but almost the whole of the Dalmatian coast-line. Most of the fighting took place on land; most of the stakes the Italians were fighting for had to do with considerations of naval strategy. The Italian coastline was ideally suited for beach life but did not boast of any suitable harbours. The mountainous eastern shore abounded with islands, inlets, bays and natural harbours. That was why the Austro-Hungarian navy was able to dominate the Adriatic despite its numerical inferiority.[21] If the Italian Admiralty wanted to turn the Adriatic into an Italian lake, it would have to take possession of Dalmatia. Except for Spalato, most of these ports had been promised to Italy. Yet, the Entente did not deliver.

There was something to be said for the slogan '*vittoria mutilata*'. Italy had to be content with a tiny foothold, the town of Zara. Both Wilson and the French were happy to support the claims of Yugoslavia. When Italian Prime Minister Orlando came back from Paris in May 1919, he commented that if he were an MP, he would vote against the treaty himself. To add insult to injury, in 1920 the Italians were also driven out of Albania which they had assumed was theirs for the taking. As a consolation prize the powers let themselves be browbeaten by Gabriele d'Annunzio's bizarre antics and the clever tactics of the despised Nitti government to turn Fiume into a free city – on the model of Danzig – that was sooner or later going to be merged with Italy.[22]

In terms of great power politics, Italy was accorded parity with France. It obviously wanted to take the place of Austria-Hungary as the great power dominating the

Balkans. The resulting antagonism with France and her clients among the Little Entente drove Italy into the arms of the revisionist powers. Rome cultivated her ties with Hungary and Bulgaria; it supported the irredentist movements of Croats and Macedonians. In colonial terms, too, Italy had won a few oases in the desert, but it had to fight hard to make good these claims. During World War I Italy had lost practically all of Libya, except for Tripoli. It took the Italians almost a decade of brutal fighting to establish their sovereignty over the rest of the country, in particular the Senussi fiefdoms along the Egyptian border.

Japan, too, had every reason to talk about a '*vittoria mutilata*'. The sudden withdrawal of the European powers after 1914 had led to an upsurge of Japanese influence in China. Japan had successfully forced the Peking government to accept its 21 demands that went a long way towards converting at least some of the northern provinces of China into a Japanese protectorate. Japan had even persuaded her old enemy Russia to condone her advance in 1916. It had used the vague threat of German subversion to bully China into a declaration of war against the Central Powers. It had sent troops to Vladivostok and along the Trans-Siberian railroad after the collapse of the Tsarist regime. Yet, at the end of the day, Japan was unable to convert these war-time windfall profits into peace dividends. The intervention in Siberia cost money to no avail. China re-asserted itself. Ten years after the end of the war Chiang Kai-shek triumphantly entered Peking, while the Japanese fell out with Chang Tso-Lin (Zhang Zuolin) their erstwhile Manchurian ally.[23]

## Germany and France

The most interesting case concerns the two perennial troublemakers of Europe, France and Germany. France had survived brilliantly as a great power. It had made the most of its potential and punched above its weight. After Verdun and the Nivelle offensive, it had husbanded its resources. Petain was rightly praised by Churchill for his leadership.[24] 'Wait for the Americans and the tanks.' France had indisputably re-gained Alsace-Lorraine (plus Togo and the Cameroons). But it had lost the cornerstone of its security policy, the Russian alliance that had for so long enabled her to keep Germany in check. Churchill as well as the Czechs had told the French that they had not really won the war unless they managed to restore the Tsarist system – or a similar friendly regime (Churchill had high hopes that Admiral Kolchak might even turn Russia into a friend of Britain).[25] But neither Britain nor Czechoslovakia would or could do much to bring about such a result. In France, Clemenceau gave the order to work for the overthrow of Lenin in September 1918 but soon shied away from the cost and effort involved. French support for the so-called 'White Russian' generals remained half-hearted, at best. In the end, Clemenceau opted for relying on a 'cordon sanitaire', consisting of the newly liberated nations in the border areas of Russia, from Finland to Georgia or maybe even Turkmenistan.[26]

The result turned out to be the worst of both worlds. Lenin had neither been defeated nor converted into an ally. As Trotsky's offer of a fourth partition of Poland in 1920, the Rapallo treaty of 1922 and the clandestine cooperation of Reichswehr and Red Army all showed, the two potentially biggest continental powers remained potential allies, with lots of common enemies, no matter what their ideological leanings might be. On top of these dangers, France had also lost her huge investments in Russia. Lenin might have been willing to cut a deal with a few favourite corporations whose know-how the Soviets needed. But he was not going to pay billions to gladden the hearts of French middle-class rentiers. Indeed, one of the more imaginative ideas for ending World War I on a compromise note read that Germany was going to be allowed to dominate the east in return for a repayment of the outstanding part of Russia's foreign debt.

That finally brings us to Germany. Of course, Germany is supposed to have lost the Great War. Embittered Germans sometimes complained that they had not been defeated in the field but stabbed in the back politically. If anything, the reverse is true. Germany had been defeated in the field during the 'hundred days' battle' following the black day of the German army at Amiens on 8 August 1918. Ludendorff duly suffered his nervous breakdown. The German government had the good sense to call it quits before the fighting actually reached the fatherland. As of November 1918, things could only get worse for Germany if it refused to acknowledge defeat. At best, the Entente could be deprived of a negotiating partner once Germany was engulfed by revolutionary chaos. For conservative Germans, that option of cutting off one's nose to spite one's face was not a very appealing one.

Defeat in World War I was a traumatic event for most Germans. As one Austrian wrote during those final days of October 1918: we take it in our stride, as we are used to bad news.[27] Germans were not. Until a few months previously, their credit rating had been fairly good. As late as June 1918, Lloyd George had debated the possibility of a French collapse in Cabinet and Sir Henry Wilson as CIGS had despaired of ever achieving victory without a second front in the east.[28] Maybe the only one who had not been properly defeated – or at least managed to persuade the public that this was so – really was Lettow-Vorbeck who returned to a hero's welcome after his successes in East Africa, entering Berlin through the Brandenburg Gate in time-honoured fashion.[29] Of course, Germany still lost all of her colonies. But then they had been 'a hostage to fortune'[30] in wartime and already a drain on resources in peacetime, anyway.

Germany lost its fleet and its overseas investments. The Damocles sword of reparations clouded the financial horizon of the post-war world. The reparations issue led to a paradoxical situation where Germany thrived on the threat of bankruptcy and prophecies of doom. Hitler's rise to power has often been blamed on the Versailles Treaty – in flagrant disregard of the fact that his party polled just 2% of the vote in 1928 but 37% at the height of the Great Depression. As Bill Clinton's advisors put it 60 years later: 'It's the economy, stupid.'

Germans were resigned to the loss of Alsace-Lorraine. Even Hitler only bothered to re-annex the area as a kind of afterthought long after having conquered it in 1940. But few Germans were willing to consent to the loss of the Polish corridor, the piece of land separating Germany from East Prussia. Most British statesmen did not regard that solution as a particularly happy one. Chamberlain (Austen, not his half-brother Neville) referred to Bismarck's celebrated statement about the Balkans when he minuted in 1925: '… the Polish Corridor (for which no British government ever will or ever can risk the bones of a British Grenadier).'[31] Some of the Prussian elite had their roots and estates in the lost territories. But did the corridor – or the coal mines of Upper Silesia – really merit such an exaggerated sense of despair and outrage?

What Germany had been afraid of before 1914 was the Russian steamroller. Once the Russians had succeeded in perfecting their system of strategic railways, Germany would no longer be able to fight a war on two fronts by defeating France before the Russians had completed their mobilisation. As it happened, that strategy turned out to be a failure in 1914 already. Yet, German industrial might – from 6" howitzers to an ample supply of ammunition – was such that Germany was able to withstand the pressure of the combined French and Russian armies. In that respect, Germany had done much better than any of the pre-war predictions gave her credit for. Without millions of British and American soldiers being sent to the Western Front, Germany could not have been defeated.

If Germany had enjoyed what many observers called a semi-hegemonial position in Europe before 1914, the logical result of her defeat should have been the redistribution of some of her resources to her closest competitor, i.e. the feared Russians. Yet, this is exactly what did not happen. Russia – together with the Habsburg and Ottoman empires – turned out to be the great loser of World War I. Germany had lost roughly 3 to 4 million people to Poland. Russia had lost roughly 15 million people to Poland, plus 6 or 7 to the newly independent Baltic States, including Finland. Even if Soviet Russia managed to re-conquer Ukraine and the area beyond the Caucasus between 1919 and 1921, it had certainly been weakened much more than Germany.

Thus, it is difficult to escape the conclusion that Germany's strategic position on the continent of Europe had actually been strengthened as a result of World War I. True, Wilson and the Czechs had taken the slogan of self-determination one step further after the armistice by applying it to Austria-Hungary, as well as Russia. Germany had lost Austria-Hungary as an ally, just as France had lost Russia. But – except for Czechoslovakia – Germany found it easy to collaborate with all the successor states of Austria-Hungary. They could always play off Italy against Yugoslavia or Hungary against Romania. Economically, too, the Balkan countries were far more dependent on German markets than on French capital.

The crucial factor in many ways turned out to be Poland. Stresemann banked on forcing a revision of the border by means of economic pressure. Germans had resigned themselves to the loss of Alsace, he told Lockhart in 1929, but from socialists to the Kaiser, none wanted to renounce the lands in the east.[32] Hitler changed tack and tried to persuade Pilsudski and his successors to join him in a crusade against Bolshevism. Whenever Göring went hunting buffaloes in the primeval forests of Bialystok, he dropped heavy hints in that direction.[33] Polish leaders, however, refused to submit to either German threats or blandishments. Nevertheless, Poland either served as a convenient buffer state against Russia or as a bait for the Russians to join in a fourth partition – an offer the western powers could never hope to top.

We are thus faced by the paradox that although Germany had certainly lost the Great War and its losses in terms of men, money and territory were far from trifling, it had nevertheless emerged as a potentially far stronger power than before. That result was partly obscured by the impression that Germany had been defeated not on the battlefield, but by means of the British blockade. That was an exercise that could easily be repeated, it seemed (unless the Germans managed to devastate the English 'home terminal areas', as they were unsentimentally described, and force a surrender by a knock-out blow from the air). As the Minister of Defence, Sir Thomas Inskip, put it in 1938: 'Pending Germany's collapse under a blockade, our other measures could be little more than gestures.'[34] In fact, German industrial might seemed to make it even more vulnerable. Oil was a particularly vital item in that respect. Iron ore from Sweden or scarce minerals might also be considered as potential bottlenecks of the German economy. This dependence on imports seemed to hand Britain a measure of deterrence, no matter what happened to the Polish corridor. In 1939, while Hitler relied on his deal with Stalin to act as a deterrent to Britain, London chose to rely on its tried and tested blockade (as radar and spitfires had taken care of the bomb scare). Neither of these deterrents turned out to be effective.

## Conclusion

What is winning? When asked why he did not favour a war against Russia, Moltke the Elder is supposed to have said: 'We don't need any extra lands, and they have not got any money.' The time-honoured idea of acquiring extra territory does not lead to any very conclusive answers in our case. The real winners in these terms were small states, carved out of Russia and Austria-Hungary, that proved unable to affect the balance of powers very much – except for Poland: Polish independence proved to be a boon for Germany, even if Germans were slow to realise that.[35] In terms of wealth and money, the US had won – but most of their gains had happened before they had entered the war – and most of their gains came at the expense of Britain. Again, the impact on the balance of power was far from clear. If anything, Britain's relative decline might be interpreted as a gain for all its potential rivals.

Maybe the important thing about winning is survival. The western powers had survived, with their political and economic systems essentially unimpaired.

Austria-Hungary and the Ottoman Empire had not survived their military defeat. Russia – and to a lesser extent, Germany – had experienced a revolution. In Germany, defeat had bred revolution; in Russia, revolution had brought about defeat. In Russia, revolution lead to the establishment of a totalitarian system in the short-term; in Germany, in the medium term, with the Great Depression acting as a catalyst or as a root cause. The two brands of totalitarianism strove to divide Europe into ideological camps – but did not shy away from deals if they benefitted them.

Simon Schama has claimed that the Napoleonic Wars made France miss out on the industrial revolution.[36] A hundred years later, after World War I, the US and Germany remained the two strongest industrial powers. The US did not 'win' anything tangible during the war. If it increased its influence in the world, it did so the way Britain won at the expense of its Dutch allies during the War of the Spanish Succession. Yet, the US were unwilling to carry any of the burdens associated with that kind of influence. Lloyd George often complained that was because Wilson had not seen fit to conduct a bi-partisan foreign policy.[37] Thus, Republicans blocked all his initiatives, from the League of Nations to the guarantee for France. In 18th-century Britain, too, the Hanoverian kings had been fairly eager to meddle in continental affairs, parliament far less so. Calvin Coolidge's famous slogan: 'The business of America is business' neatly paralleled Sir Robert Walpole's attitude towards continental commitments.[38]

Germany had won and lost at (almost) the same time. It had won in the east in the spring of 1918 and lost in the west in the autumn of 1918. Crucially, defeat in the west did not cancel the long-term benefits of winning in the east. Germans were poorer after 1918 (like almost all Europeans) and bewildered. But in terms of great power status they had lost almost nothing. The loss of the High Seas Fleet only ratified the decision already taken in 1911 that it was no use any longer competing with Britain in terms of dreadnoughts. The way the Treaty of Versailles banned conscription in Germany might have seriously curtailed the country's military potential a few decades ago. It no longer mattered that much in the high-tech style of warfare characteristic of the 1940s.

France was supposedly the great victor of 1918. But those appearances were deceptive. France had not really won at all. It had survived, gloriously maybe, but no more. It had suffered huge losses with hardly any compensating advantages. The French military tried to build a glacis in front of their borders by separating the Rhineland from Germany. None of her entente partners was willing to go along with these proposals. Even if there had been any takers, Arthur Balfour's scathing comment would still have rung true: 'No manipulation of the Rhine frontier is going to make France anything more than a second-rate Power.'[39] In terms of *securite d'abord*, France had certainly not profited from the war. The Maginot line was a symbol of that kind of rather ambivalent result. As A. J. P. Taylor acidly commented: If the Maginot line protected France, it protected Germany even better.[40]

Italy and Japan won less than they had bargained for. The peace settlement left both of them with a grievance. The nations that had won most were those that had done least fighting, like the Poles and Czechs. The Poles achieved independence from Russia, the Czechs from Austria-Hungary. Yet, it is likely that more Poles had been drafted into the Russian army than into any other; certainly more Czechs kept fighting for the Habsburgs than switched to the Legion.[41] The Czechs in particular came into their own, in lands that had not been touched by the war. But both their newly independent states were surrounded by potential enemies, with almost no reliable friends and allies.

The British Empire wound up with the most complicated balance sheet: in terms of naval strength it had improved its position. Rivalry with Germany was a thing of the past – unfortunately, it only started anew in the air, in a far most costly and even less predictable way: Germany could not send its shock troops across the Channel without a fleet. It could easily send its bombers even if they had originally been designed to strike at French or Polish targets. Britain seemed to have perfected a wonder weapon in terms of the blockade. Pre-1914 plans had already toyed with that idea.[42] During the 'Great War' it seemed to have proven its worth. But reliance on the economic deterrent without any alternatives turned out to be dangerous in the run-up to World War II. Apart from many other factors and issues, the outbreak of war in 1939 was a failure of deterrence, or as Churchill put it: 'Great wars usually come only when both sides think they have good hopes of victory.'[43]

## *Notes*

1 Stephen Roskill, *Naval Policy between the Wars, Bd. 1: The Period of Anglo-American Antagonism 1919–1929* (London: Collins, 1968), 301–325; Margaret MacMillan, *Peacemakers: The Paris Conference of 1919 and its Attempt to End War* (London: Murray, 2001), 188.

2 Ian Nish, *Alliance in Decline: A Study in Anglo-Japanese Relations 1908–23* (London: Routledge, 1972), 381; Phillips Payson O'Brien (ed), *The Anglo-Japanese alliance, 1902–1922* (London: Routledge Curzon, 2004); Sir Charles Edward Callwell, *Field Marshal Sir Henry Wilson: His Life and Diaries, Vol. 2* (London: Cassell, 1927), 307.

3 Benjamin M. Rowland (ed.), *Balance of Power or Hegemony: The Interwar Monetary System* (New York: New York University Press, 1976); D. C. Watt, *Succeeding John Bull: America in Britain's Place, 1900–1975* (Cambridge: CUP, 1984).

4 Inbal Rose, *Conservatism and Foreign Policy during the Lloyd George Coalition 1918–1922* (London: Taylor & Francis, 1999), 125.

5 Robert Blake, *The Unknown Prime Minister: The Life and Times of Andrew Bonar Law, 1858–1923* (London: Eyre & Spottiswoode, 1955), 491.

6 Elspeth Y. O'Riordan, *Britain and the Ruhr Crisis* (Basingstoke: Palgrave Macmillan, 2001); Stephen A. Schuker, *The End of French Predominance in Europe: The Financial Crisis of 1924 and the Adoption of the Dawes Plan* (Chapel Hill: University of North Carolina Press, 1976).

7 Lothar Höbelt, *Die britische Appeasementpolitik. Entspannung und Nachrüstung 1937–1939* (Wien: Bundesverlag, 1983), 86; N. H. Gibbs, *Grand Strategy, Bd. 1: Rearmament Policy* (London: Her Majesty's Stationery Office, 1976), 178.

8 George M. Trevelyan, *England under Queen Anne, Vol. 3: The Peace and the Protestant Succession* (London: Longman, Green, 1934), viii.

9 Walter C. Reid, *Empire of Sand: How Britain made the Middle East* (Edinburgh: Birlinn, 2011), 48, 252, 307; Martin Gilbert, *Winston S. Churchill, Vol. 4* (London: Heinemann, 1975), 504; Callwell, *Wilson Diaries II,* 288, 293.

10 Stephen Schroeder, *Die englisch-russische Marinekonvention. Das Deutsche Reich und die Flottenverhandlungen der Tripelentente am Vorabend des Ersten Weltkriegs* (Göttingen: Vandenhoeck and Ruprecht, 2006), 145, 262; Andreas Rose, *Zwischen Empire und Kontinent. Britische Außenpolitik vor dem Ersten Weltkrieg* (Munich: Oldenbourg Wissenschaftsverlag, 2011).

11 Gilbert, *Churchill IV,* 371, 393.

12 Richard S. Grayson, *Austen Chamberlain and the Commitment to Europe: British Foreign Policy 1924–29* (London: Taylor and Francis, 1997), 185.

13 Ewan Mawdsley, *The Russian Civil War* (Edinburgh: Birlinn 2nd Ed., 2008); Robert Service, *Stalin A Biography* (London: Macmillan, 2004), 245, 379.

14 Gilbert, *Churchill V,* 916 (14 March 1938).

15 *Documenti Diplomatici Italiani,* Series VII, vol. 7, 274 (8 Feb. 1929); vol. 8, 47 (27 Sept. 1929); Stefan Troebst, Mussolini, *Makedonien und die Mächte 1922–1930* (Cologne: Böhlau, 1987), 247 ff., 350, 354 ff.

16 Magda Adam, *The Little Entente and Europe* (Budapest: Akademiai Kiado, 1993); Gabor Batonyi, *Britain and Central Europe,* 1918–1933 (Oxford: Clarendon, 1999), 218.

17 Gilbert, *Churchill IV,* 278, 427; Adam Zamoyski, *Warsaw 1920: Lenin's Failed Conquest of Europe* (London: HarperPress, 2008).

18 Isabelle Davion, *Mon voisin, cet ennemi. La politique de securite francaise face aux relations polono-tchecoslovaques entre 1919 et 1939* (Bruxelles: Lang, 2009); Miklos Lojko, *Meddling in Middle Europe. Britain and the 'Lands Between' 1919–192* (Budapest: Central European University Press, 2006), 186–193.

19 Piotr S. Wandycz, *The Twilight of French Eastern Alliances, 1926–1936. French-Czechoslovak-Polish Relations from Locarno to the Remilitarisation of the Rhineland* (Princeton: Princeton University Press, 1988), 234; Georges-Henri Soutou, 'L'Alliance franco-polonaise 1925-1933 ou comment s'en debarasser?', in *Revue d'Histoire Diplomatique,* Vols. 2/3/4 (1981): 295–348.

20 Lothar Höbelt, "Only the Dead do not avenge themselves.' The Austro-Italian 'Negotiations' of 1914–15', in: Giovanna Orsina & Andrea Ungari (eds), *L'Italia neutrale 1914–1915* (Rome: Rodorigo, 2016), 454–468; William A. Renzi, *In the Shadow of the Sword: Italy's Neutrality and Entrance into the Great War* (New York: Peter Lang, 1987).

21 Paul G. Halpern, *The Naval War in the Mediterranean 1914–1918* (London: Allen & Unwin, 1987).

22 Paolo Pastorelli, *L'Albania nella Politica Estera Italiana* (Napoli: Jovene, 1970), 220 ff.; Giordano Bruno Guerri, *Disobbedisco. Cinquecento giorni di rivoluzione. Fiume 1919–1920* (Milano: Mondadori, 2019).

23 Yoshihisa Tak Matsusaka, *The Making of Japanese Manchuria, 1904–1932* (Cambridge, MA.: Harvard University Press, 2001), 229 ff., 259 ff.; Gavan McCormick, *Chang Tso-Lin in Northeast China, 1911–1928* (Folkstone: Dawson, 1977); Lucian W. Pye, *Warlord Politics: Conflict and Coalition in the Modernization of Republican China* (New York: Praeger, 1971).

24 Tongue-in-cheek, the manuscript pages of 'the World Crisis' on exhibit in Chartwell show exactly those pages.

25 Gilbert, *Churchill IV,* 253; Josef Kalvoda, *The Genesis of Czechoslovakia* (Boulder: Columbia University Press, 1986), 463.

26 Anne Hogenhuis-Seliverstoff, *Les Relations Franco-Sovietiques 1917–1924* (Paris: Sorbonne, 1981), 105, 150; George A. Brinkley, *The Volunteer Army and the Allied Intervention in South Russia, 1917–1921* (Notre Dame, IN: University of Notre Dame Press, 1966), xi, 86, 209, 282.

27 Diary of Princess Henriette Hohenlohe, 2 October 1918. The princess was the wife of the Austro-Hungarian Ambassador in Berlin and daughter of Archduke Friedrch, the nominal

Austro-Hungarian Army commander from 1914 to 1917. I am grateful to Herbert Fischer-Colbrie for his permission to use the diaries.

28 Elizabeth Greenhalgh, *Foch in Command* (Cambridge: CUP, 2011), 382; Gilbert, *Churchill IV*, 221.

29 Hew Strachan, *The First World War in Africa* (Oxford: OUP, 2004); Eckard Michels, Paul von Lettow-Vorbeck, *Ein preußischer Kolonialoffizier* (Paderborn: Schöningh, 2008).

30 Lord Halifax actually used that very term when discussing the possible return of German colonies (PRO, CAB 47/12, Advisory Committee on Trade Questions in Times of War, 18 March 1937).

31 Robert C. Self (Hg.), *The Austen Chamberlain Diary Letters. The Correspondence of Sir Austen Chamberlain with his Sisters Hilda and Ida, 1916–1937* (Cambridge: CUP, 1995), 270 (Chamberlain to Crewe 16.2.1925).

32 See Sir Robert Bruce Lockhart, *Retreat from Glory* (London: Putnam, 1934), 361 for his account of an interview with Stresemann.

33 Wandycz, *Twilight*, 384, 430; one of the results of that rough wooing was a translation of Pilsudski's memoirs into German with a preface by Göring.

34 PRO, CAB 47/12, Advisory Committee on Trade Questions in Times of War, 18 March 1938. A few days before, Duff Cooper had underlined the same idea in Cabinet: 'If we have to fight Germany again, blockade must be our principal weapon.' (CAB 23/92, 14 March 1938).

35 In fact, Arthur Balfour had always predicted that such would be the result of an independent Poland (Lojko, *Meddling*, 247).

36 Simon Schama, *Citizens: A Chronicle of the French Revolution* (New York: Vintage, 1989).

37 MacMillan, *Peacemakers*, 14; Rose, *Lloyd George Coalition*, 115.

38 Brendan Simms, *Three Victories and a Defeat: The Rise and Fall of the First British Empire, 1714–1783* (London: Allen Lane, 2007) provides a ringing defence in favour of the Continental Commitment even in Walpole's day. Jeremy Black in his many works on the international system in the 18th-century leans towards the opposite view.

39 MacMillan, *Peacemakers*, 205 (Balfour to Lloyd George 18 March 1919).

40 A. J. P. Taylor, *The Second World War: An Illustrated History* (London: Hamilton, 1975), 49; Wandycz, *Twilight*, 389. In the end, of course, the Maginot Line did not protect France, mainly because of its Belgian Achilles heel; see Bruno Chaix, *En Mai 1940, fallait-il entrer en Belgique? Decisions strategiques et plans operationnels de la campagne de France* (Paris: Economica, 2005).

41 Kalvoda, *Genesis*, 405 says 59,000 out of 180,000 Czech POWs in Russia signed up for the Legion.

42 Eric W. Osborne, *Britain's Economic Blockade of Germany 1914–1919* (Naval Policy and History 24, London: Routledge, 2004).

43 Robert Rhodes James, *Churchill: A Study in Failure, 1900–1939* (London: Weidenfeld & Nicolson, 1970), 245.

CHAPTER 6

# 'Winning' in the World Wars. The British Conceptions of the War-Time Leaders Lloyd George and Winston Churchill, 1914–1945

Rob Johnson

In World War I and II, Great Britain's prime ministers had distinct ideas of what constituted victory and the process of winning. They maintained these conceptions despite setbacks and sometimes contrary to the demands of allies. This chapter offers an analysis of 'victory' or 'winning' through a comparative historical case study. It examines the character of *military* victory, rather than the full spectrum of winning in diplomacy, economics, or other factors, although these elements cannot be separated entirely, not least because, as Basil Liddell Hart put it, the purpose of war is a 'better peace'.[1]

The volume, *Winning Wars,* evaluates this challenge through the idea of the utility of force, that is, appropriateness of the military instrument to achieve a strategic objective; the extent to which the ends justify the means; what policy is to be pursued; and the end state that is visualised. While both world wars were 'total' and existential for Britain, there were both broad similarities and significant differences in conceptions of force and the ends. Lloyd George's government continued the objective of dismantling Prussian militarism, begun by his predecessor, Herbert Asquith, and Churchill articulated that Nazism had to be destroyed for the sake of humanity: neither could tolerate a monstrous tyranny in a German state that sought to dominate Europe. Nevertheless, while Lloyd George laid out desiderata or war aims, Churchill refused to use these terms, instead expressing ideas that would align very closely with those of the United States: a free world of United Nations characterised by common international standards.

This chapter focusses on how the British wartime leaders conceived of 'winning', acknowledging that certain events had strategic impact and either progressed or retarded the achievement of victory. The failure of British offensives on the Western

Front between 1916 and 1917, maritime setbacks in the Atlantic (1917–18; 1941–42) or the fall of Singapore and Tobruk (February and July 1942) had some impact on the conception of victory, or at least its timings, but so too did the operational successes, such as the Allied Hundred Days in 1918, or the D-Day landings and north-western Europe campaign of 1944–45, and the retaking of Burma in 1945. Against this background, we follow the conception of aims, policies and events such as the formation of an Allied Coalition between 1917–18, the defeat of Russia in 1917 and France in 1940, and the entry of the United States into both wars. These developments had a bearing on ideas, effects, and ownership. Alongside domestic considerations, these were amongst the most significant concerns of Lloyd George and Churchill as they conceived of their respective ideas of victory. The evaluation assesses what caveats, at the conclusion of the conflicts, might be applied, in terms of price paid, loss of strategic position (in commerce, finance, influence, or power), and relative strength. It suggests that Britain was rather more successful at fulfilling its war aims than supposed, but its fundamental problem was having to wage a war on behalf of Europe, for which it was ill-prepared.

## Lloyd George and 'winning' in World War I

Before World War I, Britain had a clear idea of its position in the world, despite growing concerns about its relative strength and emergent threats. For decades, British political, commercial and financial elites shared a strong sense that a major war would be damaging for the British economy. Successive prime ministers had sought to avoid alliance obligations which might commit the country to some unforeseen crisis that did not serve Britain's interests. Its seaborne empire was global which implied that security, like that of the British Isles, depended not on foreign powers but on its own powerful navy. Imperial security threats on land were to be managed by a combination of local forces, augmented by a small British garrison that could be reinforced, if necessary, by an expeditionary force of professional volunteers.

There had been a growing sense of threat after 1900, driven in part by Germany's drive for a fleet to surpass that of Great Britain. There was also the question of German industrial and commercial rivalry which challenged Britain's economic future. Therefore in terms of the utility of force, Britain retained its faith in its maritime power. That was reinforced with a programme of modernisation and expansion of the Dreadnought class of warship after 1906. When Belgian neutrality was violated in August 1914, Prime Minister Herbert Asquith was able to unite the Cabinet by the despatch of a British Expeditionary Force (BEF), although the real reason was to protect France.

To what extent did the ends justify the means in the war that followed? The sheer costs of the war, in lives, revenue, lost markets, indebtedness, and reputation have been criticised severely since the conflict. Niall Ferguson has argued that: 'Had

Britain stood aside even for a matter of weeks continental Europe could therefore have been transformed into something not wholly unlike the European Union we know today but without the massive contraction in British overseas power entailed by the fighting of the two world wars.'[2] His damning conclusion was that, for Britain and consequently for the rest of the world, participation in the war was 'the greatest error of modern history.'[3]

Sir Michael Howard characterised the alternative interpretation. He maintained that the other battlefields of Europe offered no better prospect than the Western Front, that the British Army learned how to fight the new warfare such that, alongside its allies, it produced victory in 1918.[4] From the political perspective, Howard asked that, even if the price of preventing German hegemony in Europe is considered too high, would France and Russia have tolerated an alternative, indirect approach? Indeed, he asks, would they have survived at all if Britain had chosen an alternative way to wage war or if it had stayed out altogether?

Basil Liddell Hart, the British military writer who had helped Lloyd George compose his subsequent account of the war, posited that Britain had been wrong to abandon its traditional naval strategy in favour of a continental commitment.[5] Reginald McKenna, who was the First Lord of the Admiralty until Churchill replaced him, was strongly against a continental deployment too. Both believed it was better to keep open the sea lanes to British ports and to put into effect a blockade of Germany. British businesses would switch trade from Germany to Germany's former markets in Europe, and economic dislocation, at least in the short term, would have to be weathered. A longer war on the continent would enable Great Britain to profit from supplying France and Russia. As in the Napoleonic Wars, British banks would make available the loans and see that European armies bore the brunt of the fighting, before a British force came in at the end. There would be no wholesale re-gearing of the economy, although some adjustments were anticipated. There was a concern to ensure the passage of supplies, but price controls would be implemented only if necessary. The insurance market would be regulated to minimise the adverse effects of panic and speculation. The phrase in use was to aim for 'business as usual'.[6]

Britain's 'utility of force' was the Royal Navy, but for a continental war it needed a vast army. Building this new force took three years, and, inevitably, bloody setbacks. Equally, Britain's war policy appeared in a piecemeal way. Asquith had stated in November 1914, that Britain's aim was the liberation of Belgium, the security of France, and, crucially, that Prussian militarism was to be erased from Germany. In April 1916, Asquith had reiterated this desire to see 'the destruction of the military domination of Prussia.'[7] The practicalities of this aspiration were not discussed. In 1917, Arthur Balfour noted: 'We entered this War ... with little in our minds besides the necessity of defending Belgium and the necessity of preventing France from being crushed before our eyes.'[8] There were no plans for Germany. Sir Edward Grey, the Foreign Secretary, had suggested in 1915 that, had Germany offered to evacuate Belgium, Britain might be prepared to make peace. Nevertheless, in a Cabinet Memorandum that year, he had also insisted that Germany must first suffer a 'crushing defeat', and that 'the military hierarchy in Germany should be dethroned.'[9]

The emissary of President Woodrow Wilson, Colonel Edward House, had looked favourably on these proposals in 1916 and intimated that, if a peace conference was called and Germany acted unreasonably, the Americans would join the war on the side of the Entente and enforce Germany's evacuation of Belgium, among other measures.[10] This House-Grey Memorandum was kept secret, and the government War Committee was led to believe that the priority remained the defeat of the German army and the dismantling of Prussian militarism. There was still faith in the utility of the force now at Britain's disposal to achieve victory. Britain believed it had the opportunity to expunge the litany of setbacks it had suffered between 1914 and 1916. It now possessed a larger army, with conscripts pouring in. There were 1.4 million British and Commonwealth troops, in 53 divisions, in France. Its war industries were beginning to reach the production levels required for a modern war. Its ministries were now organised and the armed forces' leaders were more confident of victory. Moreover, it was thought that American participation at this stage, even as peace negotiators, might give them disproportionate influence just at the moment of British military success. The Americans might insist on concessions from Britain that would negate its efforts in the war. This dilemma of when to make peace, when a military solution still seemed open, was to afflict Germany in 1918 in similar ways.

The Entente powers and their Italian allies had agreed on coordinated offensives in the summer of 1916, but by November, each of the Allied offensives had drawn to a close without success.[11] In Britain, long-standing disappointments with Asquith's lethargic war leadership now reached their peak. The pugnacious and energetic David Lloyd George, then serving as Secretary of State for War, supported by the Conservative opposition, replaced Asquith. He created a five-man executive to govern, supported by co-opted members of the Cabinet, including Winston Churchill. The establishment of this War Council did two things: one, it centralised power and led to the more single-minded prosecution of the war. Two, it reasserted civilian primacy in war policy. Out with Asquith went Grey, the Foreign Secretary, who had bungled the formation of a coalition with Italy, Greece, Bulgaria and Serbia, and alienated the United States.

Lloyd George's objectives in the war were very simple. He wanted to win. He insisted on 'a fight … to the finish – to a knock out.'[12] Paul Guinn was critical of Lloyd George for siding with the generals, on every policy, with the implication he did so for political reasons, but the alignment was really because the generals, like him, were motivated by the desire for military victory.[13]

Soon after Lloyd George took office, the Central Powers made a peace overture, offering vague conditions that would ensure 'a lasting peace'. The Entente dismissed the proposal as insincere. Then the Americans called the belligerents to put forward their respective terms for peace, because the United States was growing weary of British demands for more credit and the damage being done to global commerce.[14] There was even talk in America that the United States might take counter-measures against the British blockade of the Central Powers. The British proposals for peace were therefore listed, as requested. There was no mention of the dismantling of Prussian militarism. Instead the emphasis was on liberation and restoration, including the populations under Ottoman tyranny in the Middle East.

Woodrow Wilson's objectives and motives have been subjected to intense scrutiny, but can be summed up as a combination of a pragmatic desire to protect American national interests, a deep conviction to avoid war where possible, and a moral aspiration to see international institutions created to prevent war. His idealism permitted him to believe that democracy offered a better chance for peace, and he shared the general American distaste for European colonialism. There was no sympathy for British interests and the motive for military intervention, in the end, was driven by the German decision to launch a campaign of unrestricted U-Boat warfare, a last ditch effort by Berlin to strangle the British and French into submission.

America's entry into the war on the side of the Entente was welcomed, but the need to build an American army, almost from scratch, imposed pressure on the Entente. Indeed, shipping losses in the first months of 1917 were so catastrophic that the Cabinet was warned that, unless the trend was reversed urgently, their war effort was unsustainable. To make matters worse, Russia collapsed into revolutionary paralysis, the Italians were defeated at Caporetto, and the much-vaunted French offensive in Champagne proved such a disaster that mutinies broke out. Industrial unrest was on the rise at home too, with socialist agendas more in evidence.

Lloyd George was compelled to address the situation with his characteristic energy and zeal. In a series of public speeches, he insisted that Germany must evacuate Belgium and occupied portions of France. He noted that, 'No German statesman has ever said they would restore the independence of Belgium.'[15] No peace would be possible, he repeated, until that guarantee could be given. Jan Smuts, a member of the War Cabinet, agreed but also suggested that the Kaiser would be deposed. Back channels to Germany were opened via the Vatican, to elicit a response, but there was no German reply. This was, in part, because, with Russia defeated, Berlin believed it could win the war militarily.

The paradox was, that while public representatives in assemblies in Russia, Germany, and France discussed peace terms, with some caveats, British parliamentarians, realising Britain's leadership of the alliance for the time being, were in no mood to compromise. MPs voted overwhelmingly against peace overtures or terms.[16] For them, too many lives had been lost and too much treasure expended to consider peace without victory. Sir Edward Carson, inside the War Cabinet, and Asquith, insisted that Britain could never make peace before a significant military victory.[17]

Lloyd George acknowledged the mood. The former *Daily Mail* editor summed up the government's position in the Commons: 'The war aim of the Allies is perfectly well known ... we were out to beat the German to his knees and then skin him alive.' Lloyd George applauded the sentiment, if not the content of the speech.[18]

New measures were introduced by Lloyd George to achieve a military victory prior to substantive talks, including the convoy system to protect vulnerable merchant shipping from submarine attacks. The War Cabinet also examined solutions to the supply crisis by clearing the coasts, and thus increase the ranges at which U-Boats would have to operate. A vast offensive had already been planned in November 1916.[19] It was thought that it should start at a point close to the railway junctions and ports so that a continuous stream of men and munitions could be brought to bear. In April and June 1917, there had been successes near Arras and at Messines, facilitated by a greater concentration of British artillery than at any time previously in the war. The plan was to replicate this saturating bombardment at Ypres through the dry summer months, aided by larger numbers of armoured vehicles and orchestrated in a combined arms operation. The ultimate objective was a decisive battle that would go some way towards the liberation of Belgium, and, at the very least, inflict defeat on the German army. Sir Douglas Haig, commanding the British forces in France, was more sanguine and saw the need to consider the Western Front as one continuous line of engagements, all of which contributed to the defeat of the German armies by destroying German reserves.

The Third Battle of Ypres (July–November 1917) did not live up to expectations for a variety of reasons, including prolonged wet weather which dashed any hope of manoeuvre. German resistance proved resilient, and the losses to their divisions was matched by heavy casualties on the British and Commonwealth side. An armoured offensive late in the year at Cambrai, while more successful, could not be reinforced, given the exhausting battle at Ypres. Germany therefore held on to its Belgian territory. The result was that Lloyd George lost faith in the British army's promises of the deliverance of Belgium. His plan was to await the arrival of American reinforcements in large numbers.[20] The Admiralty argued that Britain could not wait that long, given the losses it was taking in merchant shipping. The forthright Chief of the Imperial General Staff (CIGS), Sir William Robertson, deplored Lloyd George's idea of pausing operations on the Western Front in favour of peripheral campaigns in the Middle East. Robertson spent a great deal of his time trying to persuade the British government that the centre of gravity of the entire war was Germany. It was on the Western Front, he argued, that the war would be won or lost.[21]

Lloyd George was so exasperated by Robertson's objections that he tried to have him replaced as CIGS, but when that looked unlikely to succeed, he tried to give effective command of British forces in France to the French.[22] He even tried to take control of a large-scale strategic reserve, which the army desperately needed to stave off an imminent German offensive. This looming attack, involving many divisions that would be coming from the termination of the campaign against Russia, was regarded as a very serious threat by General Headquarters in France, but Lloyd George was more concerned that the army did not seem to grasp that British lives were not limitless.[23] He was thinking of the future of the country and the empire.

The German offensive broke over the British and French armies in March 1918. It was only with considerable difficulty that the attacks were slowed and then halted in the summer. With growing American assistance, greater numbers of armoured vehicles, and refinements in all arms tactics, the Allies drove back the increasingly exhausted German forces. The last 100 days proved decisive. German logistics broke down, their manpower reserves spent, and the civilian population angry and hungry.[24] As the British army and its allies pushed across north-eastern France, it became increasingly clear that a military victory was at hand. To stave off the inevitable, the German authorities appealed for an armistice, and, perhaps to forestall even greater American influence over the outcome, Britain and France agreed, but with strict terms.

In his memoirs, Lloyd George asked: 'Ought we to have interfered in the realm of strategy?'[25] His answer was that military personnel regarded the conduct of war as their exclusive domain and that he had respected that, but this war, as a total war, had also required 'high politics' to ensure a coalition was formed, to guarantee the co-operation of the Commonwealth and the Allies, to furnish manpower, to sustain public morale, to arrange the provision of food to the population, and to organise industry to supplies munitions and equipment. This was undoubtedly true, as the military could not have won the war without the assistance of the apparatus of the government. Nevertheless, on the repeated assaults against the fortifications against the Western Front, Lloyd George asked: 'Should governments intervene or leave the decision entirely to the soldiers?'[26] He seemed to think the failure of the Asquith government to restrain the soldiers and insist on campaigns in other theatres was a 'strategical blunder' but he exonerated himself from the decision not to halt the Third Ypres offensive in 1917, even though he claimed he always knew it would fail and result in heavy losses, on the grounds that he would have been overruled by his Cabinet colleagues. He surmised that 'the fighting of a battle is mainly the decision for the generals.' This is one of the least convincing conclusions of his entire memoir.

Lloyd George knew that, as the end of the war approached, he had to contend with the peace principles advanced by Woodrow Wilson, while insisting on the terms that Britain had sought consistently as the best way to ensure a lasting settlement. Belgium and Serbia were to be independent, territorial restorations were to be made to France, and territorial awards would be made to other allies. The Ottoman dominions would be broken up into independent Arab entities under British and French supervision, while Palestine would be internationalised as a remedy to jealous religious factions. In 1918, General Allenby had been encouraged by Lloyd George to mount a significant offensive in the Near East to complement the success of General Sir Stanley Maude in Mesopotamia. The result was that Baghdad and Jerusalem were in British hands, and, by September1918, the Ottoman armies in the Middle East had been broken. Britain was free to make changes to the region as it had imagined, although France became a significant obstacle as Parisian imperialist factions insisted on Syria being treated as a colony.[27] There were other Asian problems to resolve, not least the future relationship with Russia, now in Bolshevik hands, and the possibility of a resurgent and aggressive Turkey. The primary concern, however, was still Germany.[28]

Despite the common criticism of the costs of the war and the alleged inadequacy of the peace settlement of 1918–19, British war aims were achieved. A military victory had preceded a set of negotiations, exactly as the British government had demanded in 1916. As an end state, Britain insisted that Belgium and parts of France that had been occupied were liberated and restored. They were. Although Britain had feared the interference of the United States, its role as global peacemaker was relatively short-lived. Britain and France were to be the leading powers of the new League of Nations. The French and British empires remained intact too, and, for all its losses, Britain emerged with greater territorial responsibilities, not less. The British had desired the end of Prussian militarism and they got it. The Kaiser was deposed by his own people, and a German republic was founded on democratic principles. By the mid-1920s, it seemed as if the British desire to see Germany demilitarised had, remarkably, come to pass. The new scourge of Bolshevism, while menacing in Asia, was, in fact, contained in Europe thanks to the efforts of the Polish army and German counter-revolutionaries.

Britain also achieved the maritime security it had wanted. The Dardanelles and the Black Sea were in British hands. The German fleet was impounded in Scapa Flow, then scuttled itself, terminating that maritime surface threat for good.

In the Middle East, by 1920, Britain's Arab partners had independence or the guarantee of it, and elsewhere in the empire, the war had initiated a new benign spirit of guardianship in place of the excesses of the Victorian era. For the Dominions, there would be no going back to the old colonial relationship, and Britain acknowledged their new-found independence.[29]

The British had visualised an end state, and, despite the tortuous need to mobilise fully for a continental war, they had, despite all the setbacks, achieved their objectives. Whether that decision to pursue a continental commitment had been right at all is a different matter, but, based on the course the government set itself, its sense of ownership, its intervention for strategic effect, and the ideas it fought for, namely liberation, upholding a set of international standards, and security, the British authorities can be said to have achieved their victory.

## Churchill and 'winning' in World War II

In 1939, Britain's national interests were identical to those of 1914. It desired peace for commercial stability, security in its colonies, and maritime dominance to protect its sea lanes. There was no continental commitment, since France and its Eastern European partners were, collectively, thought strong enough. Italy was supposed to assist in keeping Germany contained.

At the outbreak of war in 1939, few in the British government wanted significant American involvement. Not only was there an expectation that the American public would oppose intervention, there was also a strong sense that Washington would demand the dismantling of the British Empire as the price for its support. President Roosevelt was known to be a critic of Britain's colonial apparatus. The hope was that Washington would opt for a benevolent neutrality, and this, in turn would prevent Japan from joining the war. However, there were serious concerns that American financial demands would amount to a price the United Kingdom could not pay. The Americans might demand naval bases, draw Australia and New Zealand into the American orbit, and replace the British commercial system which would make a continuation of imperial rule impossible.

Alongside these fears, Neville Chamberlain and his Foreign Secretary, Lord Halifax, imagined their war aims were to fight for peace, freedom, security and the sanctity of treaties. Chamberlain claimed, naively, his desire was 'to establish a new Europe ... with a new spirit in which the nations that inhabit it will approach their difficulties with goodwill and mutual tolerance.'[30] The public, with good reason, did not accept such high-sounding rhetoric. The defeat of France, Scandinavia and the Atlantic in 1939–40 indicated what two decades of optimism in international forums and disarmament had produced: humiliation and retreat.

On becoming prime minister, amidst the crisis, Churchill expressed his aim was military victory.[31] His famous address to the House of Commons in 1940 made that clear: 'You ask, "what is our policy?" I can say: It is to wage war, by sea, land and air, with all our might and with all the strength that God can give us; to wage war against a monstrous tyranny, never surpassed in the dark, lamentable catalogue of human crime. That is our policy.'[32] He then added: 'You ask, "what is our aim?" I can answer in one word: It is *victory*, victory at all costs, victory in spite of all terror, victory, however long and hard the road may be; for without victory, there is no survival. Let that be realised; no survival for the British Empire, no survival for all that the British Empire has stood for, no survival for the urge and impulse of the ages, that mankind will move forward towards its goal.' Churchill knew that the formidable armed forces of Nazi Germany would have to be defeated, and, given the alignment of the Soviets with Hitler in 1940, and the occupation of most of Europe by the spring of 1941, the British would have to draw on their imperial resources, and the support of the United States, to have any chance of victory. His intervention for effect was therefore not only to deprive the enemy of their goals, but to ensure a global coalition was formed of which Britain was the vanguard.

Churchill refused to be drawn on the specifics of any war aims. He knew this might prove counter-productive as the war unfolded. Victor Rothwell argues that, since domestic agendas of reform would have to be stipulated, Churchill's inclination was to avoid the subject and lay out more general principles.[33] The general sense was that Britain would struggle to meet the costs of a welfare programme, new housing, reconstruction, and job creation. Nevertheless, an equally plausible explanation was that Churchill feared specific war aims would bring up the question of the colonies, particularly amongst the Americans. In November 1942, Churchill had growled: 'I have not become the King's First Minister to preside over the liquidation of the British Empire.'[34] In 1943, when the Soviets requested a discussion on war aims, Churchill again refused. This was in part, of course, because Churchill and most of his colleagues wanted to prevent Germany and the rest of Europe falling into the hands of the communist government in Moscow. This had been the consistent line of British foreign policy through the inter war years and was resumed in 1945. The co-operation with the USSR after 1941 was a marriage of convenience. The best that could be hoped for was that co-operation might moderate Soviet ambitions and methods after the war.

The short term concern was to survive the onslaught of Nazi legions, then to defeat Germany militarily, and finally to find some mechanism to prevent Berlin from causing another war. A war of revenge was a particular anxiety.[35] There was also a concern that, while Hitler was the bellowing catalyst of belligerence, there might be something innately aggressive about the modern German. It seemed as though German foreign policy had always been expansionist. The concept of *Lebensraum* had characterised German elites of the 1910s, including the Kaiser, as much as it did Hitler and the Nazis.[36] The sheer ferocity of the German air offensives against cities, their ruthless ground operations, reprisals against civilian resistance, and then the wholesale mass murder of minorities, angered Churchill. He argued, in 1942, that Germany should be punished severely and it should endure significant casualties, a contrast from a more moderate opinion he had held the previous year. Yet he still hoped to avoid reparations which were widely held to have caused a German desire for revenge in the inter war years. He also remained consistent in refusing to see Germany reduced to permanent ruin. One American suggestion that Germany might be rendered an agrarian space, he rejected utterly.[37] Churchill knew that Germany would have to be restored, especially if it was to resist communism, but that did not preclude the wartime needs of defeating Germany militarily and destroying its material means of resistance.

Although there were committees, established in 1943, to discuss what to do with Germany, Churchill refused to be committed to policies which might have to be overturned later. The prevailing view of the committee staffs was that Germany represented a permanent and long-term threat to Britain's security. Clement Attlee, as Chairman of the Cabinet Committee on Armistice and Post War Matters, hoped that, after 50 years of democratic re-education, Germany's aggressive tendencies might be neutralised.[38] In the short term, the objective was to inflict very large scale casualties to reduce the military manpower potential for some time. Churchill rejected Stalin's proposal to execute tens of thousands of German officers held prisoner, which was an illegitimate proposition.[39] Attlee was unconcerned if many Germans were used as slave labourers by the Soviets, particularly if it could ensure rapid reconstruction in the immediate aftermath of the conflict. He disagreed with Eden's more lenient tendencies.[40]

Churchill was more concerned with reducing Germany's future capacity to threaten Britain by seeing the country partitioned along a north-south axis. The Protestant north would be separated from the Catholic pro-Austrian south, and possibly a third state, based on the industrial Ruhr and Rhine Valley would lean westwards. Churchill imagined a monarchical restoration, with the south under Habsburgs or the Bavarian Wittelsbachs. He put these ideas to Stalin in October 1944.[41] The plan was dropped because it was felt that democratic states would not fight to enforce such a territorial division and nationalistic Germans would demand reunification.[42] The British government was nevertheless keen to ensure that Austria was restored as a separate state, and that the *Anschluss*, greeted with apparent enthusiasm in 1938, would be reversed. The proposal was accepted at the Allied conference at Tehran.

At the end of the war, with Germany's defeat, Britain's proposal for a detailed settlement was rejected by the United States and the Soviets, both of which favoured unconditional surrender. Britain was concerned to terminate its continental commitment as rapidly as possible, and to return to its historic association with the Commonwealth.[43] The problem was that the USSR was changing facts on the ground which prevented a British disengagement with Europe. The Soviets had already repudiated old agreements over Poland and Czechoslovakia. British hopes for an Eastern European bloc, including Romania and Finland, to resist German and Russian power, proved hopeless. The Foreign Office remained concerned that German proclivities to admire power might turn them towards Stalin, or that weak Balkan states might tempt Germany to reassert itself there and imagine a new *Lebensraum* project.[44]

These fears were unrealised in 1944–45 and were instead replaced by concerns about zones of occupation. Britain was eager that it should control the north of Germany, to occupy naval bases, industrial zones, and maintain proximity to its Scandinavian and seaboard allies. The Americans disputed this, but London was afraid the US would enter another period of isolation and simply give away whatever it took control of in 1945. Roosevelt had warned that he expected all US forces to have been withdrawn from Europe within two years of the end of the conflict there.[45] At Yalta, the Soviets seemed determined to partition Germany, and their commitment to retain all their gains only made the British more nervous.

There seemed a strong possibility that liberated Western European countries would fall under Soviet sponsored communist movements, and much effort would be required to prevent this. Churchill felt that the solution lay in influencing Stalin and dissuading him for this course. Attlee contemplated an invasion of Spain to depose General Franco but Churchill condemned the idea, arguing it would foster antagonism that would spread to France and Italy. There was a more robust approach with Greece, however. British planners regarded Greece as too important, and too close to its Eastern Mediterranean interests, to be allowed to fall to the communists. In October 1944, Stalin had agreed to Churchill's demand not to intervene just as communist partisans and their government enemies clashed in Greece's civil war. The temporary Treaty of Varkiza in 1945 appeared to have concluded the issue, but a new wave of violence from 1946 drew in support from the United States, against Yugoslavian-backed communists.[46]

Churchill could not adopt an anti-Soviet position in the war: he needed their military power against Germany in Europe, and he needed public backing, which would not have tolerated such a position. The British public would also not have accepted a long-term subjugation of Germany, a position widely regarded as having caused the war. That said, the mood in April and May 1945 was decidedly anti-German, as news of the death camps was revealed.[47]

If Churchill avoided specific war aims, he was remarkably prescient in how the war should be conducted. His memoranda on is stages, drawn up in December 1941 under the headings of 'Mediterranean', '1943', and 'Pacific' were accurate forecasts.[48] His approach to victory was to weather the air-sea onslaught of 1940–41, seek every opportunity to strike back at the Axis powers, including a campaign to 'set Europe ablaze' with partisan warfare. He sought to contain Germany by contesting the Mediterranean and North Africa, and prioritised the European theatre, even to the extent of assisting the Soviets, over the defence of East Asia.[49]

Above all, he cultivated the relationship with the United States. He knew there would be criticisms of the British Empire, but he also knew Roosevelt appreciated the greater cause at stake was the defeat of Nazism. Churchill's view of the British Empire was, in any case, that it was a fundamentally beneficent and civilizing influence in the world. It was not, like Nazism, not an edifice of race and power alone.[50] Churchill felt Britain should be treated with parity by the United States because it had been the sole power offering resistance in 1940–41 and because its forces were still fighting in large numbers in every theatre. The Americans nevertheless felt Britain should make way for US leadership because of America's greater military strength, economic power, logistical support to the alliance, and because of its strategy. There were tensions, over finance, over the future of the empire, and over the higher direction of the war. Critics note that this vital alliance into which Churchill entered to save Britain, including its empire, contributed ultimately to its decline and disappearance. After 1945, Britain's power was sharply curtailed. Its loss of prestige, power, share of global commerce and domestic wealth were considerable and the United States was compelled to assist.

Churchill ensured that he did not antagonise the Soviets so to ensure that the little amount of trust between east and west did not break down entirely, and also to ensure that sympathies in Britain for the Soviets' defence were not squandered. In the course of the war there were disagreements, not least over the timing and location of a 'second front' and the shape of the post-war world, but these were resolved, and Churchill was the architect of Britain's victory, even though it was dependent on the United States.[51] After 1945, Britain's global power, relative to the new Superpowers, was evidently diminished. As many observers have pointed out, the paradox of World War II was that, in winning the conflict, Britain lost much of the subsequent peace.

## Comparative analysis

This chapter puts forward three unifying ideas to measure success: ownership, intervention for effect, and fighting for ideas. 'Ownership' indicates the possessions for which a war is fought. In World War I, Lloyd George was eager to acquire territory that could be used to ensure a stronger negotiating position at the end of the conflict, and this included the liberation of Belgian and French soil, overrunning the German colonies in Africa and Asia, and controlling Arabia and Mesopotamia under British protection in place of the Ottoman Empire. The result was that Britain possessed even greater territorial responsibilities in 1918 than in 1914, renewing an older debate on the cost and consequence of imperial rule. While Lloyd George could assert that Germany was deprived of colonies and Belgium and France had been liberated, there was no denying that the price Britain paid was considerable and the war resulted in a significant loss in the share of global commerce, measured in terms of shipping and manufacturing. Moreover, the British sterling was replaced as the global currency by the dollar.

Winston Churchill was far more sanguine about 'ownership', seeking to preserve the British Empire, while avoiding unnecessary responsibilities. His approach was more concerned with the existential questions of defence and a moral position which would emphasise the cohesion of the alliance. Churchill had to manage a series of setbacks throughout the war and the compromise of accepting American and Soviet interests even where they collided with Britain's.[52] Nevertheless, by 1943, Churchill could see the need for a strong post-war Anglo-American partnership, while he warned of the perils of Soviet territorial ambition and its ideological potency: a consistent position regarding communism he held throughout his career.

The second criterion is 'intervention for effect', which might only mean the defeat of a threat rather than the acquisition of some vital resource. Lloyd George modified the intention of his predecessor as Prime Minister, namely to liberate Belgium and to dismantle German militarism. Instead, he developed the idea of ensuring that Germany could not pose a future threat to France and Belgium, by post-war penalties, but not so comprehensively that Germany would cease to exist. This placed him at odds with his French allies, but closer to the Americans. Lloyd George was nevertheless consistent in wanting to see the warlike means of Germany reduced and removed, a view he expressed from at least 1912 and during the peace negotiations in Paris. Moreover, he secured the impounding of the German navy in 1919 and insisted on the demilitarisation of the German state.

For Churchill, the 'intervention for effect' was even clearer. The nature of the Nazi regime and its aspirations meant there could be no negotiated settlement with Germany as there had been in 1918–19, and his policy was to wage war and achieve military victory.[53] If the purpose was clear, the execution was not. The war was, wrote Churchill, 'the same kind of problem as unfolding a long, sustained, interlocked argument.'[54] He referred to 'a proposition which, whether of few or numberless parts, is commanded by a single unity of conception'. He noted there must be adaptation to the context, and there must be room for creativity, but above all one had to appreciate how each operational element contributed to the overall context or the 'big picture'. As Eliot Cohen notes: 'Churchill believed that the formulation of strategy in war did not consist merely in drawing up state documents sketching out a comprehensive view of how the war would be won, but a host of detailed activities which, when united and dominated by a central conception, would form a comprehensive picture.'[55]

The final and related element is the notion of 'fighting for ideas.' In World War I, Lloyd George embraced the existing rhetoric of a Great War for civilisation. The Allied cause, for public consumption, was a worthy, just war against German brutality, occupation, and abrogation of law. For evidence, Lloyd George pointed to the German invasion of neutral states, illegal technologies, and unrestricted submarine warfare. It was also an objection to German actions which had led to the death of civilians, through air bombardment, the sinking of passenger and merchant ships, or by reprisals. The British prime minister, a long-standing radical politician, also believed that the war must result in 'a land fit for heroes to live in', that is, a domestic agenda that alleviated poverty and want. He was suspicious of class prejudice, borne of his modest Welsh, non-conformist background, and believed the justification for British colonial rule was not some innate racial advantage but a guardianship. He belonged to the liberal tradition that saw imperial governance as temporary and only justified if it enabled peoples, in time, to govern themselves. It was for this reason he had preceded Woodrow Wilson's '14 points', but deplored the haste of the American agenda, which, he reasoned, could only result in chaos and more conflict.

Churchill mobilised ideas in his speeches to the British public, but also, indirectly, to the occupied populations of Europe and the people of America. Even in 1940, facing invasion, he told his colleagues: 'I take up my task with buoyancy and hope. I feel sure that our cause will not be suffered to fail among men. At this time, I feel entitled to claim the aid of all, and I say, "come then, let us go forward together with our united strength."'[56] He expressed the military mission of Britain to be the destruction of Nazism and the liberation of Europe, and he ensured that his narrative aligned with the United States. This was no easy task. Roosevelt, like many Americans were suspicious of British motives, especially in regard to imperialism. In October 1942, *Life* magazine produced an open letter to the British people and reminded them the United States was not fighting to 'hold the British Empire together.'[57] Critics have subsequently condemned Churchillian rhetoric as misleading, concealing the evident failure and eclipse of British power.[58] Supporters point to the strategic dilemma Churchill faced. He knew there was a moral choice to make in favour of the future of civilization that was more important than British economic prosperity. He would thus engage in this fight, regardless of costs, for the greater good, but pragmatically preserve whatever he could in the process.

### *Notes*

1 Basil Liddell Hart, *The Remaking of Modern Armies* (London: John Murray, 1927), 92; Basil Liddell Hart, *Paris, or the Future of War* (London: Kegan Paul, Trench, Trubner & Co. Ltd, 1925), 20.
2 Niall Ferguson, *The Pity of War* (London: The Penguin Press, 1998), 459.
3 *Ibid.*, 462.
4 Sir Michael Howard, *The Continental Commitment* (London: Temple Smith, 1972), 57–8.
5 David French, *British Strategy and War Aims, 1914–1916* (London: Allen and Unwin, 1986), xi.
6 French, *British Strategy and War Aims*, 14–16.
7 Asquith, speech to French Senators and Deputies, 10 April 1916, in *War Speeches by British Prime Ministers* (London: HMSO, 1917), 104–109, cited in Paul Guinn, *British Strategy and Politics* (Oxford: Clarendon, 1965), 122.
8 *Parliamentary Debates*, House of Commons, xcvi, 30 July 1917, cols. 1847–55.
9 Foreign Office Memorandum on War Aims, August 1916, referenced by David Lloyd George in *The Truth About Peace Treaties*, Vol. I (London: Victor Gollanz, 1938), 131–50.
10 Arthur Link, *Woodrow Wilson and the Progressive Era, 1910–1917* (New York: Harper Brothers, 1954), 161, 198; Charles Seymour, *The Intimate Papers of Colonel House*, Vol. 1 (Boston: Houghton Mifflin, 1926), 347–8, 353–4, 380; David R. Woodward, *Trial by Friendship* (Lexington, KY: University of Kentucky Press, 1993), 12–15.
11 Cyril Falls, *The Great War* (New York: Perigree Trade, 1961), 83.
12 John Grigg, *Lloyd George: War Leader* (London: Allen Lane, 2002), 239. Actually, Lloyd George insisted the phrase was used in an interview with the American press in reference to Germany's choice–'a fight to the finish with England'–and he had no doubt Britain would meet the challenge successfully. Lloyd George, *War Memoirs*, I, 508.
13 Guinn, *British Strategy and Politics*, 180, 184.
14 Ross A. Kennedy, *The Will to Believe: Woodrow Wilson, World War I, and America's Strategy for Peace and Security* (Kent: Kent State University Press, 2009), 68–74, 86–89, 94–98.
15 Guinn, *British Strategy and Politics*, 238.

16 *Parliamentary Debates*, House of Commons, xciii, 16 May 1917, cols. 1625–1734; xcvi, 26 July 1917, cols. 1479–1590; G. Lowes Dickinson, *Documents and Statements Relating to Peace Proposals and War Aims* (December 1916 to November 1918), xvi, 145.

17 Guinn, *British Strategy and Politics,* 240–1.

18 Lloyd George, *War Memoirs*, II, 1121–39; *Parliamentary Debates*, House of Commons, cvii, 16 August 1917, cols 1465–71.

19 Lloyd George, *War Memoirs*, II, 1250–2; *Military Operations: France and Belgium*, 1917, I, edited by Captain Cyril Falls (London: Macmillan, 1940), 21 and II, 8–10.

20 C. R. M. F. Cruttwell, *The Role of British Strategy in the Great War* (Cambridge: Cambridge University Press, 1936), 66.

21 David R. Woodward, *Field Marshal Sir William Robertson* (London: Praeger, 1998), 60–2.

22 Guinn, *British Strategy and Politics,* 286–7; C. E. Callwell, *Field Marshal Sir Henry Wilson: His Life and Diaries*, II (London: Cassell, 1927), 32, 38.

23 Lloyd George, *War Memoirs*, II, 2032.

24 Lloyd George, *War Memoirs*, II, 2033.

25 Lloyd George, *War Memoirs*, II, 2033–5.

26 Lloyd George, *War Memoirs*, II, 2035.

27 Minute, 22 Jan 1919, FO 608/97/403. TNA ; de St Quentin to Millerand, 'Visées anglaises sur la Syrie', 28 July 1915 in Vincent Cloarec, *La France et la question de Syrie, 1914–1918* (Paris: CNRS Éditions, 2002), 126.

28 Margaret Macmillan, *Peacemakers* (London: John Murray, 2001), 473, 476.

29 Howard, *Continental Commitment*, 64.

30 Ivan Maisky, *Memoirs of a Soviet Ambassador: The War 1939–43* (London: Hutchinson, 1967), 22.

31 Eliot A. Cohen, *Supreme Command: Soldiers, Statesmen and Leaders in Wartime* (New York: The Free Press, 2002), 94–96.

32 *Hansard*, 13 May 1940.

33 Victor Rothwell, *War Aims in the Second World War: The War Aims of the Belligerents, 1939–45* (Edinburgh: Edinburgh University Press, 2005), 65.

34 *Hansard*, November 1942.

35 Anthony Eden, *Freedom and Order: Selected Speeches, 1939–46* (London: Faber, 1947), 110.

36 The overreaction against Fritz Fischer in the 1960s over this same question indicated just how sensitive German historians were to this same question. J. Jenkins, 'Fritz Fischer's "Programme for Revolution": Implications for a Global History of Germany in the First World War,' *Journal of Contemporary History*, Vol. 56, No. 3 (2013). See also Dieter Fricke, 'Zur Rolle des Militarismus nach innen in Deutschland vor dem ersten Weltkrieg', *Zietschrift fur Geschichtswissenschaft*, Vol. 6 (1958): 1303–05.

37 Rothwell, *War Aims*, 67.

38 Minutes, 11 August 1943, FO 371/34461, no. 12405, The National Archives.

39 It is striking how many German prisoners died in Soviet captivity. Official Soviet statistics are that 381,067 German Wehrmacht prisoners died in Soviet camps, but Rüdiger Overmans believes this figure represents only a third of the actual numbers who perished. Rüdiger Overmans, *Soldaten hinter Stacheldraht. Deutsche Kriegsgefangene des Zweiten Weltkriegs* (Ullstein: Propyläen, 2000), 246.

40 Sir Llewellyn Woodward, *British Foreign Policy in the Second World War*, vol. V (London: HMSO, 1970), 202.

41 'Anglo-Soviet Political Conversations at Moscow, October 9–17, 1944' Record of Meetings, Ismay Papers, V1/10, King's College London.

42 War Cabinet Minutes, 5 October 1943, CAB 65/40, TNA.

43 Rothwell, *War Aims*, 72.

44 Memorandum of the German Department of the Foreign Office, 15 July 1944, FO 371/39116, no. 9330. TNA.

45 Rothwell, *War Aims*, 8.

46 Christina J. M. Goulter, 'The Greek Civil War: A National Army's Counter-insurgency Triumph,' *Journal of Military History*, Vol. 78, No. 3 (July 2014):1017–55. Stalin changed his mind several times over the Greek conflict.

47 Armistice and Post-War Committee, Minutes, 17 May 1945, CAB 87/69, TNA.

48 CHAR 23/10, 16 December 1941 (Churchill Papers, Cambridge); Martin Gilbert, *Road to Victory: Winston Churchill, 1941–45* (London: Heinemann, 1986), 9ff.

49 Brian Bond, 'Alanbrooke and Britain's Mediterranean Strategy, 1942–44' in Lawrence Freedman, Paul Hayes and Robert O'Neill, (eds), *War, Strategy and International Politics* (Oxford: Clarendon, 1992), 175.

50 Kirk Emmert, *Winston S. Churchill on Empire* (Durham, NC: Carolina Academic Press, 1989).

51 Andrew Roberts, *Masters and Commanders* (London: Penguin, 2008), 69.

52 David Stafford, *Roosevelt and Churchill: Men of Secrets* (London: Little Brown and Company, 1999), xvii–xviii; Alex Danchev, 'Britain: The Indirect Strategy' in David Reynolds, Warren Kimball, A. O. Chubarian, (eds), *Allies at War* (New York: St Martin's, 1994), 1–26.

53 There were critics of his methods, see Field Marshal Lord Alanbrooke, *War Diaries, 1939–1945*, edited by Alex Danchev and Daniel Todman, (London: Phoenix, 2002 edn.), entry for 10 September 1944, 590.

54 Churchill, cited in Cohen, *Supreme Command*, 108.

55 Eliot Cohen, 'Churchill at War', *Commentary*, Vol. 83, No. 5 (May 1987): 40–49; Cohen, *Supreme Command*, 110.

56 Ibid.

57 Niall Ferguson, 14 March 2003, http://www.niallferguson.com/journalism/history/whats-really-in-it-for-britain, accessed November 2019.

58 Peter Clarke, *The Last Thousand Days of the British Empire: Churchill, Roosevelt, and the Birth of the Pax Americana* (New York: Bloomsbury Press, 2008), 507.

CHAPTER 7

# 'Winning' in the Cold War and the Nuclear Age, 1945–1990

Jonathon Riley

The British diplomat and historian Harold Nicolson remarked in the introduction to his book *The Congress of Vienna*, that, 'The basis of any alliance, or coalition, is an agreement between two or more sovereign states to subordinate their separate interest to a single purpose.'[1] His book was published at the close of World War II, a war that had been won by a coalition of nations who were ideologically completely opposed, but who were united in the face of a common threat from Nazi Germany. As with most other coalitions of this kind, the disappearance of that unifying purpose led to its break-up: winning in a coalition, ironically, is often the harbinger of the re-emergence of tensions and competitions that were there before the coalition was formed, were suppressed, but can now re-emerge. So it was at the end of World War II.

Even before the defeat of Japan, in July 1945, tensions emerged at the Potsdam Conference where the Soviets pressed for reparations from Germany, which they would achieve by removing industry and materiel from their agreed zone of occupation; the Allies, by contrast, were thinking in terms of reconstruction under what became known as the Marshall Plan. It was, however, at this conference that President Truman told Stalin that he possessed an Atomic bomb; a week later, Truman demonstrated his weapon at Hiroshima. Stalin reacted calmly at Potsdam but later called it: 'superbarbarity ... the balance has been destroyed ... That cannot be.'[2] For Truman and the western allies, however, the bomb provided the means to counterbalance the overwhelming military power of the USSR in Europe – around 300 divisions in 1945 as against less than one-third of that total from the west, and that imbalance growing as the west rapidly demobilised.

So began what became the Cold War. During 1946 and 1947, the Soviets consolidated their control over East Germany, Poland, Hungary, Romania, Bulgaria and Czechoslovakia; and they extended their influence into Albania and Yugoslavia. In response, the United States began a programme of containment of Soviet power, which began with the publication of *The Sources of Soviet Conduct*, a highly influential article written by George F. Kennan under the pseudonym 'Mr. X' and published in *Foreign Affairs* magazine in July 1947. Kennan had been the Deputy Chief of Mission of the United States to the USSR from 1944 to 1946 and was under no illusions as to the nature of Soviet power. In 1947, the programme dramatically widened in nature to become US policy, now generally referred to as the Truman Doctrine. Truman's avowed purpose, and his definition of winning, became to counter Soviet, and indeed all communist, expansion. Truman announced his intentions to Congress in March 1947,[3] and developed his theme further on 4 July when he pledged to contain Soviet threats to Greece, by providing aid to the Royalists in the Greek Civil War, and to Turkey. Truman told Congress that: 'it must be the policy of the United States to support free peoples who are resisting attempted subjugation by armed minorities or by outside pressures.'[4] Winning for the west was therefore to halt the forward progress of communist expansion, but not necessarily to push it back or to win the battle of ideas.

Direct use of American military force was to be avoided unless absolutely necessary, but Congress appropriated funds immediately to support the governments of Greece and Turkey.[5] More generally, the Truman Doctrine implied American support for other nations threatened by communism. It became the cornerstone of American foreign policy for the next 45 years, leading almost immediately, in 1949, to the formation the North Atlantic Treaty Organisation, or NATO. The Soviets responded by creating the Warsaw Treaty Organisation (WTO) and thus the superpower confrontation in Europe was born. The Cold War's first near-miss in terms of military competition was the Berlin Blockade of 1948–1949 and although they could not out-face the west on that occasion, the Soviets received a boost through the victory of communism in the Chinese Civil War in 1949. Already, it seemed clear that winning in terms of conventional military confrontation in Europe was not a realistic prospect for either side, for the costs would be too high. Winning meant the successful extension, and consolidation, of one ideology over another: in this case, communism and capitalism. Marxist–Leninist ideology viewed its own triumph as inevitable; Truman set out to show they were wrong.

However, in August 1949, communism received yet another boost when the Soviets detonated their own nuclear bomb, having been working on a programme since 1942 which was reinforced after the discovery of the Nazi's material and personnel at the end of World War II.[6] Although both sides now had nuclear weapons, they were short on the means of delivery. Initial work focused on developing strategic bombing from the air, following the pattern of World War II. Long-range, very high-altitude aircraft like the US Convair B-36 bomber and the Soviet Tupolev Tu-95 were the result. In terms of the policy and doctrine to guide the employment of such weapons, the US adopted 'massive retaliation' in response to any Soviet conventional, chemical or nuclear attack – a phrase first used by Secretary of State John Foster Dulles in 1954:

> We need allies and collective security. Our purpose is to make these relations more effective, less costly. This can be done by placing more reliance on deterrent power and less dependence on local defensive power … But there is no local defense which alone will contain the mighty landpower of the Communist world. Local defenses must be reinforced by the further deterrent of massive retaliatory power. A potential aggressor must know that he cannot always prescribe battle conditions that suit him.[7]

Although there could be no worthwhile or even meaningful victory by military means, and little ability to disrupt or destroy the opposing political system by force in Europe, fighting and winning was possible outside Europe. This could be either by direct involvement beyond the shadow of nuclear weapons, or by the use of proxies, or even in limited operations within one's own sphere of influence provided this was below the threshold of confrontation with the other side. This was brutally demonstrated during the Korean War from 1950 to 1953, which ended in stalemate and the formal creation of the two states in the Korean peninsula, one under communist control and one capitalist. But the Americans and the Soviets also competed for control in Central and South America, in Africa and in South-East Asia. The Soviets suppressed internal dissent as well, in East Germany and Hungary during the 1950s. This spreading competition gave rise to further crises such as Suez in 1956, the Berlin Wall in 1961 and, perhaps the closest that the major rivals ever came to outright war, the Cuban Missile Crisis of 1962. Following the humiliating climb-down by the Soviets during the crisis, and the consequent fall of Nikita Krushchev as Soviet leader, a new phase of the Cold War began. The Soviet Union and China, formerly close partners, split over ideological differences and this split affected the whole communist sphere. On the other side, some American allies, particularly France after the US failure to back the Suez intervention, began to distance themselves from the US. In the case of France, this eventually led to their withdrawal from the NATO integrated command structure and the removal of NATO's major military headquarters from Versailles to Mons. The Soviets continued to repress dissent with impunity inside their own sphere of influence, crushing the 1968 Prague Spring in Czechoslovakia. Impunity, not only because the balance of nuclear power allowed them to do so, but because US conventional military power was tied down for more than a decade in the biggest of the proxy wars, that in Vietnam which ended not in the restoration of the *status quo ante bellum* as in Korea, but with the triumph of communism in Vietnam and Cambodia. By the close of the 1970s, it seemed that it was communism that was winning both the physical battles and the battles of ideas.

The doctrine which gave rise to this bipolar world, characterised by stalemate in Europe and conflict beyond, was that of Mutually Assured Destruction – or, appropriately, MAD. As the nuclear arsenals of the two sides grew it became clear that the full-scale use of strategic weapons – which could now be delivered by aircraft, by land-based missiles, and by missiles from submarines – would inevitably mean the destruction of both an attacker and a defender.[8] From 1955 onwards, for example, the United States Strategic Air Command (SAC) kept one-third of its bomber crews at 15 minutes notice for take-off, to attack targets inside the Soviet Union and destroy them with nuclear bombs in the event of a Soviet first-strike attack on the United States. In 1961, President John F. Kennedy increased funding for this programme and raised the commitment to 50%. MAD was derived from deterrence theory, which proposed that the threat of extreme, destructive retaliation by one side, in response to their use by another, would prevent the use of those weapons. Neither side would have any incentive to use these weapons as the price of victory would be too high – not only in terms of the loss of life and destruction of civilisation in Europe, but also the effects on continental US. and the sacred Russian *narod,* or motherland. The term was first used by Donald Brennan, a strategist at the Hudson Institute, in 1962.[9]

MAD can be described under game theory[10] as a 'Nash equilibrium', named after the mathematician John Forbes Nash Jr. This equilibrium is a proposed solution of a non-cooperative game involving two or more players, in which each player is assumed to know the strategies of the other players. This is achieved through a rational process of elimination, i.e. that strategy A is inferior to strategy B when I choose C, so my opponent will choose B – his best choice. No player therefore stands to gain by changing only their own strategy, rather than that of both or all players.[11] In terms of game theory, if each player has chosen a strategy, and no player can benefit by changing their own strategy while the other players maintain theirs, then the set of all current, in-play strategy choices and their corresponding benefits constitutes a Nash equilibrium. Game theorists have used the Nash equilibrium concept to analyse the likely results of proposed geopolitical, strategic, decisions and predicting outcomes since it is chiefly concerned with best responses to a set of circumstances.

Some caution is required, however, since the type of game theory used here is more explicitly referred to as 'one-shot Nash equilibrium', because a nuclear contest, if viewed as a game, will be played only once and there are no chances for future retaliation. Nuclear stand-off, without actual use, however, is *not* a one-shot game and so it may be appropriate to describe the situation of the Cold War as an infinite game in which there was no end, simply because in a one-shot game, the chances of one side launching a nuclear strike were considerably higher, as consequences were non-existent should they launch. It should also be understood that the Nash equilibrium assumes that each actor responds to a challenge without knowing what the other side is doing, or else they act immediately when threatened.[12]

MAD, and the modelling provided by game theory, taught that each side in a nuclear confrontation would have enough nuclear weaponry to destroy the other side. Game theory could not inherently predict the outcome of a MAD scenario, but only show responses to one side's action by the other side. For example, side

A chooses not to launch, but side B chooses to do so. This yields A dying and B surviving. However, if A had a backup plan on death to kill B, game theory is unable to factor that information in based on the responses given. That said, under MAD, either side, if attacked by the other in a first strike, would suffer significant damage to its nuclear arsenal, but through dispersion or concealment could retain enough power to respond with the same or even greater force in a second strike. The result would inevitably be an irreversible escalation of violence which would end in mutual, total, and assured destruction. The doctrine was predicated on the assumption that neither side would construct massive systems of shelters, since doing so would destabilise the equation by lessening the effects of a second strike – and would probably be far too expensive and time-consuming anyway.[13]

The doctrine further assumed that neither side would dare launch a first strike because the other side would have enough visibility of the battlespace to see the attack coming, and would launch a response as a result of a warning, rather than on impact, as game theory suggested. The result would still be that of unacceptable losses for both sides. Thus, winning was defined by a stand-off resulting in a tense but stable peace. Winning could also be defined, it can be argued, by the results of the nuclear arms race that developed between the Soviets on one side, and the US, Britain and France on the other. It was a race paralleled between India and Pakistan with China also developing its own nuclear arsenal. In this race, both sides invested huge sums to keep nuclear parity, or at least retain their second-strike capability.

Some of the assumptions of the doctrine stand examination, for example, that a launch by the other side would be correctly identified and that false readings, or mistakes, would not lead to a full-scale exchange; and that launches could be correctly attributed – for example, a launch against the US from the Russian–Chinese border or a launch from a nuclear submarine. Game theory cannot predict these issues, but it is not the point of game theory to do so! It also assumes that the launch process will not fall victim to a rogue commander; one famous example of this was the near-escape during the Cuban Missile Crisis when an argument broke out aboard a Soviet nuclear missile submarine with no radio communications to the outside world. The executive officer, Vasili Arkhipov, refused to launch a missile in spite of an order from Captain Savitsky to do so.[14] It also assumes that every leader with a nuclear capability cares about the survival of their subjects. Winston Churchill warned that no strategy could: 'cover the case of lunatics or dictators in the mood of Hitler when he found himself in his final dugout.'[15] It also assumed no development of anti-missile technology, an assumption which was to be breached in the closing stages of the Cold War.

There were, of course, huge risks in the MAD equation, emanating from the notes of caution described in the Nash equilibrium. Risk is usually defined as the likelihood that something will go wrong; under MAD, that likelihood was the product of possible miscalculation and/or misunderstanding of a particular perceived threat. Miscalculation or misunderstanding could lead to a launch based on faulty or partial information and to mitigate the risk, therefore, several measures were necessary. The first was layered intelligence from multiple sources, to bear down on uncertainty and therefore misunderstanding. The second was a dual or triple lock mechanism on launch codes. The third was the ability for the two superpowers to speak directly to each other at an appropriately high level, should this be required.

The strategy of MAD was fully developed by the early 1960s, primarily by US Secretary of Defense Robert McNamara, who announced his ideas in a speech at Ann Arbor, Michigan, in February 1965.[16] According to McNamara, there was the very real danger that any opponent of the US that possessed nuclear weapons could try to eliminate America's forces with a surprise, devastating first strike and thus theoretically win a nuclear war relatively unharmed. A true second-strike capability could only be achieved when the nation had a guaranteed ability to survive a first strike and then fully retaliate.[17]

The US had achieved an early second-strike capability through its high-readiness strategic nuclear bombers, mentioned above; the potency of these bombers was further increased by maintaining a percentage actually in the air, on patrol, at all times. This meant that even before the development of other delivery systems, the US could still retaliate, even after a highly damaging first-strike attack. It was, however, expensive and as the disparity in numbers of intercontinental ballistic missiles (ICBMs) widened between the US and the Soviets, priority was increasingly given to the development, deployment and readiness of ICBMs over bombers. It was with the advent of ballistic missile submarines of the George Washington class in 1959, 18 of which could be kept at sea for long periods and were undetectable, that a genuinely survivable nuclear force became possible and a retaliatory second strike capability guaranteed.

There was, however, an alternative view, put forward as many in the US began to turn against Massive Retaliation – not least because the doctrine was not able to stop the tide of communist-inspired revolutions in Africa, Asia and South America. It was, quite simply, too blunt an instrument and in addition, it was not credible: no-one believed that a massive nuclear strike would be launched against Moscow in retaliation for a communist take-over in Cambodia. It was felt by some that a more flexible approach was needed to address the sorts of competitive conflicts that were being fought between the ideologies outside the Central European theatre. The term 'flexible response', which described the use of a wide range of diplomatic, intelligence, military and economic tools (the DIME), to deter or defeat and an enemy, first appeared in US General Maxwell D. Taylor's book *The Uncertain Trumpet*, in 1960, which adversely criticised US national security policy.

Immediately after assuming the presidency in 1961, even as MAD reached its fullness, John F. Kennedy instructed his advisers to begin drafting a new strategy based on the belief that the United States should adopt this policy of a wide variety of military and non-military options for responding to communist advances. Kennedy presented to Congress the first draft of what became known as Flexible Deterrent Options in March 1961: the programme was rapidly adopted as official national security policy in the United States; NATO – many of whose member states were deeply worried by the implications of Massive Retaliation – officially adopted the strategy in 1967.[19]

Flexible Response gave NATO, and the US in particular, the ability to meet any hostile act with a proportional response. Economic aid, trade deals and arms supplies were among the chief hard responses; softer responses included changes in diplomatic relations, or media campaigns. It also, however, required Kennedy to modernise the US missile arsenal, increase conventional forces by land, sea and air. And increase the size, reach, capability and mandate of US Special Forces.

All ideas of deterrence of course rested on its being credible to both sides, and credibility was vital: both the Soviets and the west had therefore to invest significantly in defence. During the 1970s, NATO figures indicated that in the European theatre there was a rough balance in the numbers of strategic bombers (major missile systems, submarines and multiple warheads), but an imbalance of conventional forces: 58 WTO divisions immediately available as against 27 NATO, and total divisions in national orders of battle being 152 WTO and 115 NATO; 19,000 versus 6,100 tanks; and 3,410 versus 1,700 aircraft. The Soviets also enjoyed the advantage of interior lines of communication, but were at a disadvantage in terms of quality.[20] The need for credibility also meant that both sides had to protect their nuclear assets by diversification and dispersion, and by the hardening of missile silos. The cost of this sort of investment was high: figures for the United Kingdom indicate that at the time of the Korean War, 11.2% of GDP was spent on defence. By 1959 this had fallen to 7% and by 1970, to 5.4%. This was higher than any other European NATO nation, but less than that of the US.[21] By contrast the Soviets were spending 15–17% of GDP on defence throughout the 1970s and early 1980s.[22] Of course, western GDP was far greater than that in the communist world and the strain of maintaining parity, or superiority, was telling on the Soviet economy.

To NATO nations, the conventional force imbalance did not matter since at this time there was no intention of fighting a symmetrical war against a Soviet attack in Europe – under Massive Retaliation, and underpinned by the credibility of the doctrine and the means to implement it, any Soviet attack of any kind would inevitably invoke a nuclear response. For most of the period of the Cold War, ammunition stocks, for example, reflected this thinking: the British Army of the Rhine maintained no more than five to seven days' of supply depending on the intensity of combat and consumption.[23] In the event of a Soviet conventional attack on Western Europe, NATO planned in the first instance to use tactical nuclear weapons, before escalating if necessary to intermediate and then strategic weapons. The Soviets answered by issuing a statement that any use of nuclear weapons, whether tactical or otherwise, against Soviet forces would be met by a full-scale Soviet retaliatory strike, adopting the language of massive retaliation. Actually, the Soviet position was rather more extreme, for they planned to *begin* any operations in Europe with chemical and nuclear weapons, rather than escalate to them later, something not fully understood in the west.[24]

By the 1970s, however, both NATO and the Soviets began to consider compromises order to create a more stable and less tense international system, ushering in a new relationship known as détente. The chief fruit of this was the Strategic Arms Limitation Talks (SALT), begun in 1969, and the opening of US diplomatic relations, including formal recognition, with the People's Republic of China; to some this was mere cynicism, exploiting the ideological split between China and the Soviets rather than a genuine move towards mutual toleration between the opposing ideologies. SALT I was signed in May 1972 by US President Richard Nixon and Soviet Premier Leonid Brezhnev; SALT II in 1979.[25] Détente, however, collapsed soon afterwards with the Soviet invasion of Afghanistan at Christmas 1979.[26] In spite of this, the original MAD doctrine was modified on 25 July 1980, with US President Jimmy Carter's adoption of the 'Countervailing Strategy' under Presidential Directive 59. According to its architect, Secretary of Defense Harold Brown, countervailing strategy stressed that the planned response to a Soviet attack was no longer to bomb Soviet cities, killing large numbers of civilians; but first to kill the Soviet leadership, then attack military targets. Brown hoped to win through a Soviet surrender before total the destruction of the Soviet Union – and probably also the United States.[27] This was a step towards the development of a war-fighting doctrine under the nuclear cloak, while still maintaining the possibility of assured destruction for at least one protagonist.[28]

Following the collapse of détente, the 1980s were, therefore, a period of renewed high tension. The election of Ronald Reagan as president of the US in the wake of the humiliating years under Jimmy Carter, and the election of Margaret Thatcher as prime minister of Great Britain, brought an escalation of the Cold War confrontation led by these two close allies and marked the end of containment as victory: from now on, the west would push communism back. The Reagan Administration implemented a new policy towards the Soviet Union through National Security Decisions Directive 32 which was designed to confront the Soviets in three key areas. First, Soviet access to modern technologies would be curtailed, enabling the west to overmatch them in terms of capability rather than numbers. Secondly, Soviet resources would be eroded by depressing the value of Soviet commodities in world markets. Thirdly, the US in particular, and NATO, more generally, would markedly increase defence spending both to strengthen their position in any future negotiations, but also to force the Soviets to devote more of their already-diminished economic resources to defence, to a point where the Soviet command economy would simply be unable to function and would collapse. A massive increase in US conventional forces was quickly apparent, along with a new doctrine, 'Air-Land Battle',[29] which was predicated on conventional war fighting rather than a defence leading inevitably to a nuclear response. Reagan also revived the B-1 bomber program in 1981, an initiative which had been cancelled by his predecessor, and also continued secret development of the Northrop B-2 Spirit or Stealth bomber. Directive 32 therefore redefined what winning could mean: first, tactical victory might be possible without recourse to strategic nuclear weapons, so long as the quality of the force was able to overmatch its opponent even in the face of a quantitative disadvantage. Secondly, that strategic victory could be realised through the threat of force and the effect of over-heating the economy of the opponent – just as it had through the application of blockade against Germany during the Great War.

The multiple independently targetable re-entry vehicle, or MIRV, was another weapons system originally designed specifically to contribute to the MAD doctrine of nuclear deterrence; it was now brought into the new equation by Reagan. With a multiple payload, a single ICBM could carry many separate and, as the name describes, separately targetable, warheads. MIRVs were created to counter the Soviet A-35 anti-ballistic missile systems around Moscow. Since each A-35 could destroy only one incoming missile, then increasing the number of projectiles would also require an increase in the defensive missile numbers. This made defence against MIRV attacks costly and difficult. The Peacekeeper, produced under the Reagan administration, could carry up to ten warheads, each with a yield of around 300 kilotons of TNT or together, an explosive payload equivalent to 230 Hiroshima-type bombs.[30]

Next, in response to Soviet deployment of the RSD-10 Pioneer intermediate-range mobile missile system, the US administration deployed the Pershing II intermediate missiles in West Germany as well as tactical cruise missile systems in Germany and Britain. The objective of this deployment was to engineer the removal by both sides of this entire class of nuclear weapons and was in accordance with NATO's double-track decision, which was to offer the Soviets and their allies a mutually agreed limitation of medium-range ballistic missiles and intermediate-range ballistic missiles, combined with the threat that in case of a failure to agree, NATO would deploy more middle-range nuclear weapons in Western Europe.[31] The Strategic Arms Reduction Talks II were to eliminate this class of weapons but no agreement was ratified.

Reagan also proposed the Strategic Defense Initiative, or SDI. This was to be a space-based missile defence system that could make nuclear war impossible and thus deliver victory. However, the technology was immature and its opponents cruelly dubbed it 'Star Wars', after the George Lucas films which had recently been released. Critics also thought that its deployment would be likely to accelerate the arms race (something that they failed to understand was Reagan's objective anyway) and that the huge expenditures required would cripple other programmes. However, Soviet leaders Brezhnev and Gorbachev certainly became seriously worried.[32]

Reagan also returned to the business of confronting communism around the world. Nearly one-third of the world's population lived under communist rule at this time and under a policy known as the 'Reagan Doctrine'[33] his administration promised aid and military assistance to push back left-wing dominance and suppress communist insurgencies. Inevitably this meant propping up some highly repressive right-wing regimes, such as that of Marcos in the Philippines, or the white apartheid government in South Africa; and aiding guerrilla movements opposing governments linked to the Soviet Union, such as the Contras in Nicaragua, the Mujahideen in Afghanistan, and the UNITA in Angola.[34] Winning was once more about rolling back the opposing ideology rather than containing it.

Reagan and Thatcher's strategy worked and worked more quickly than anyone expected. Under Soviet President Mikhail Gorbachev, moves to restructure the failing economy, *Perestroika*, were put in hand along with a policy of openness to the west – *Glasnost*. However, these had unintended consequences. Nationalism and pressures for independence gathered pace in Central and Eastern Europe, especially Poland, where resistance to the communist regime had been under way for some years, focused on the city of Gdansk. Gorbachev, in contrast to his predecessors, refused to use Soviet troops to prop up the Warsaw Pact regimes as they had in Hungary, Czechoslovakia and East Germany in the past. The end came in 1989 in a wave of generally peaceful revolutions, with the exception of that in Romania. The communist regimes of Central and Eastern Europe collapsed; the Communist Party of the Soviet Union even lost control and was banned following an abortive coup attempt against Gorbachev in August 1991. This in turn led to the formal dissolution of the USSR in December 1991 and the collapse of communist regimes in other countries as far away as Mongolia, Cambodia, Albania, Yugoslavia and South Yemen. Today only North Korea, Vietnam, Cuba and China remain communist and arguably, economic communism remains only in Cuba and North Korea, prompting Francis Fukuyama famously, and wrongly, to declare 'the end of history':[35] history, that is, as struggle, not as chronology. The west, and, in particular, the US, won the Cold War. It was won not by fighting, except through proxies, but through winning the battle of ideas: eventually, the desire for individual liberty, accountable government and high living standards offered by the west, captured the loyalty of people living under communism. But it was also won by the west outspending the enemy in every area of defence and security. This was only possible because the Soviets viewed NATO, under US leadership, as a credible military threat which had to be countered. The fact that we now seem to have entered a phase of international relations that could be described as a new Cold War shows that nothing – including winning – is finite. We would do well to bear this in mind today.

## Notes

1 Harold Nicholson, The *Congress of Vienna: A Study in Allied Unity: 1812–1822* (London: Phoenix, 1946), 49.

2 Cited in John L. Gaddis, *The Cold War: A New History* (London: Penguin, 2005), 24–26.

3 David McCullough, *Truman* (New York: Simon and Schuster, 1992), 547–549.

4 Michael Beschloss, *Our Documents: 100 Milestone Documents from the National Archives* (Oxford: OUP, 2006), 194–199.

5 Barın Kayaoğlu, 'Strategic imperatives, Democratic rhetoric: The United States and Turkey, 1945–52.' *Cold War History*, Vol. 9 No. 3, (August 2009): 321–345; *André Gerolymatos, An International Civil War: Greece, 1943–1949* (London: Yale University Press, 2017), 100–111.

6 David Holloway, *Stalin and the Bomb: the Soviet Union and Atomic Energy* (London: Yale University Press, 1994), 421; Jeffrey Strickland, *Weird Scientists: the Creators of Quantum Physics* (New York: Lulu.com, 2011), 549.

7 John Foster Dulles, 'The Evolution of Foreign Policy', Department of State press release 81, 12 January 1954, http://nuclearfiles.org, accessed 21 January 2020.

8 See, for example, Colonel Alan J. Parrington, U.S.A.F., 'Mutually Assured Destruction Revisited' in *Airpower Journal*, (Winter 1997); http://archive.org/wayback, accessed 22 January 2020.

9 See Daniel Deudney, *Whole Earth Security: the Geopolitics of Peace* (Washington: Worldwatch Institute, 1983), 80.

10 Game theory is defined as 'the study of the ways in which interacting choices of economic agents produce outcomes with respect to the preferences (or utilities) of those agents, where the outcomes in question might have been intended by none of the agents.' https://plato.stanford.edu/entries/game-theory/, accessed 21 January 2020.

11 Martin J. Osborne and Ariel Rubenstein, *A Course in Game Theory* (Cambridge, MA: MIT Press, 1994), 14 *et seq*.

12 Information provided by the Department of Politics, Lancaster University, January 2020.

13 Glenn A. Kent, Randall J. de Valk and David E. Thaler, *A Calculus of First-Strike Stability* (RAND note N-2526-AF, June 1988).

14 Wolfganag Panofsky, 'The Mutual-Hostage Relations between America and Russia' in *Foreign Affairs Journal*, Vol. 52, No. 1 (October 1973): 118.

15 Robert Rhodes James, *Winston Churchill: His Complete Speeches 1897—1963* (New York: Chelsea House Publishers, 1974), 8628–8630.

16 John T. Correll, 'The Making of MAD' in *Air Force Magazine*, 27 July 2018.

17 Alain Enthoven and K. Wayne Smith, *How Much is Enough?* (New York: Harper and Row, 1971), 207–210.

18 Dan Alex, 'USS George Washington', www.militaryfactory.com/ships, accessed 20 January 2020.

19 Francis J. Gavin, 'The Myth of Flexible Response: United States Strategy in Europe during the 1960s', *The International History Review*, Vol. 23, No. 4 (December 2001): 847–875.

20 www.nato.int, accessed 23 January 2020.

21 www.ukpublicdata.org/ukdefencespending, accessed 23 January 2020.

22 http://fas.org/nuke, accessed 23 January 2020.

23 Author's interview with Field Marshal Lord Brammall, 10 June 2014.

24 Lothar Rühl, 'Noch 1990 zielte die NVA Richtung Westdeutschland und Benelux', *Die Welt*, 31 July 1991.

25 Encyclopaedia Brittanica, https://www.britannica.com/event/Strategic-Arms-Limitation-Talks, accessed 25 January 2020.

26 Office of the Historian, U.S. Department of State, 'The Soviet Invasion of Afghanistan and the U.S Response, 1978–1982', https://history.state.gov/milestones/1977-1980/soviet-invasion-afghanistan, accessed 24 January 2020.
27 See the *U.S. Department for Defense Annual Review for the Fiscal Year 1980* (Washington D.C., 25 January 1979), 61.
28 Milton J. Leitenberg, 'Presidential Directive (P. D.) 59: United States Nuclear Weapon Targeting Policy', *Journal of Peace Research* Vol. 18, No. 4 (1981): 309–317.
29 Department of the Army *Field Manual (FM) 100-5, Operations,* 4 September 1981.
30 Brent Scowcroft, John Deutch and R. James Woolsey, 'A Small, Survivbale, Mobile ICBM' in *The Washington Post*, 26 December 1986, A-23.
31 NATO e-library, Special meeting of foreign and defence ministers, 12 December 1979. https://www.nato.int/cps/en/natolive/official_texts_27040.htm, accessed 23 January 2020.
32 Atomic Heritage Foundation, 'Strategic Defense Initiative', 18 July 2018, http://www.atomicheritage.org/history/strategic-defense-initiative-sdi, accessed 18 January 2020.
33 Office of the Historian, U.S. department of State, 'Reagan Doctrine' https://history.state.gov/milestones/Reagan-doctrine, accessed 24 January 2020.
34 Victor Bondi (ed), 'Government and Politics: The Cold War: Third World Woes', in *American Decades: 1980–1989* (Detroit: Gale Research, 1996), 267.
35 Francis Fukuyama, *The End of History and the Last Man* (London: Free Press, 1992).

CHAPTER 8

# Western Strategic Goals and 'Winning' in Iraq and Afghanistan since 2001

Daniel Marston and Carter Malkasian

The purpose of this chapter is to examine US and UK strategy in the wars in Afghanistan and Iraq.[1] We explore the idea of winning and impediments to strategy formulation. Our main observation is that policy goals repeatedly changed. 'Winning' in the traditional sense of defeating an opponent on the battlefield or attaining pre-war goals was not applicable. Effective strategy depended on the ability of civilian and defence leaders to adapt to changing goals. What emerges is that clearly defined policy goals are a helpful component of strategy but perhaps less so than recognising that strategy can be a process of muddling through.

In this chapter we unpack four core themes regarding winning and strategy in the Afghan and Iraq wars: how policy goals changed; how lack of forethought impeded creation of strategy suited to these changing goals; how lack of robust and honest dialogue within the civil-military realm obstructed forethought; and how these issues skew strategy formulation for allies. We had the opportunity to work in the wars with commanders at all levels, and to engage with senior leaders, civilian and military, in the US, UK, and within NATO. Our findings are informed by that experience.

The first section of this chapter will review the idea of winning in war and general perceptions of the failure of US strategy in Afghanistan and Iraq. The second section will describe how policy goals changed. The third section will explain why the United States had difficulty designing strategy that could accommodate changing goals. The fourth section examines how civil–military relations affected how strategy was formed. The fifth section looks at the creation of strategy formulation in the Coalition context. The sixth section is the conclusion in which we offer a few recommendations for civilian and military defence leaders.

## The idea of winning in war

What is winning in the context of the wars in Afghanistan and Iraq? The answer is complicated by the fact that winning itself is an ambiguous concept.

Today, strategists often interpret winning as attaining policy goals that secure national interests. Those could involve seeking a negotiated peace, turning the war over to a partner, or even withdrawal. Although the traditional interpretation of winning can still pertain at the operational and tactical levels of war, it has become much less helpful to think about strategy in terms of 'winning', rather than in terms of attaining policy goals.

A starting point of military strategy is to identify the policy goals. Generals and civilian leaders should strive to know in advance how they wish a war to end before they start. As Hew Strachan, doyen of historians of war, instructs, 'Strategy implies that the government has a policy and that the strategy flows from the policy: it is an attempt to make concrete a set of objectives through the application of military force to a particular case.'[2] Or, in military terms: Allenby's advice to 'think to the finish'; in political or more philosophical terms: Carl von Clausewitz writes in Book VIII of *On War*, 'No one starts a war – or rather, no one in his senses ought to do so – without first being clear in his mind what he intends to achieve by that war and how he intends to conduct it.'[3] Whether in political or military terms, this is all about the underpinning logic of 'what for' or the 'in order to'.

Trying to assess whether the United States and United Kingdom met their policy goals in Afghanistan and Iraq cannot be undertaken at this stage (in 2020), as the war continues in both countries. What is apparent, even now, is that the Americans and the British failed in creating viable and coherent strategies.[4] Strachan criticised: '"The long war" is about shaping military means to political objectives, but conceptually it remains devoid of strategic insight and of political context. Strategy needs to recognise war's nature; this one uses a policy statement as a substitute for defining war .... Strategy's absence is not an exclusively American problem. The same point can be made about Britain.'[5]

In Strachan's estimation:

> Strategy has been absent throughout the wars in Iraq and Afghanistan. In part that is because the political objects have been unclear, or variable, or defined in terms too broad to be deliverable in strategic terms. Because there has been no clear relationship between the ends and the limited (and often inappropriate) means, strategy is simply not possible. The result has often been war shaped by platoon and company commanders, a series of ill-coordinated tactical actions, where killing and casualties define success.[6]

His criticism represents the general consensus that strategy in Afghanistan and Iraq was poorly conceived.

Many American and British policy makers and the military chiefs failed to ask in 2001 and 2002 'to what end?' in relation to the use of force in Afghanistan and Iraq. There were many planners in both countries who attempted to debate these fundamental questions as the US and UK planned the wars in both Afghanistan and Iraq. However, any objections were overruled by policy makers and military commanders who did not want to debate such fundamental points.[7]

## Changing goals

Though identifying goals is a useful first step in forming strategy, goals should not be viewed as static. In Afghanistan and Iraq, goals changed significantly over time. In general, they became more and more limited. This occurrence is far from unprecedented; conflicts frequently witness changing policy goals on the part of one or both sides. Changing US goals during the Vietnam War and changing British goals during the Napoleonic Wars readily come to mind. Furthermore, goals can expand – the dreaded 'mission creep' – as well as contract. At a glance, goals seem especially likely to change during longer conflicts. International, domestic, and economic factors change with time, influencing the goals of ongoing wars. But goals can also change during relatively short conflicts, such as the change of US goals from restoration of the union to ending slavery over the course of the Civil War or the change of goals from complete victory to a negotiated armistice during the Korean War. Any notion that militaries are the instrument of a single policy goal from beginning to end of a war would seem ahistorical.

When the Afghan and Iraq wars started, the administration of President George W. Bush believed it could win in the traditional sense. The adversaries would be permanently defeated, and at some point the United States could leave both countries. The strategy played out first in Afghanistan and then in Iraq.

On 7 October 2001, the United States invaded Afghanistan in retaliation for the attacks of 11 September. The goals were to defeat al-Qaeda, overthrow the Taliban, and set up a new democratic government. The combination of US airpower, special operations forces, intelligence operatives, and the northern alliance chased out al-Qaeda and upended Mullah Omar's Taliban regime. Osama bin Laden fled from Tora Bora to Pakistan.

Rapid success left the original goals in place. Bush kept a light military footprint to complete the defeat of al-Qaeda and the Taliban and help set up a new democracy that could prevent terrorists from coming back.[8] The idea was to withdraw eventually, but no plan was composed to lay out how to make that happen, other than the default of killing or capturing al-Qaeda and Taliban leaders. The relative peace and quiet between 2002 and 2006 relieved any pressure to adjust the goals. Although unstated, goals informally expanded around nation-building – economic development and improving human rights. They were never formally stated but resources were slowly placed against them. In the meantime, the United States invaded Iraq.

Bush and his team went into Iraq flush from the success of Afghanistan. Besides removing weapons of mass destruction, the goals were to overturn the regime of Saddam Hussein and establish a new democratic government. At first, success was as impressive as in Afghanistan. It turned out to be ephemeral, and an insurgency erupted before the year was out.

The pressure of the insurgency caused the administration, largely pressed forward by Secretary of Defense Donald Rumsfeld, to alter its goals. Rumsfeld was sensitive to the costs of staying in Iraq with a large number of forces in an internal conflict. He wanted 'to get this thing wrapped up' and, having authority over operations in Iraq, in September 2003, directed that power be handed over to the Iraqis as soon as possible so that the United States could get out.[9] A new strategy was emplaced to reduce US forces and turn over the war to the new democratic Iraqi government and its armed forces. The strategy was sometimes called 'Iraqization,' after the similar 'Vietnamization' of the Vietnam War. General George Casey executed it. It is unclear how much Bush himself bought into turning the war over to the Iraqis short of victory; as the insurgency heightened, Bush was at times sceptical of the strategy.[10]

The Iraqization strategy persisted into 2006 when rising sectarian violence showed it was unlikely to work. The Iraqi army was unable to carry out security duties, and it became increasingly clear that the Iraqi government would fail if the United States continued with Iraqization. US means and policy goals then changed in the famed Iraq surge. Bush removed Rumsfeld. Goals reverted to defeating the insurgency – winning. US troops would stay until that goal was accomplished. Bush remembers telling his team: 'We must succeed .... If [the Iraqis] can't do it, we will .... We have to make damn sure we do not fail.'[11] Bush was similarly resolute when violence returned to Afghanistan in 2006. He remained determined to defeat the Taliban and score, in his words, 'a victory for the forces of liberty.'[12]

The election of President Barack Obama in 2008 led to a dramatic change of goals in Afghanistan and Iraq. America's domestic political environment had changed from 2001–2003. The great recession was damaging the US economy. Obama needed to focus on recovery at home rather than two expensive foreign wars. At the same time, the terrorist threat, though decreased, was still a concern, 11 September a visceral memory. Consequently, Obama wanted to shift strategy away from winning but could not yet responsibly withdraw. Winning did not return as an objective, at least its old definition. Generals sometimes wrote guidance entitled 'What Winning Looks Like', but they did not mean winning in any traditional sense. Winning was redefined to mean handing the war over to a capable government that could carry on the fight on its own.

In Afghanistan, Obama surged but at the same time, cautious of investing too much, refocused the goals of the war on removing the terrorist threat to the homeland and reducing the chances a terrorist safe haven would re-emerge in Afghanistan. Gone was Bush's intent to defeat the Taliban no matter what. The United States would deny al-Qaeda safe haven, reverse the Taliban's momentum, and, in Obama's words from his 2 December 2009 West Point speech, 'strengthen the capacity of Afghanistan's security forces and government, so that they can take lead responsibility for Afghanistan's future.'[13] Supporting the Afghan government against the Taliban was solely a means to 'keep pressure on al-Qaeda.' The United

States was to hand off to the Afghan government, beginning the drawdown of the surge in July 2011.

The strategy and its goals evolved over the next years. At first, it was not clear whether the US military would withdraw from Afghanistan completely. Then, the outcome of the surge signalled to Obama the unsustainability of the whole Afghan endeavour, especially as he tried to reduce the budget deficit in the wake of the recession. Meanwhile, the death of bin Laden on 1 May 2011 reduced the terrorist threat. In January 2009, a Gallup poll had found that 66% of Americans agreed with the decision to send military forces to Afghanistan.[14] Days after bin Laden's death, a Gallup poll showed 59% of Americans believed the US mission in Afghanistan had been accomplished.[15] Planning to withdraw accelerated. 'America, it is time to focus on nation building here at home,' Obama announced in his June 2011 address on the drawdown.[16] Over a series of announcements between 2010 and 2014, Obama laid out a schedule to draw down US military forces to zero (excluding a small embassy presence) by the end of 2016. Thereafter, the Afghan government was to fight on its own, with US and allied funding of its army and police.

In Iraq, of course, Obama pursued a more abrupt withdrawal strategy. Bush had signed a status of forces agreement with Prime Minister Maliki in December 2008 that stipulated that US forces would withdraw by the end of 2011. Obama, opposed to the Iraq war, immediately pursued the goal of drawing down but did not commit to going to zero and attempted to arrange a bilateral security agreement that would allow a few thousand US troops to stay. Nevertheless, the impetus to stay waned. By 2011, the threat of terrorism was deemed sufficiently low to get out while domestic pressure to reduce spending rendered an Iraq presence less palatable.[17] When Maliki balked at signing a bilateral security agreement, which included the difficult task of getting the Iraqi parliament to formally confirm legal immunity for US troops, Obama chose not to look for another diplomatic option to keep a small presence. At the end of the year, complete withdrawal took place.

In 2014, an unexpected event disrupted the two withdrawal strategies and altered US goals once more. Just as the wreck of al-Qaeda was finally drifting into the distance, the rise of the Islamic State in 2014 resurrected the threat and its political repercussions. Its rise had been assisted by the sectarian policies of the Maliki government. In six short months, the Islamic State had swept up half of Iraq and nearly a third of Syria. Over the following year they conducted or inspired multiple terrorist attacks in the United States and Europe. The terrorist threat that had seemed crippled after the death of bin Laden reappeared, changing decision calculus. The post-bin Laden relief turned into renewed domestic concern over terrorism. The percentage of Americans who feared they could be a victim of a terrorist attack jumped from 36% in 2011 to 49% in 2015.[18]

In Afghanistan, the Islamic State threat was accompanied by widespread Taliban battlefield successes as the US drawdown approached its endpoint. In light of the setbacks of 2015, the intelligence community assessed that if the drawdown went forward on schedule, security could deteriorate to the point that terrorist groups could form a safe haven in Afghanistan.

Obama shifted his strategy to deal with the changing circumstances and accepted, with revisions, strategies to fight limited, long-term conflicts in Iraq and Afghanistan with minimum troops and heavy use of air strikes, which he had previously placed under stringent restrictions.[19]

In Iraq, Obama had to send back in thousands of troops and authorise a withering air campaign. The war against the Islamic State was larger than Iraq, extending into Syria. Obama's goals were first to degrade and then to defeat the Islamic State. The definition of defeat was not the destruction of the Islamic State, but their removal of them from cities and elimination of their ability to conduct or inspire external attacks. US forces were deployed in minimal numbers and were to avoid combat. A cleaner definition of winning would have demanded larger numbers of US forces in direct combat, something Obama opposed.

In Afghanistan, the new strategy remained limited but the idea of turning over to the host government was cast aside. The virtue of Obama's drawdown timeline is that he had budgeted to adjust course. Obama suspended the Afghanistan drawdown – and also granted his commanders greater authority to use air strikes – in order to prevent the Afghan government from falling. He wanted to use the minimum resources possible and was not interested in winning.

By this point, in both Iraq and Afghanistan there was no exit plan. Obama had settled on strategies that deployed small numbers of forces with no end date. Afghanistan and Iraq were indefinite limited wars. The belief that the host government could stand on its own in the foreseeable future was gone; in its stead was a determination to keep a sustainable, minimum number of forces in place. This de facto strategy endured for five years.

The Trump administration brought further turbulence to US goals. President Donald Trump left Obama's strategy in place while debating openly over whether to pull out all troops. He eventually extracted large numbers from Syria but not from Iraq. In Afghanistan, the administration stuck with Obama's strategy until February 2020 when it signed a conditional agreement with the Taliban. The agreement stipulated that the military would withdraw within 14 months if the Taliban met certain conditions. Goals seemed to have changed.

Thus we can see how much policy goals fluctuated in Iraq and Afghanistan. It should be noted that goals were deeply affected by the host governments. The respective war efforts depended upon the host governments for access, resources, military forces, local popular acceptance, and international legitimacy. Yet, as the well-known principal–agent model describes, the Iraqi and Afghan political leaders had different goals from the United States and its Coalition allies and partners. Maliki had a sectarian agenda. Karzai and late Ashraf Ghani prioritised political survival. The differing goals of these leaders obstructed attainment of US goals. In turn, US presidents contracted their own goals. The host governments were a reason that US policy goals changed.

The fact that goals change should not be surprising. 'War itself shapes and changes policy', writes Strachan.[20] Or as Lawrence Freedman states in his magisterial *Strategy*, 'The picture of strategy that should emerge ... is one that is fluid and flexible, governed by the starting point and not the end point.'[21] It seems intuitive that lengthy wars will see changing goals, and that winning may likely be deemed unattainable or not worthwhile. And even if victory is not impossible, the political circumstances surrounding a conflict naturally change over time. War goals will be pressured by these changing circumstances. Why strategies are so often insensitive to the possibility of changing goals and inflexible while political circumstances change is the question to which we now turn.

## Forethought

In Iraq and Afghanistan, all three US administrations set strategies that neglected how policy goals could change in the future. Military officers are fond of saying that 'no plan survives first contact' but that statement generally applies to how ways and means have to adapt to meet *static* ends. What we are asking here is: why did strategy have difficulty accounting for the possibility of *changing* goals? The best answer is a lack of forethought. US leaders clung too long to their own preconceived notions of how the war would play out and neglected options that did not fit their biases. They undervalued how goals can change with circumstances and were frequently not on the lookout for the unexpected.

In the Bush administration, lack of forethought can be seen in the form of an overconfidence that persisted through its early years. In Afghanistan, the 2001 victory left them supremely confident: the Taliban were presumed defeated, and there was no reason to change their goals. Rumsfeld announced at a news conference in early 2003: 'We clearly have moved from major combat activity to a period of stability and stabilization and reconstruction activities .... The bulk of the country today is permissive, it's secure.'[22] The ease of the 2001 success carried away sensibility. Bush and his team discarded opportunities to bring the Taliban into the peace process and to build a stronger Afghan army and police. The White House disregarded arguments by President Hamid Karzai and Ambassador Zalmay Khalilzad, and later Lieutenant General Karl Eikenberry and many others, that insurgents were staging a comeback. Bush and his team turned to Iraq and glided forward in Afghanistan without much deliberation over what could happen if their assumptions were off.

In the runup to the Iraq invasion, the successes of Afghanistan encouraged using military force and setting maximalist goals. As is well known, Bush and his team created a strategy to overthrow the Saddam regime. They neglected planning for stabilization operations – 'Phase Four operations' – and outright dismissed consideration of the outbreak of an insurgency. Vice Chief of the Army General Jack Keane who was privy to the planning recalls 'we never even considered an insurgency as a reasonable option.'[23] When the insurgency broke out, Rumsfeld refused to recognise it for months. Later, he stuck to his Iraqization plan with its goals for a drawdown well into the outbreak of sectarian violence. The plan and timetable had become a kind of buffer against consideration of new goals and new options. One of the main findings of the recent US Army official history of the war in Iraq is that: 'military and civil leaders who embark upon war have a shared responsibility to ensure that war aims are achievable .... It is also incumbent upon them to ensure ... the organizational capacity to execute in a sufficiently coherent way, then adapt quickly enough as change happens.'[24]

If lack of forethought comes across clearly during the Bush administration, it is also present in the Obama administration, seen in the Afghan surge and the Iraq and Afghanistan withdrawal decisions.

In the creation of the Afghan surge, strategic discussions narrowed around one view of the threat and one paired option – counterinsurgency and a surge – to address it. Generals David Petraeus and Stanley McChrystal believed that the terrorist threat to the United States and worsening battlefield situation in Afghanistan warranted escalation. McChrystal was emphatic that, if no new strategy were implemented, the war would likely end in failure because the insurgency could gain such momentum that within a few years the government might be defeated.[25] Other possible futures were inadvertently marginalised.

Petraeus and McChrystal had witnessed counterinsurgency and unswerving resolve turn the tide in Iraq, and thought the same could be done again. Many other generals shared their outlook. The turnaround in Iraq encouraged officers to see the impossible as possible. 'You have to recognise also that I don't think you can win this war. I think you keep fighting .... This is the kind of fight we're in for the rest of our lives and probably our kids' lives,' Bob Woodward quotes Petraeus as saying.[26] Counterinsurgency grew into a wider plan and a set theory of victory with a set vision of the future. The same goals would be pursued for years. The plan was ill-suited to change. McChrystal and Petraeus tended to believe that political and economic circumstances would prevail to allow long-term counterinsurgency. In Petraeus's calculus, successes could convince Obama to extend deadlines to start withdrawing troops. 'Put time back on the clock' was a favourite saying. He thought politics would bend to the strategy.

During the surge debate, the generals' case that defeat was on the horizon and their overconfidence in counterinsurgency crowded out the practical alternatives of sending fewer reinforcements or foregoing them entirely. Vice President Joe Biden created one option with his counter-terrorism-plus plan but, instead of opening up the debate to a discussion of a variety of options, his idea polarised the debate into two rival and inflexible camps. Biden could never overcome the single-minded focus of the military. Impressed with McChrystal's assessment, Gates dismissed a longer-term, no-surge strategy outright, arguing 'standing pat, middling options, muddling through are not the right path forward and put our kids at risk for no good purpose.'[27] Nothing shows this better than Obama's West Point speech in which holding steady spelled failure: 'There are those who … suggest that we go forward with the troops that we have. But this would simply maintain a status quo in which we muddle through, and permit a slow deterioration of conditions there. It would ultimately prove more costly and prolong our stay in Afghanistan, because we would never be able to generate the conditions needed to train Afghan security forces and give them the space to take over.'[28] Events would show that surging had nothing on muddling.

The Iraq and Afghanistan withdrawal decisions were similarly narrow-minded. In the process of creating strategy, Obama and his team concentrated on getting out, regardless of the unexpected. The Obama administration discounted the return of a terrorist threat – in the form of the Islamic State or Taliban battlefield successes (which could destabilise Afghanistan) – and the domestic political imperative to deal with the threat. Low-cost ways and means to mitigate the damage from unexpected yet reasonably possible contingencies were rejected lest they anchor the United States in the region. Timetables for withdrawal blinkered decision-makers from the possibility that their own policy goals could change in the future.

By 2009, Al-Qaeda in Iraq was widely assessed as wiped out, even though cells and leaders survived in hiding. Once the drawdown was complete, these cells regrouped into the Islamic State. The administration disdained warnings of a reconstituted threat. Iraqi Prime Minister Nouri al-Maliki was rebuffed when came to Washington in October 2013 to ask for help. Increasing warnings from the US intelligence community also had no traction. 'This was not something that people were not being made aware of in terms of implications,' recalls Secretary of Defense Leon Panetta.[29]

The Afghanistan story follows the same line. The military and intelligence community widely predicted that the Afghan armed forces would have trouble holding back the Taliban without US advisors and air strikes. The White House agreed with the assessment but pressed forward regardless under the assumption that it was time for the Afghans to make do on their own and that the United States would not be compelled to return and help them. The Taliban were not al-Qaeda and the terrorist threat from al-Qaeda or any other group was deemed quite low. At this time, Obama was willing to bear the risk of defeat. Thinking about what might have to be done if the terrorist threat proved greater than expected did not happen.

In its withdrawal decisions, the Obama administration neglected options to keep small numbers of troops, provide reasonable authorities to conduct air strikes or use special operations forces to fight insurgents, or consider creative diplomatic options such as finding an informal alternative to a bilateral security agreement. If Bush's error was overconfidence, Obama's error was too little insurance on the withdrawal policy. When the unexpected happened, he was underinsured.

Good strategy should be alert to biases. It should consider a range of futures and a range of options. In this sense, we can see that US leaders were averse to forethought. US presidents and generals repeatedly saw their plans fall short when reality defied their expectation: for Bush, when the Taliban turned out not to be defeated and Iraq became a quagmire; for Petraeus and McChrystal, when the Afghan surge proved unsustainable; for Obama, when the terrorist threat returned; for Trump, when the political costs of leaving loomed steeper than assumed. If they had thought more about the different ways things could play out, then opportunities might not have been missed, options might not have been discarded, and America may have experienced less costly wars.

Forethought is not a new idea in strategy. It is implicit in historian John Lewis Gaddis's use of Isaiah Berlin's description of the fox versus the hedgehog to illuminate grand strategy, in which foxes 'pursue many ends, often unrelated and even contradictory, connected if at all, only in some *de facto* way.'[30] Gaddis warns of how too much precision and decisiveness in strategy can entrap a leader, cutting off an ability to adapt to changing goals. The environment surrounding strategy and the goals that guide it naturally shift: 'Assuming stability is one of the ways ruins get made. Resilience accommodates the unexpected.'[31]

## Civil–military relations

The creation of options and consideration of different futures was affected by civil–military relations. Lack of robust dialogue within civil/military relations was an area of significant tension.[32]

During the early years of the wars, the military voice was constricted. A senior officer relayed a story from his involvement in the planning for the invasion of Iraq. He was in the room while a senior policy maker was briefing the CENTCOM planning staff on the policy aims of the Iraq invasion. In his hearing, one senior military officer turned to another and stated: 'We are f*****.' The second officer said, 'So tell the policy maker what you think and why we are f*****.' The senior officer refused, saying it was not his place to do so. It was his responsibility to get the military to Baghdad; the policy makers could figure out the rest.[33] This exchange was characteristic of the lack of robust dialogue. It opened the way for Bush and his team to pursue infeasible military options and obstructed the kind of debate that could have better revealed the dangers that lay ahead.[34]

Later in the wars, the voice of the US military was too loud, partly in overcompensation for the constriction of the past. Advocacy of a surge and counterinsurgency strategy in Afghanistan famously 'boxed in' President Obama. The generals' voice carried political weight that was difficult for a politician to override. The experience generated its own civilian reaction in Obama's middle years when military advice was disregarded. Civilians tended to distrust the generals. US generals did not help matters in Iraq or Afghanistan during this period by focusing on their own preferred options, so that strategic debate became a contest between rival factions with hardened concepts instead of a consideration of different options to address an uncertain future. Things might have been different had the generals presented a larger range of options and been open to any of them. It was only late in his administration that a balance emerged in which debates yielded a fluid strategy that kept options open while pursuing politically acceptable limited aims.

Overall, frequent breakdowns in civil–military dialogue occurred, not only in discussions about the nature of war,[35] but also in strategy development and strategy's links to coherent and viable policy.

Across military and civilian leaders in the United States and the United Kingdom, there is a general consensus that civil–military relations should be managed in a way that prevents the military from making high-level policy decisions on war, lest it impinge upon civilian control of the military. At the same time, the military is expected to play a role in the execution of a war and advise the high-level decisions. Hew Strachan describes this consensus:

> The challenge for the United States and for the United Kingdom was – and is – the link between the policy of its administration and the operational designs of its military agents. In the ideal model of civil–military relations, the democratic head of state sets out his or her policy, and the armed forces coordinate the means to enable its achievement. The reality [one hopes] is that this process – a process called strategy – is iterative, a dialogue where ends also reflect means, and where the result – also called strategy – is a compromise between the ends of policy and the military means available to implement it.[36]

The interpretation of Samuel Huntington's *Soldier and the State* has cast a long shadow over the current tensions within the civil–military domain on both sides of the Atlantic. The central theme from *Soldier and the State* that is most directly related to the discussion in this chapter is the term of 'objective civilian control'. The thesis states a clear delineation between the roles of civilian and military decision makers. The civilian side of decision making deals with policy and politics, while the military focuses their expertise on the management of violence.[37] Huntington asserts that 'Politics is beyond the scope of military competence, and the participation of military officers in politics undermines their professionalism, curtailing their professional competence .... [T]he military officer must remain neutral politically.'[38]

On the role of the military in forming strategy, Huntington writes: 'When required in his executive capacity to make decisions involving both military and political elements, the military man ideally should formulate his military solution first and then alter it as needs be on the advice of his political advisors.'[39] This quote has been widely interpreted to mean that the military will not, and should not, counter the advice or views of the political policymakers; rather, they should adjust their own plans to fit the directives of the civilian leadership. They should certainly not ask the difficult question, 'To what end?'

Huntington's guidance was reinforced in the education of many officers who fought in Iraq and Afghanistan. Major General William Rapp, who commanded at the US War College until 2017, declared that 'Military leaders are taught civilian leaders will clearly articulate the ends of policy and military advice should be limited to matters of ways, means and risk.'[40]

A variety of scholars and senior officers consider the 'Huntington model' less relevant now than 50 years ago.[41] They highlight the need for robust dialogue between civilian and military leadership regarding strategy formulation and the interaction of war and policy. Hew Strachan exemplifies the challenge to the Huntington model:

> The principal purpose of effective civil–military relations is national security: its output is strategy. Democracies tend to forget that. They have come to address civil–military relations as a means to an end, not as a way of making the state more efficient in its use of military power, but as an end in itself. Instead the principal objective, to which others become secondary, has been the subordination of the armed forces to civil control.[42]

Professor Risa Brooks, notes that:

> Huntingtonian norms generate a reluctance among military leaders to engage in debate about factors in strategic assessment that bear on the civilian's policy choices or political objectives. Strategy – or the theory for how to use force to achieve political goals – sits at the intersection of the political and military spheres. Therefore, the military's role in formulating strategy is ambiguous in the objective control model.[43]

Robust dialogue between military and civilian leaders was inconsistent through the Afghan and Iraq wars. In 2014, a report on strategic competence by a team at RAND led by Linda Robinson, found that 'policymakers and military leaders may not see strategy as essentially as adaptive art for coping with the uncertainties of war and the lack of perfect knowledge.'[44] Even more telling are the findings in the realm of civil–military relations, which indicate that policymakers and military leaders were divided into three groups regarding the role of dialogue and debate in strategy formulation. One group followed a very traditional Huntington model of linear process, where civilian leadership presents the objectives to military leadership, who then craft a strategy to achieve those objectives. There is no mention of a dialogue. Another group expected decisions to be made using the presidential style of leadership and committee. The most interesting response came from the remaining group, who demonstrated a more nuanced understanding of policy, strategy, and war. They specifically stated that 'the line between policy and strategy is actually very blurry, and civilians and the military need to engage in an iterative and dynamic dialogue to inform policy decisions and subsequent strategy.'[45]

We saw plenty of examples of the lack of robust dialogue. In one example, a civilian intelligence officer in a meeting in Baghdad in the summer of 2007 asked difficult questions in response to a PowerPoint slide that showed positive tactical statistics. He then continued to argue that the 'political and sectarian Iraqi government' was not in alignment with the recent tactical progress, and was actually playing the long game against US tactical success and political vision. His views were dismissed by some in the room, while others countered with comments along the lines of 'there is no political guidance from DC, so we need to create a breathing space.'[46] This bottom-up approach which was not tied to the strategic level, nor was it part of a wider dialogue regarding shifting policy aims that were consistent with the reality of the battlefield and what was possible in the end.

In a variety of other examples, senior civilian policymaker visits in Afghanistan and Iraq pushed aside strategic issues in favour of discussing tactical topics. During such visits, military officers sometimes asked higher level questions, seeking answers from the delegations, regarding policy aims from DC and how they may or may not be linked to operations on the ground. Too often, they were rebuffed. In one meeting, a policymaker turned to a senior officer and stated he did not need to worry about strategy, as his mission was hunting and killing on the battlefield.[47] Linda Robinson's report presented this last issue as a chief finding: 'policymakers have a tendency to eschew strategy and focus on tactical issues.'[48]

One of the most eye-opening examples of the absence of strategic dialogue occurred in 2016. A former US ISAF commander admitted to a small audience of military officers, under Chatham House Rules, that his number one failure was that he should have asked the president of the United States 'to what end' in Afghanistan. Instead, he saluted, thanked the president, and told him that he would achieve progress.[49] Both civilians and military officers felt constrained at times to from debating and offering their opinions, due to the perceived boundaries of 'strategy formulation'.

Good strategic dialogue has certain characteristics. First, military leaders should ask questions about policy goals, but not set them. They must recognise that their voice has political weight and creates political costs for a president to override. Second, the military and civilians together should consider a variety of options to attain goals. Third, dialogue should be iterative because goals change in wars and expectations on the effectiveness of various ways and means need to be updated by experience on the ground. Fourth, the military should treat the dialogue as sensitive and not share their opinions in public unless permitted by the civilian leadership or required by the legislature. Simply talking in public politicises.

## Lack of dialogue in coalition relations

Another area of significant tensions is a notable lack of robust dialogue within the coalition, in particular between the British and Americans.[50] Lack of robust dialogue in civil–military relations has not been an issue in the US only; it has been a problem in the UK as well,[51] characterised by a common lack of effective communication between policy makers and senior military officials. The British Iraq Inquiry pointedly noted that 'One manifestation … was failure to give weight to the candid analysis that was regularly supplied by the JIC [Joint Intelligence Committee], by some commanders in theatre, and by others that things were going wrong.'[52] The pervasive issue was further compounded by lack of robust dialogue between Coalition partners. As a British senior officer noted in 2007 in Iraq, 'it is critically important that the ends, ways and means are in balance, and that the realities of what he [British generals] can do are understood nationally and within the theatre chain of command.'[53] The communication gap between the two key allies was even cited as a major strategic lesson in the US Army Official History on Iraq:

> Coalition warfare was largely unsuccessful for several reasons … with both the US command in Baghdad and the British command in Basra operating as though events in central Iraq and far southern Iraq were unrelated. … These differences illustrated problems that stemmed from assuming the coalition … held a common understanding and commitment to the strategic objective. They did not.[54]

How did this come to be?

On the eve of the war in Iraq, the British government was not prepared to create a viable and coherent strategy for the war. The Ministry of Defence (MOD) attempted to create policy for the use of force in Iraq, but they should not have been the lone driver in that area, let alone strategy. The Foreign and Commonwealth Office (FCO) was to be involved in policy and strategy making, but there were no clear lines of responsibility or linkages with MOD.[55] This was further exacerbated by the lack of properly organised lines of effort in the military domain. The Permanent Joint Headquarters (PJHQ) in Northwood (north-west London) and its three-star commander were not always directly linked to the Chief of Defence in London.[56] PJHQ worked closely in the initial war planning for Iraq with the US Central Command, and were relied upon for the best understanding of the American decision making and personalities. Other senior military leaders in the UK, meanwhile, were somewhat left out of the picture on some key issues and debates.[57] As Hew Strachan recalled, 'the gap which opened between the Ministry of Defence, fighting inter-departmental battles in Whitehall, and the operational direction of the war in Iraq and Afghanistan, was one of the key deficits exposed in Britain's conduct of the war after 2003.'[58]

Attempts to create policy and a coherent strategy were undermined by the reality, largely unacknowledged and never transparently communicated, that the UK's primary goal in joining in the Iraq campaign was to maintain its special relationship with the US. The UK failed to develop an independent strategy for Iraq because it lacked the wherewithal to develop or implement one. Although UK forces managed to take responsibility for the Basra area, they had little idea of what to do with it, other than figure out a way to hand over responsibility and withdraw as soon as possible. Their primary objective – successful early victory against Saddam Hussein – had already been achieved, and leadership did not undertake a serious attempt to develop or articulate another one.[59]

The lack of dialogue between the US and UK developed early in the war. By late 2003, debates had already arisen regarding force levels in the British sector, MND (SE). In early 2004, even with the worsening security situation in Iraq, General Walker, Chief of the Defence Staff (CDS), made it clear that there was to be no increase in troops. At the same time, Her Majesty's Government was determining that Iraq was a war that the UK needed to leave as quickly as possible,[60] and subsequently made a decision in July 2004 that the British would gradually hand over security to the Iraqi security forces, so that withdrawal would be complete by 2006.[61] This mission and timetable were to be the overarching plan for the British war in Iraq. Meanwhile, much of the reporting to the Americans within MND SE, with the support of Whitehall, was to paint a rosy picture to allow the British to withdraw. This set a pattern of fragmented discourse between American and British personnel that was not rectified until it was almost too late in 2007. By then, there were major tensions between the two allies over perceived miscommunication of intent.

An early example of this disconnect and lack of robust debate occurred in Iraq at the start of 2005. The outgoing, deputy commander MNF-I,[62] Lieutenant General Fry, produced a challenging assessment of the British campaign in southern Iraq. He called for more troops and resources to try to secure the British position in and around Basra. He presented three options for proceeding, the last of which specifically noted accepting failure in the mission. This assessment was passed along to the Chief of Defence Staff, General Walker, who passed along a different version of the assessment to the prime minister – one which avoided asking for more troops to get the campaign back on track, and did not discuss possible failure.[63] By mid-2005 the British military had decided that they had done all they could in MND (SE) and that shifting to the war in Afghanistan should be a priority. Officially Iraq was still listed as the main effort, but from the summer of 2005 on any decisions regarding MND (SE) were influenced by the new UK effort in southern Afghanistan[64] – a consideration that would not be fully communicated to the Americans.

The security situation in Iraq generally and in MND (SE) in particular continued to deteriorate throughout 2005. In October, the British Chiefs of Staff gave a stark assessment of the situation, reporting to HMG that 'Ministers needed to be clear that the campaign could potentially be heading for 'strategic failure', with grave national and international consequences if the appropriate actions [are] not taken.'[65] Nothing changed, as a result of this, however, and movement to 'exit' proceeded. As the Iraq Inquiry stated, 'US and UK in strategies for Iraq had in effect been on different courses ... while [the UK] had come to see its main task in Basra as one of keeping the situation calm while building the case for drawdown.'[66]

With the security situation in Iraq becoming even worse in 2006, the British remained focused on exit and reinforcing Afghanistan, while sticking to positive narratives. When the new Chief of Defence Staff, Air Chief Marshal Sir Jock Stirrup, visited Iraq, he reported back to the MOD that there was to be no change in the plan: the UK would continue to hand over to Iraqi security forces, and focus on further support operations in Afghanistan. He did identify that the withdrawal before the war was complete could be seen as 'Strategic Failure',[67] and advised HMG that 'astute conditioning of the UK public may be necessary'[68] to deliver such a message. He failed to address how HMG or the British military should approach delivering a similar message to their US counterparts.

The GOC MND (SE), Major General Richard Shirreff (mid-2006 to early 2007), was very candid about how he viewed the situation, telling the Iraq Inquiry that 'we had a strategy that involved extraction rather than necessarily achieving mission success. It was, in a sense, an exit strategy rather than a winning strategy.'[69] He was even more candid in a letter to the British Prime Minister in early 2007, where he pulled no punches in his assessment of British policy and strategy development since late 2003:

> Impertinent as it might be for a tactical commander to comment on the strategic level and appear to second guess those who must make the hard strategic choices about Main Effort and the need to balance risk between Iraq and Afghanistan, I do so because the impact of not getting our strategic thinking right is felt most acutely on the ground. I am not convinced that we have correctly identified the strategic Main Effort and resourced it appropriately to avoid potentially disastrous failure in Iraq. We have no choice but to achieve mission success in Iraq. This means concentrating our resources in pursuit of national interests: clouting, not dribbling ... [W]e have left it very late and depended on a willing and generous ally to make good our shortfalls ... [W]e can achieve mission success here, however, complex and intractable the problems facing us.[70]

With the arrival of General Petraeus and the 'Surge' of 2007, the war in Iraq was set to head in a different operational direction, and tensions and lack of communication between the two countries become even more pronounced. The new British GOC, Major General Jonathan Shaw, was upfront about differences of opinion about British planning; in his closed-session testimony to the Iraq Inquiry, he described his meeting with the British Chief of Defence Staff, Sir Jock Stirrup, in which he asked about the US surge plan and whether this impacted on his orders for OP ZENITH [exit strategy]. He reported the CDS' response as: 'the surge has nothing to do with us ... I have cleared this with General Casey and he is very happy.' General Shaw also recalled that General Casey had made his displeasure about OP ZENITH clear in their very first meeting, telling Shaw bluntly that 'I am not happy with your plans.'[71]

It was clear to people in London that neither US military command in Baghdad nor policymakers in DC were pleased with the UK's new approach. The British Embassy in DC had picked up on concerns about British plans, and shared them with Whitehall. Tony Blair read the reports and asked, 'What is going on in Basra?'[72] Despite questions and concerns raised at 10 Downing Street and elsewhere, PJHQ and Whitehall carried forward their plans for withdrawal, first to the COB and later out of Iraq altogether. This clearly diverged from the stated aims of MNF-I and MNC-I, but PJHQ and the UK Chiefs remained firm on the need to withdraw assets and forces from Iraq to support the operations in Afghanistan. On 1 February, a British cabinet paper discussing the transition in southern Iraq was distributed, in anticipation of an official announcement by the PM on 7 February. The paper covered many issues that would plague the British over the coming 14 months, and hinted at Iraqi Army and Police problems with maintaining security inside the city. It also referenced tensions with the US military, and their concerns about withdrawal plans. These tensions would continue throughout 2007 and early 2008.[73]

Many Americans felt that the British were not being honest in their reporting, and consequently were dubious about accepting statistics relevant to security. Many American and British officers felt that reporting from MND SE was at least partially suspect, and this truly soured the Coalition relationship. One thing I will add from my own experience: I made it clear to many of my UK contacts that it was hard to understand how British leadership could fail to see the damage that was being done, and the impact it would have on operations in Afghanistan, where British forces would still serve under US command. How could the British military expect to be trusted in Afghanistan, if they had left Iraq without seeing it through? It was astonishing and perplexing to watch.[74]

Working within a coalition is not new, but the complexity of the situation in Iraq, combined with the role of new national command structures, tied to shifting national interests, added a new set of complications. There were some divisional and many lower-level commanders who were pushing for more robust debates in theatre with the American command, as well as within MOD and Whitehall.[75] The Iraq Inquiry was even more damning:

> Ground truth is vital. Over-optimistic assessments lead to bad decisions. Senior decision-makers – Ministers, Chiefs of Staff, senior officials – must have a flow of accurate and frank reporting. A 'can do' attitude is laudably ingrained in the UK Armed Forces – a determination to get on with the job, however difficult the circumstances – but this can prevent ground truth from reaching senior ears. At times, in Iraq, the bearers of bad tidings were not heard. On several occasions, decision-makers visiting Iraq (including the Prime Minister, the Foreign Secretary and the Chief of the General Staff) found the situation on the ground to be much worse than had been reported to them. Effective audit mechanisms need to be used to counter optimism bias, whether through changes in the culture of reporting, use of multiple channels of information – internal and external – or use of visits.[76]

One thing I would have added to this conclusion is the need for openness and frankness within the Coalition. Brigadier Julian Free's[77] assessment is key: '[I]n all dealings with the US military it is best to be completely honest as to what one can do, what one will do, what one cannot do and what one will not do.'[78]

## Conclusion

So, perhaps at the heart of this book is a message that this is not just an academic exploration for its own sake, but a really vital aspect (both morally and practically) of the political and military business of the application of force. In short, know that things are not likely to play out as you expect. You are not entitled to clear or enduring policy goals. From that starting point, three takeaways stand out.

First, defence leaders should expect goals to change, especially in a long war. Even if goals start as total, a variety of other international, political, and cultural factors can whittle them down. And if goals start out as minimal, the same factors, perhaps combined with unexpected opportunity or ill-advised overconfidence, can expand them. When planning, it might be best to accommodate the possibility of change, rather than consistency of goals over the long term. Defence leaders themselves have very limited ability to affect policy goals and might do better by being adaptable and flexible. Long-term plans may not survive shifting goals. In this regard, exit strategies may be an ideal rarely to be fulfilled. Changing goals are likely to render them overtaken by events.

Second, defence leaders should consider different ways the future may play out and factor in the unexpected. They should provide a range of options that are all realistic to civilian policy makers. The defence leaders themselves should try to abstain from preferring any single option so as not to bias the decision-making process.

Third, to feed such forethought, robust dialogue between civilian and military leaders should be undertaken. Marginalising the generals or boxing in civilians' limits consideration of future possibilities and options. Generals need to hear policy goals. Civilians need understand what is feasible. Neither should try to limit options.

This all points to the virtue of strategies that muddle through rather than seek a decisive outcome. A strategy that muddles through eschews high goals that will be difficult to attain. It errs on the side of managing rather than solving a problem. Exploration of many possibilities and options is a virtue. Timetables and overcommitting to a plan, which can turn into a straitjacket, are pitfalls. Trying different ways and means is expected. A strategy that muddles through is wary of assuming the same goals will be pursued years into the future, cognizant that the farther we peer ahead, the more likely policy goals will change. Because goals and the environment change, strategy should buy insurance. Options should be reserved for the possibility things go worse than expected. Resources or authorities should not be relinquished until necessary. Dithering is tolerated. Making a decision early abdicates options and surrenders time for the environment to change under our feet. Waiting preserves options and the maximum flexibility to adjust as long as possible.

A great danger of a strategy of muddling through is mission creep. The answer to mission creep is to recognise its potential in the first place. Understand at the outset of an intervention or small war that goals can expand. Treat that as a natural risk, meaning that the endeavour may be longer and more costly than expected. For a wise strategist, the chance of mission creep should be an added reason not to get involved. Staying out of wars is usually the best strategy of all, if one cannot define national interests.

## *Notes*

1 We would like to thank Sir Lawrence Freedman and Sir Hew Strachan for taking the time to read and offer comments for this chapter. Any mistakes are ours.

2 Hew Strachan, *Directions of War* (Cambridge: Cambridge University Press, 2013), 64–5. See also the many works by Strachan, Lawrence Freedman, Frank Hoffman, Colin Gray, and Risa Brooks which deal with the failure of strategy making in the early 21st century.

3 Michael Howard and Peter Paret, trans. *On War* (Princeton: Princeton University Press, 1976), 579.

4 Colin Gray noted that 'there is a black hole where American strategy ought to reside.' *Another Bloody Century: Future Warfare* (London: Phoenix, 2005), 111.

5 Hew Strachan, *Directions of War*, 64–5. See also the many works by Strachan, Lawrence Freedman, Frank Hoffman, Colin Gray, and Risa Brooks which deal with the fundamental failure of strategy making in the early 21st century.

6 Strachan, *Directions of War*, 218.

7 Strachan, *Directions of War*, 68–9.

8 Afghanistan Papers, *Washington Post*, interview with former National Security advisor Stephen Hadley, 16 September 2015.

9 Frontline, 'Losing Iraq', television documentary, Public Broadcasting Service (PBS), July 2014.

10 Andrew Payne, 'Presidents, Politics, and Military Strategy: Electoral Constraints during the Iraq War', *International Security*, Vol. 44, No.3 (Winter 2019/2020): 165–80.

11 Ibid.

12 https://georgewbush-whitehouse.archives.gov/news/releases/2007/02/20070215-1.html. See also 2008 State of the Union, https://www.theatlantic.com/international/archive/2018/01/afghanistan-at-the-sotu/551855/.

13 Stanley A. McChrystal, *My Share of the Task* (London: Penguin, 2013), 359; President Barack Obama, 'West Point Speech', www.cnn.com, 2 December 2009.

14 Robert Gates, *Duty* (New York: Knopf, 2014), 492.

15 Gallup polls on Afghanistan, http://www.gallup.com/poll/116233/afghanistan.aspx.

16 President Obama on the Way Forward in Afghanistan, White House, https://obamawhitehouse.archives.gov/blog/2011/06/22/president-obama-way-forward-afghanistan, 22 June 2011.

17 Payne, 'Presidents, Politics, and Military Strategy: Electoral Constraints during the Iraq War', 180–200.

18 Gallup Poll, 2–7 June 2015, www.gallup.com.

19 Phil Stewart, 'Obama approves broader role for U.S. forces in Afghanistan', *Reuters*, 10 June 2016.

20 Strachan, *Directions of War*, 54–5.

21 Lawrence Freedman, *Strategy* (Oxford: Oxford University Press, 2013), xi.

22 Peter Tomsen, *The Wars of Afghanistan* (New York: Public Affairs, 2011), 631.
23 Frontline, 'Losing Iraq', television documentary, Public Broadcasting Station (PBS), July 2014.
24 US Government, *The U.S. Army in the Iraq War Volume 2: Surge and Withdrawal 2007–2011* Vol II (Carlisle, PA: Army War College, 2019), 621–2.
25 Bob Woodward, *Obama's Wars* (New York: Simon & Schuster, 2010), 161.
26 Ibid., 332–3.
27 Gates, *Duty*, 375.
28 President Barack Obama, 'West Point Speech', www.cnn.com, 2 December 2009.
29 Frontline, 'The Rise of ISIS', television documentary, Public Broadcasting Station (PBS), October 2014.
30 John Lewis Gaddis, *On Grand Strategy* (Princeton: Princeton University Press, 2018), 4.
31 Ibid., 147, 155.
32 See the CJCS Study, *Decade of War, Volume I: Enduring Lessons from the Past Decade of Operations*, 2012.
33 Story told to Daniel Marston in spring 2004.
34 The following articles present the most up-to-date debates: Risa Brooks, 'Paradoxes of Professionalism: Rethinking Civil-Military Relations in the United States', *International Security*, Vol. 44, No. 4 (Spring 2020): 7–44; James Golby and Mara Karlin, 'Why 'Best Military Advice' Is Bad for the Military—and Worse for Civilians', *Orbis*, Vol. 68, No. 1 (Winter 2018): 137–53; John Kiszely, 'The political-military dynamic in the conduct of strategy', *Journal of Strategic Studies*, Vol. 42, No. 2 (2019): 235–58; Hew Strachan, 'Strategy in theory; strategy in practice', *Journal of Strategic Studies*, Vol. 42, No. 2 (2019): 171–90; Hew Strachan, 'Strategy and Democracy', *Survival*, Vol. 62, No. 2 (2020): 51–82; issues related to lack of robust and honest dialogue across the coalition was recently covered in Daniel Marston, 'Operation TELIC VIII to XI: Difficulties of Twenty-First-Century Command', *Journal of Strategic Studies* (2019); and earlier works, including Frank Hoffman, 'Dereliction of Duty Redux?: Post-Iraq American Civil-Military Relations', *Orbis*, Vol. 52, No. 2 (Spring 2008): 217–35 and Linda Robinson, *Improving Strategic Competence* (Washington, DC: Rand), 2014.
35 Clausewitz, *On War*, 'War is thus an act of force to compel our enemy to do our will', 75. This last point it highlights is that war is a human activity and thus, debates are needed.
36 Strachan, *Directions of War*, 45.
37 See Brooks, 'Paradoxes of Professionalism: Rethinking Civil-Military Relations in the United States', 9–17 and Eliot A. Cohen, *Supreme Command: Soldiers, Statesmen, and Leadership in Wartime* (New York: Simon and Schuster, 2002), 225–51, for more detailed analysis of Huntington.
38 Samuel Huntington, *The Soldier and the State: The Theory and Politics of Civil-Military Relations* (Cambridge: Harvard University Press, 1957), 456.
39 Ibid., 73.
40 Quoted in Brooks, 'Paradoxes of Professionalism: Rethinking Civil-Military Relations in the United States', 26.
41 There are many who have critiqued Huntington's conclusions; notable examples include Peter D. Feaver and Richard H. Kohn (eds), *Soldiers and Civilians: The Civil-Military Gap and American National Security* (Cambridge, MA: MIT Press, 2001); H. R. McMaster, *Dereliction of Duty: Lyndon Johnson, Robert McNamara, the Joint Chiefs of Staff, and the Lies That Led to Vietnam* (New York: HarperPerennial, 1997); and Cohen, *Supreme Command: Soldiers, Statesmen, and Leadership in Wartime.*
42 Strachan, *Directions of War*, 76. See discussion in chapters 1–4 on the impact of General Colin Powell's analysis and interpretation of *On War* on US strategy development.
43 Brooks, 'Paradoxes of Professionalism: Rethinking Civil-Military Relations in the United States', 35.

44 Robinson, *Improving Strategic Competence*, 33.
45 Ibid.
46 Notes from my notebooks, 2007.
47 There were some military commanders who were looking for a dialogue on questions of 'to what end' versus how many soldiers were fighting in the streets of Ramadi or in the canals of Helmand Province. There were some senior officers who did feel, even as late as 2017, that it was not their place to ask such questions. Notes from Afghanistan and Iraq, 2006–2017.
48 Robinson, *Improving Strategic Competence*, 33.
49 2016 briefing by a senior US general on his top ten mistakes as an officer.
50 See the CJCS Study, *Decade of War, Volume I: Enduring Lessons from the Past Decade of Operations*, 2012. This work claims to highlight strategic lessons; I would argue that the eleven lessons provided, as well as the way they are framed, are better classed as operational rather than strategic lessons.
51 See Hew Strachan, 'Strategy and Democracy', 71–2: 'In 2011, Cameron (Prime Minister) told the Chief of Defense Staff, David Richards, that it was Richards's job to do the fighting while Cameron himself would do the talking.'
52 *Iraq Inquiry* (IQ), Section 9.8, para 184. It added more depth to the issue: 'Senior decision makers—Ministers, Chiefs of Staff, senior officials—must have a flow of accurate and frank reporting. A "can-do" attitude is laudably ingrained in the UK Armed Forces.' See also the Executive Summary; paragraph 197; paragraph 715; and the assessment submitted by Lt General Fry to the Chief of Defence Staff, which specifically posited potential failure in the British mission in MND SE. This frank assessment was not passed along to the Prime Minister by the CDS, but deleted from his own report.
53 Lt General Bill Rollo, 'Campaigning and Generalship: Iraq 2008' in Jonathan Bailey, Richard Iron, and Hew Strachan (eds), *British Generals in Blair's Wars* (Farnham: Ashgate, 2013), 185. See also Marston for a more detailed analysis of the British Iraq case study.
54 US Army Iraq Official History, Vol. II, 616–7.
55 Strachan, *Directions of War*, 43–4.
56 Strachan, *Directions of War*, 71.
57 Lawrence Freedman, 'Political Impatience and Military Caution', *Journal of Strategic Studies* (2019): 11–2. This was an issue in the US, as the Chairman of the Joint Chiefs (CJCS), was not as tied into the planning and debates being run by CENTCOM.
58 Strachan, *Directions of War*, 75.
59 Correspondence with Sir Lawrence Freedman, June 2020.
60 IQ, Executive Summary, para 720.
61 See IQ, Executive Summary, paras 693–712 for more detail.
62 Always a British Lt Gen.
63 IQ, Executive Summary, para 715.
64 IQ, Executive Summary, paras 721–731.
65 IQ, Executive Summary, para 740.
66 IQ, Executive Summary, para 746.
67 IQ, Executive Summary, para 757.
68 IQ, Executive Summary, para 757.
69 IQ, Lt General Richard Shirreff interview, para 7.
70 IQ, 9.5 July 2006–June 2007, paras 448 and 449. See also Shirreff's testimony, 6–7, for more details.
71 IQ, Shaw Testimony, CLOSED, 32–3. See General Shaw's chapter in *Blair's Generals*, as it paints a different picture from his formal testimony.

72 IQ, 9.5 July 2006–June 2007; see paras 476 and 477 for more details. There are other paras in the high 400s that highlight that there was some manipulation going on within the British government to justify withdrawal. The British deputy general in MNF-I, LTG Lamb, commented to the Assistant Private Secretary that there was apprehension within the US military regarding the British plans. See para 508 and Marston for more details.

73 Things shifted a bit with the Operation Charge of the Knights in March/April 2008. See Marston, 20–5 for more details.

74 See Marston for a much more in-depth discussion and evidence relevant to this tension. This tension was confirmed over a series of meetings in 2007 and 2008 in Iraq.

75 Marston, 26.

76 IQ, Section 9.8 Conclusions, para 197. In para 177, the committee reported, 'UK strategies [not strategies as one might define it] tended to focus on describing the end state rather than how it would be reached. On none of the 20 occasions when UK strategy was reconsidered was a robust plan for implementation produced.'

77 The 4th Infantry Brigade commander during the Charge of the Knights Operation in 2008.

78 Reports and interviews with Brigadier Free and Brigade Staff 2008 and 2009. This officer stated this after the Charge of Knights Operation in 2008, which highlighted the lack of communications between the British and the Americans in Iraq. Brigadier Free viewed honest and robust dialogue with the US military as extremely important. See Marston for a more detailed analysis of this case study.

CHAPTER 9

# A Hollow Victory? Assad's Regime and 'Winning' the Conflict in Syria since 2001

Richard Kuno

## Introduction

Before we can deal with the question of 'what winning means for Assad', a few technical terms and research assumptions must be made clear. We need common ground for the following analysis. Only from this common ground can the journey start and put the phenomenon of 50 years of Assad rule in its regional, cultural, historical and strategic context. We can only understand why conditions in Syria are as they are when we leave our western point of view and dive deep into a culturally and historically completely different environment.

What is the Middle East?

To make the term Middle East clear for the purpose of the present analysis, we use a very simple explanation: the Middle East is called the Middle East because it's in the middle.

If we look at where Syria is located on a world map, we can see that it is one of the central fields of the global chess board. At precisely this point the three largest and culturally most influential worlds of the Islamic cosmos meet with each other – or often clash together: The Arab world, the Persian world and the Turkic world. All three of them have expanded their empires over Syria several times. Sometimes for decades, sometimes for centuries, and each empire left its trace and legacy: pockets of religious or ethnic groups, administrative traditions, trade connections and so on. Like a changing of the guard, Persian, Arab and Turkic empires followed each other in ruling Syria and the Levante, the eastern shores of the Mediterranean.

But our definition of the Middle East is not yet complete.

If we open our triangle of the Arab, Turkic and Persian worlds towards the west, then we finally get a four-sided arena. This box-like arena is where the western world meets the east. Syria was the zone of influence, battlefield and marching ground for western powers. All western expansion or power projection to the east was aimed at or needed to make use of this land bridge between Europe, Asia and Africa.[1]

The Middle East is also the cradle of civilisation between Mesopotamia and Phoenicia. Not only is the Ugarit alphabet from northern Syria one of the oldest alphabetic scriptures, but Damascus is the oldest continuously inhabited capital of the world, looking back on around three and a half thousand years of uninterrupted history. Being part of the oldest cosmos of city states and ancient empires – Hittites, Babylonians, Egyptians, Medes, Israelites, to name just a few – one can argue that if politics is the art of achieving, increasing and preserving power, then it was invented right there: in the earliest city states and empires of mankind – even before the Greeks gave politics its name, derived from the word πόλις, which means 'city'.

What is the principal source of political loyalty in the Middle East: constitution, race, religion, language, or nation state?

The whole construct of the 'nation state' has grown incrementally over centuries of European and American history. It is deeply rooted in the west's socio-political DNA and therefore the nation state is the angle from which westerners not only see themselves or their European neighbours but the global society. After World War II, 'nation states' have been created that have populations with vastly different historical experiences. This is especially true in the Middle East where not a single sovereign country exists that is older than 1946 and that consists of a mono-ethnic or mono-lingual or mono-cultural population. In the Middle East the nation state is an imposed shell that cuts through ethnic and cultural ties of regional populations. It has not developed into an effective source of political loyalty.[2]

## A hollow victory? What winning means for Assad

Western military leaders', politicians', media's and common people's view on winning is guided by a western perception of politics, warfare and history. But for the Syrian regime, winning is something completely different, which is why we have difficulties understanding and especially acknowledging it. A perfect example for our skewed comprehension of what winning means for Assad is what *The Economist* wrote on its frontpage in September 2019 – 'Assad's Hollow Victory' – depicting him standing lonely and phlegmatic on a heap of rubble. The real question is whether Assad's victory is hollow or stunning. To anticipate an answer: it is stunning. *The Economist*'s misunderstanding comes from a clash of concepts, the western concept of what winning is and a Middle Eastern concept of what winning is. In the Syrian case the latter is very much driven by Assad and his regime's very specific social and historical background.

Before we can devote ourselves to the question of what winning means for Assad or the Syrian regime, we have to examine the nature of the ruling system that has brought him into his position and that has proven itself, even after years of conflict, as very robust and resilient, now having been in place for half a century.[3]

First, it is important to remember that the current president, Bashar al-Assad, was not the designated successor of his father Hafiz al-Assad. As the second son and thereby second in line, he had been systematically kept away from government affairs. Keeping away the second successor has been a usual procedure for centuries, and not only in Syria, because power transition is the most delicate task to handle in an autocratic system or monarchy. In order to prevent Bashar from becoming – even unintentionally – a potential alternative to the designated heir of Hafiz, he had to pursue a professional career that was as far away from power and politics as possible: that of a medical doctor.

Hence the Syrian regime ran into serious trouble when Bassil al-Assad, the designated successor, died in a car crash in 1994, and his father was suffering from leukaemia so it was evident that his power had to be handed over to the next generation within a few years. So, Bashar was immediately ordered back to Damascus and had to undergo a presidential crash course. Bashar did not have a specific profile or strong image. Bassil had been systematically built up by his father to inherit his ruling system, a perfectly smooth-running machine, and enjoyed respect, if not fear, even among the highest ranks of his father's old guards in politics, administration, secret services and the armed forces. When he died, a life-long project of intense preparation died with him.

As Hafiz was already sick with cancer when his heir suddenly disappeared from the equation, it was clear that his replacement would need much more support from the old representatives of the system than Bassil had needed – Bassil who had been prepared to uncompromisingly dominate even his father's strongest fellows. Hafiz' old cronies were very powerful figures with lots of experience in a personally and politically very dangerous environment. When Bashar finally came into office in the year 2000, everybody understood that Bashar was the figurehead of his father's ruling system rather than a 'dictator'. The strongest figures around him were the directors of the most powerful secret services, a handful of ministers and generals and his brother-in-law Asif Shawkat, who were, with very few exceptions, all members of the Alawite community.

## Who are the Alawites?

The Syrian population is very fragmented, not only in terms of religious groups and denominations but also language-wise and ethnically. Of course, Syria is regarded to be an Arab country, but looking closely at the ethnicities present in Syria, we see that there are not only Arabs; there are Kurds, Armenians, Turks, Arameans, Circassians, all with their different languages. When looking at religious communities, the fragmentation of the Syrian society becomes even more evident. Beside the Sunnite majority of around 70% of the Syrian population, there are three larger minority groups of around 10% each: Alawites, Druze and Christians, not to mention the smaller communities like the Twelver Shiites, Ismailis and Yazidis. Regarding the geographical distribution of the groups we see two main constants: the Sunnites dominated the cities and plains while the minorities originally dwelled in the mountainous and remote regions. The Alawites in particular lived under very harsh conditions. Influenced by Ibn Taymiyya's Alawite fatwa, which said that 'Alawites did more harm to Islam than Christians and Jews', they had to face numerous brutal extinction campaigns throughout history and were repeatedly expulsed from their refuges.[4]

So, how did the most deprived and most under-privileged of all groups in Syria take over and established a ruling system that withstood all enemies – inside and outside the country – up until today? There was only one short historical period when this could take place. After World War I, and even more so after World War II, modern western political ideas and ideologies spread across the Middle East. People were looking at Europe or the Soviet Union and tried to overcome the 400 years of Ottoman colonisation that was associated with corruption, standstill, exploitation and cultural oppression. Communist, liberal, socialist, nationalist, pan-Arabist, or even Nazi parties were founded and struggled for power. The head scarf disappeared from the streets of Beirut, Damascus, Cairo and Baghdad, and religiosity became an attribute of old-fashioned people. The idea of an Arab nation or race became increasingly important, either in the apparel of Pan-Arabism (Baathism) or as a form of regional nationalism, like Syrian nationalism. For the first time in history it became more important to be an Arab or 'Syrian' than to be a Muslim, Christian or Alawite.[5]

Besides this beneficial zeitgeist, there was another helpful circumstance that helped the Alawites to break up the social box they were in. The founding of the new Syrian state in 1946 meant the young nation needed modern armed forces and police. Service in the military or police was not regarded as prestigious in the Syrian society. For the traditional Sunnite and Christian, joining the military or police was something for the lower classes. So, while the representatives of the old elites avoided the armed services, for the Alawites it was their ticket to social progress. From the very beginnings of the Syrian army, police and security forces, their ranks were over-proportionally filled by Alawites.

After the failing of the United Arab Republic between Syria and Egypt in 1961 and a period of political instability and turmoil in the 1960s, Hafiz al-Assad finally prevailed and took over in a military coup in 1970. The elections shortly after granted him the office of the president of the republic with 99% of the votes. Hafiz was commander-in-chief of the air force, former defence minister, representative of the military-pragmatic faction within the Baath party – and Alawite.[6]

## The Assad system

Hafiz al-Assad was clear that he had to implement a system for his rule that was different to those of his predecessors. His new system of rule was based on three principal pillars: surveillance, protection and corruption.

The first – and most common – pillar of Hafiz' plan represents the hard edge of an autocratic system: a Soviet-style weave of mutually monitoring secret services to oppress any kind of political resistance, be it in civil society or inside the security forces themselves, implemented by Soviet intelligence advisors.

The second pillar was protection, or to be more precise, freedom of religion and proportional participation. Despite the fact that the Baath ideology is in principle a secular one, religious belonging became overwhelmingly important, especially for the Alawite community. As Hafiz was more interested in a solid power base than in promotion of the 'best and brightest', the country's key positions were filled with members of his own denomination. At the same time all other minority communities became beneficiaries of a policy of general protection of religious practices, places and festivals, so that, for example, Christians could (and still can) celebrate Easter on the streets with timpani and trumpets. In this manner the minority rule of the Alawites became the guarantee of religious freedom for all the other minor religious groups, like Christians, Druze, Shiites, Ismailis, Yazidis and so on, who were (and still are) in fear of an overwhelming majority of Sunnis. Discrimination and pogroms are part of the collective memory of all non-Sunni communities.[7]

Finally, the third pillar: corruption. While our common understanding of corruption is that of a sign of social decay, for Hafiz it became a method of domination. When we observe corruption as a pandemic plague, we imagine a poor country that is not able to pay its servants a decent income, so that they are forced into corruption and bribery by poverty. In this sense corruption is a bottom-up phenomenon induced by bad economic circumstances. In Syria, the opposite was the case. When Hafiz became the head of government it was paramount for him to secure his rule. He very quickly started to bring his trusted men into the key positions of the state, many of them Alawites but all of them loyal solely to him in person. In order to ensure that his selectees were loyal only to him, he interwove professional success and career with corruption – even if that sounds like a paradox. When Hafiz offered a certain candidate a political, administrative or military career, he offered an opportunity for self-enrichment too, and he expected his beneficiary to make use of it. This demonstrates the outstanding strategic talent and cunning of Hafiz, because with his two-sided offer he strengthened his position in Syria. The offer was a litmus test: if you want unconditional personal loyalty from your liegeman, you need to make sure that he is neither a true political idealist nor a law-abiding person. Also, someone who is busy with filling his pockets does not complain about bad leadership anymore – he is part of it. In effect, Assad created an environment of yes-men. Thirdly, the moment anyone accepted the offer, the secret service opened his file and each and every single misdeed was precisely recorded. So, whenever anyone became a problem, all the evidence of his misbehaviour was readily at hand for extortion or 'legal' prosecution. As the years passed, being part of the ruling system became increasingly a stigma. Not only because of political dissatisfaction but because of the rampant and ever growing corruption, which pervaded deeper into society. This triggered hatred and frustration among the population. If Hafiz' rule ever ended, many would lose wealth and could face vengeance.

The original Assad-system, inherited by Bashar, is a self-enforcing mechanism of domination by entanglement.

## Real power is a ball

Often the concepts of power, government or rule are depicted as a pyramid: the ruler is on top and the ruled are at the bottom. In between, the degree of power of an individual decreases the lower it is on the pyramid structure. This depiction reflects an idealised mechanistic imagination of power which is influenced by western statehood and bureaucracy. A formal view of the Syrian political system would suggest the same model of power, maybe even stronger than in any western country, showing the president as the paramount leader at the very top and then the classical republican hierarchy with ministries, military echelons, and administrative bodies from nation level down to local communities.

The real nature of power in Syria is difficult to recognise, and to see it as a pyramid is totally misleading. This is why western observers and decision makers had difficulties finding the right approach to the Syrian crisis when it emerged in 2012. Western politicians and media quickly drew the conclusion that in order to solve the crisis and to 'change the regime' it would be necessary for Assad to step down from presidency. This demand is understandable but was far from achievable. Even if Bashar had wanted to, he could not step down. To understand why, we need to imagine a different model of power, which comes much closer to the reality in Syria: a ball.

In this alternative model the 'centre of power' is literally the centre of a ball. The crucial point is that the ruler does not stand alone and vulnerable at the top of a pyramidal hierarchy but he is in the centre of the ball and thereby protected by the many layers of social groups around him – cushions of fearful minorities, and legions of corrupt administrators and entangled military and interior security officers – for whom the survival of the leader has become their own survival. So, every layer has the highest intrinsic incentive to actively protect the power centre out of pure self-interest. This is the highest form of security that an authoritarian leader can achieve, and this is what Hafiz prepared for himself and his successor. When the leader in a pyramid system is changed, the tip of the pyramid can easily be cut off and being replaced, while the pyramid is still a pyramid. This means the basic functions of state can keep working and can guarantee a minimum degree of public order and stability.

As a rule of thumb, when a member of the majority rules, we can observe a true pyramid. When a member of a minority rules, a 'ball system' is in effect. Recent examples where a member of the majority ruled are Tunisia[8] and Egypt. In both countries power transitions were manageable without total breakdown of statehood. Even after decades of Mubarak dictatorship, the Egyptian president was even changed twice within a few years.[9]

In a ball system there is no regime change possible without destroying the ball, which has become statehood itself. If you want to extract the centre of power, you need to break up the protective layers. With the breaking of the ball, which is cutting through the genetic make-up of the ruling system, statehood itself is broken.

A sad example of a very strong ball system has been Saddam Hussein's Iraq. Toppling Saddam Hussein, who had installed a similar system to Hafiz', was breaking up Sunnite minority domination.[10] The end of Saddam's rule led straight into a total breakdown of statehood and a decade-long Sunni-Shiite civil war. The result was a de-facto trisection of Iraq into a Sunnite, a Shiite and a Kurdish part.

The same is true with Syria: Syrian statehood is based on Alawite domination of all key positions. Toppling Bashar would mean toppling the Alawites from all key positions. Ousting the system's Alawite powerbase from their leadership positions in the security forces would most likely lead to retaliation pogroms. So, no matter how overwhelmingly the international community and the majority of the Syrian people want Assad out, the old system would not easily go but fight. This is exactly what we have observed since 2012.

Indeed, even today Bashar is a beneficiary of the system that his father once designed and installed strategically. It protects him and his power out of pure self-interest. On the other hand, he simply cannot resign, even if he wanted to, because the layer that benefits the most from him being the head of state is his own group of origin, the Alawites. The backbone of his rule is made up of his own Alawite brothers and sisters. For him, resignation would mean putting to death his own origin.

## Assad's strategy for survival and 'winning'

All alarm bells in Damascus rang on 10 April 2003. All over the world newspapers headlined with a picture of Saddam's toppled statue in Baghdad. The west had wiped out Saddam's military within weeks – the strongest, best-equipped and most experienced armed forces of the Arab world. That Damascus could be the next on the list for a 'regime change' was intensively discussed in Syria, the entire region and beyond. There were serious reasons to fear a coming attack by the west or a western-led alliance. Syria was regarded by many analysts as a latent threat to Israel and an obstacle to a solution to the Palestine conflict and the Lebanese crisis, not speaking of Syria's strong ties to Iran. Additionally, it was an old ally of the former Soviet Union and it would not easily fit into a new and more pro-western Middle Eastern order. Russia was still not back on its feet but military ties with Syria remained intact on a low but stable level. Furthermore, Syria was an occupational force in Lebanon, which created a lot of enmity among the Lebanese population and among some Arab gulf states who supported Sunnite political movements in opposition to the Syrian presence. All in all, many factors were in favour of an expulsion of the Assad system and the regime was well aware of that.

The Syrian armed forces, as well as the entire society, had already gone through decades of organised corruption which led to a total run down of the military, aggravated by the fall of the Soviet Union and the loss of substantial defence support from there. The Syrian regime understood that it would not make sense to ramp up its arsenals in order to reach a level of credible self-defence capabilities against the US. The gap was simply too big.[11]

As direct confrontation equalled defeat, an indirect approach was needed. At this point Middle Eastern strategy design at its best began.

The first step that the system had to take was to make sure that every member of the regime's inner circle would stay loyal and hold the line. In order to keep the west out of the theatre in 2012, when the turmoil of the Arab Spring reached Syria, certain red lines had to be respected in strategy design – at least superficially and for a limited but crucial timeframe.

When in 2012 the situation started to become difficult for the Assad system with the mass protests on the streets in Damascus and throughout the country, especially in the city of Dara'a, the immediate reflex would have been to use military force. This is what some of the old guards from the days of Hafiz would have pledged for. They had seen the Muslim brotherhood insurgency in the 1980s and learned their lesson when it was finally crushed with the massacre of Hama in 1982. But a second Hama was forbidden – especially when Barack Obama said in public that military intervention against civilian protesters would trigger immediate intervention. Thereby, paradoxically the hard-liners became as much a threat to the survival of the system as the demonstrators themselves.

Other members of the inner circle may have been tempted to go to exile, to enjoy the deposits of their offshore accounts. However, the entanglement of the Alawite population at large made this kind of exit-strategy impossible for Assad or other leadership figures who were tempted to leave the sinking ship.

Some members of the ruling elite simply became too powerful, were too independent, acted autonomous or were too obscure in their actions and intentions – a risk that Assad and his closest advisors could not tolerate anymore. It became clear by July 2012 that the time had come to close the ranks – and to neutralise potential risk factors. Indeed, we can see that the number of decision makers – especially those deeply rooted in the system with an old track record – was abruptly and significantly reduced from July 2012 onwards, either by violent incidents or by defection. As a result, consistent strategy execution became much easier.

An advantage was that allies were already at hand at the beginning of the conflict – Iran and Russia. They had not only been Syria's closest allies for decades, but they were also in a certain way 'natural enemies' of the west. Thereby, Syria, Russia and Iran are bound together by more than friendship – by mutual interests.

For Russia, Syria in 2011 was its only Middle Eastern ally left and it played a crucial role as a base for Russian power projection towards the Mediterranean and the wider Middle East.[12] Additionally, it was proof of Russian reliability and trustworthiness for allies far beyond Syria. It became a test range for new tactics and technologies, and it brought Russia back on eye-level with the US when it came to crisis management and influence in a geo-strategically critical theatre. From 2015, Syria became even more important as part of Russia's integrated geo-strategy. By gaining influence in the Syrian conflict, Russia was able to open a 'second front' against the west, threatening the EU's soft belly in the south-eastern flank of NATO. This could only work when Moscow's ally Assad stayed in place. Whenever NATO's or the EU's pressure on Russia in the Ukraine/Crimea crisis would become too high, Russia could disbalance Europe's centre of gravity by escalating warfare in that region, where most refugees of 2015 came from – a vital threat that made European decision makers accessible for any future settlement: on Assad, as well as on Crimea.

For Iran, since the Islamic revolution, Syria had changing roles but an ever-growing importance. From Iran's fight against Israel in the early days of the Islamic Revolution to its struggle with Saudi Arabia today, Syria has always been crucial in Iran's Middle East strategy. Syria was Iran's force enabler in its build-up of Hizballah; Syria condemned Saddam's aggression and stood by Iran for eight years during the first Gulf war; Syria fought against spreading Sunnite Islamism as part of the genetics of its regime and thereby became an ideal partner in Iran's power-struggle with Saudi Arabia. In short, in almost every major conflict both partners were united. Both being 'pariah states', their alliance became even stronger when western pressure on the Syrian regime increased after 2011. Iran helped in mobilising military support for the regime in Damascus, with an appetite for the destruction of Sunni Jihadists, like Al-Nusrah Front and ISIS. The defence of Shiite holy sites, like the tomb of Imam Ali's daughter Zaynab in Damascus, could be used as a call-to-arms for Shiite volunteers. For Iran, a non-Sunni regime in Damascus was and will be vital for its Middle East strategy as a whole and for its confrontation strategy with Riyadh and Israel in particular.[13]

Just 20 kilometres to the north of Damascus, there is the notorious military prison of Sednaya. The proportion of Islamist detainees started to grow in the 1980s with the Muslim Brotherhood insurgency. Additionally, due to the insurgencies in Afghanistan and Iraq since 2001 and 2003, the prisoners increasingly came from the ranks of experienced fighters from both theatres, many of them with sophisticated knowledge of tactics, leadership, and IED technology. Despite being a declared enemy of Al-Qaeda and other Islamist movements, Syria became attached to the 'axis of evil' in 2003. From then on, when Damascus started to prepare for an upcoming existential challenge, the prison of Sednaya served two different purposes. The first purpose was to lock-up dangerous people, many of them returnees from the Jihadi battlefields. But the second and much more important purpose was to keep ready a critical mass of fighters who were able to organise an insurgency and to execute a maximum of terror and turmoil instantly when a chance appeared – and when needed by the government. At first this sounds contradictory. This critical mass of fighters was the fiercest enemy of the regime itself. But their 'storage' and deliberate release had the potential to become useful. The regime needed a fire accelerant when power finally slipped out of its hands, there nothing to resort to anymore other than fighting fire with fire. When the riots in Dara'a got out of control and the protests on the streets in Damascus attracted international attention, it was too late to quell the protests with conventional riot police and secret service terror. In 2011 the Syrian government was close to collapsing: the armed forces began to disintegrate, police stations were burned down, hundreds of thousands of protesters called for an end to Assad's rule and the west threatened with intervention. So, in May 2011 Assad declared a general amnesty that included the inmates of Sednaya military prison. This declaration was sold by the regime (and misunderstood by international observers and by the Syrian public) as a concession of the regime in face of the mass protests. In reality it was pouring oil into fire.[14]

The wilful release of the detained Jihadists from Sednaya represents the practical start of the regime's final survival strategy and simultaneously the beginning of the intended destruction of public order in Syria, executed by the regime's own fiercest enemies. In game theory this strategy is called 'brinkmanship' and stems from the Cold War with the possibility of a nuclear Armageddon: the credible threat of mutual destruction, including self-destruction. This strategy associates one's own fate with the enemy's fate: *if you kill me – you kill yourself.*[15]

The full outbreak of armed violence and destruction was started by independent Jihadist splinter groups (many formed by ex-convicts) which quickly formed loose alliances. Then the Al-Nusrah Front entered the scene and finally ISIS emerged and brought propagandistic atrocities and public terror to never-before-seen levels.

This 'Islamist Weapon of Mass Destruction' was aimed at several targets simultaneously: the dissatisfied Syrian population would be forced back to preferring security over freedom, and external enemies would realise that chaos in Syria triggered a mass exodus of refugees towards safe neighbour states which would potentially destabilise the entire region. In sum, the intended regime change would start to produce collateral damage that would not stand in relation to its gains anymore. Assad would be revived as the 'lesser of two evils'.

The direct impact of the Syrian crisis reached Europe's heartland by 2015 with the refugee wave and millions of immigrants. For almost every political leader in Europe and in the region (which by 2011 demanded regime change), stability now became more important than change. After 2015, voices calling for Assad to step down became quieter, instead, voices calling for an 'end to the violence' in Syria became more prominent.

With the release of the Islamists from Sednaya, the regime killed several birds with one stone: Damascus followed national and international demands for amnesty and reform at an early stage of the crisis – and could wash its hands of any following responsibility. Damascus proved it was right (even in its methods) when fighting Islamism for decades, much too often being unheard when warning the west against this threat. The regime showed its own population and international decision makers what would be the alternative if Assad lost power: widespread Islamist terror, potentially the emergence of a second Afghanistan. With this brinkmanship strategy it forced back interior and international support that it had lost in 2011.

Against Recep Tayyip Erdogan, the prime minister of Turkey, who was at first a fierce enemy and one of the most aggressive interventionists on behalf of a regime change, Damascus had a trump card – the Kurdish population in northern Syria along the Turkish borders. Turkey was a serious opponent: a NATO member, the region's strongest military power and under Erdogan's leadership increasingly fundamentalist. Additionally, Syria was under Turkish (Ottoman) rule for more than 400 years, and Erdogan stressed the Ottoman legacy as part of Turkey's claim for regional leadership. When in 2011 the regime took the strategic path of brinkmanship, the Kurds were the decisive factor to neutralise the Turkish threat. At the time that public order collapsed, that the army became ineffective, and that police and security forces were consumed with fighting for their own survival, Turkey was supporting rebel groups with logistics and infrastructure. The strategic answer was another 'release', namely that of the Kurds into a de-facto autonomy. By its strategic retreat from the north the regime left them with the responsibility for their own survival and did not have to invest further efforts in defence or oppression of a region, where it would have been facing superior opponents anyway. Additionally, Assad created a serious threat for Turkey when he allowed the potential emergence of a sovereign Kurdish state. At that point they became a threat to Erdogan and Turkey's national integrity, because with a Syrian state in pieces, the Kurds could potentially take the chance to proclaim an independent Kurdish state and a new surge of the separatist insurgency in Turkey's south-east became imaginable.[16] As a result, Syria's territorial integrity under a strong centralist leadership turned into Erdogan's own interest. Assad was revived again as the guarantor of the desired status-quo. What makes this strategy even more remarkable is the Kurds themselves later called back Assad to protect them from Turkish occupation. This happened in 2019, when the Syrian Arab Army returned to the Kurdish provinces and was welcomed by the Kurdish population like liberators.

## Conclusion: What is winning for Assad's Syria

Often western strategy design is torn between a dichotomy of political and military winning conditions which contradict each other. Western strategists are confronted with manifold conflicts of military, economic, political, humanitarian and legal objectives. For a non-western ruler (not only for a Middle Eastern or a Syrian one) strategy design can be much easier, because there is not always a dichotomy between military and political objectives, therefore, strategy can be more integrated and ruled by a single objective, which is survival.

For Damascus, strategy design is an ongoing process without time constraint: it starts when a problem emerges (and this normally is long before an average western observer would realise it) and it's dynamically adjusted and developed before, during and after a conflict, with a clear objective. Because the winning condition (survival) is undisputable, it's easier for Assad to distinguish between different options. One can only assess a new situation as beneficial or unfavourable when the objective can serve as a constant. Under western strategy design conditions (where objectives are manifold and often contradictory), it's much more difficult to take a decision – especially a 'right' one.

So, what makes Assad 'win'? There are several factors that decide this. First, the Syrian conflict is not decided in Syria. Powers around the theatre decide. This paradigm can be regarded as the basic principle of any Middle Eastern strategy design per se. Dealing with allies, proxies and outside enemies was the day-to-day business of any leader in the Middle East in history. Today, the power brokers in Syria are Russia and Iran, on which the survival of the regime is very much dependent. The first years of unrest after 2011 saw many outside powers who jumped on the wagon of turmoil: Turkey tried to re-establish hegemony over Syria by supporting protesters, the Free Syrian Army, and Islamist groups; Gulf states engaged with money and logistics when they saw a chance to block Iranian influence and to break up the 'Shiite crescent'. The west supported oppositional groups and the Kurds, who seemed to be the only bulwark against the advance of ISIS for a while. Most outside powers have withdrawn or changed their engagement, at least for now. The price for further endeavours in Syria became too high: Turkey couldn't risk a free Kurdistan, which loomed in northern Syria; Saudi-Arabia meanwhile is too much consumed with its engagement in Yemen and suffers under the financial burden of its military ventures, aggravated by historically low oil revenues; Europe desperately needs to avoid another refugee wave like in 2015; for the EU, stability and reconstruction have become the paramount objectives; the US have finally decided to scale down their engagement after almost two decades of military endeavours in the region, frustrated by results that have stayed far away from their intended outcomes. Even now, after the strategic retreat of many international opponents of Assad, the re-establishment of his rule over every corner of Syria is hard to imagine without Russian and Iranian help.

Secondly, Assad is the centre of a ball, not the tip of a pyramid. The core factor why the Assad regime is so resilient is that it represents a minority-rules-majority system. This is also the reason why (western) demands for Assad stepping down were unrealistic. In the 1970s, from day one of his rule, Hafiz, Bashar's father, managed the transformation of Syria into a totally Alawite-dominated machinery which is based on entanglement. Like Saddam in Iraq, Hafiz had to turn Syria upside down by installing someone he could personally trust in every single key position. The most natural way to staff trustworthy people in a Middle Eastern context was (and still is) to rely on wider family and brothers of faith, tribe or ethnicity. For Hafiz this general rule came in combination with several Alawite-specific advantages: the Alawites share the collective memory of poverty and of being systematically excluded and brutally oppressed – often by their own Syrian compatriots. In 1970, with Hafiz grasping power, they all had a unique chance in history to rise to the top of the Syrian society, administration and wealth. During the following decades of Assad rule, their entanglement with corruption and oppressive violence grew. Thereby, for the entire Alawite community it became literally a question of physical survival to keep Assad in power. If Assad is toppled, the whole Alawite community would go with him, and this despite not every Alawite being a supporter of Assad.

Here, a factor of mutuality becomes effective: asking Assad to step down meant demanding him to put thousands of his brothers and sisters to death. He could never fulfil the demand to step down from presidency, even if he wanted to. For Assad and for his Alawite constituency, stepping down was never an option. It would mean collective punishment, brutal retaliation, perhaps genocide.

Thirdly, the enemy of my enemy is my friend. Power politics in its purest form. In a political arena that is ruled by the assumption that winning is a zero-sum game, politics and strategy is totally focused on the very own benefit. In such an environment it is quite simple to categorise the political players in terms of beneficial or unbeneficial for their own interests. So, when militant Islamists, who are the Assad system's worst enemies, are at the same time the worst enemies of an even stronger enemy, the west, instrumentalising these Islamists as enemies of the west, makes them for this particular task Assad's 'friends': the west had to be made to understand that Assad is the lesser of two evils. When the west decided to engage the militant Islamist groups in Syria militarily, it had the unintended consequence of paving the way for Assad's revival. Assad also knew that the Kurds became a natural threat to Turkish national integrity when they were released into autonomy. Erdogan – a declared and active enemy of Assad – preferred him to stay in power instead of risking the disintegration of Syria. Assad (again the lesser of two evils) had to regain control over his Kurdish compatriots to prevent the emergence of a Kurdish state. The Kurds themselves had to welcome back Assad, because without him they would face Turkish occupation. The Assad system could arrange its survival strategy according to the weak points of each actor.

When the so-called Arab Spring reached the Syrian streets in 2011, the regime was clear that the protests could not be oppressed with military force anymore. The west had already intervened militarily in Libya not long before, which led to the overthrow of Gaddafi. With a small core constituency of just 10% of the population, the upcoming conflict couldn't be managed with their own forces. There was an urgent need for proxies and allies. These came quickly in the form of Russian and Iranian military advisors, Hizballah, Shiite militias of Afghan and Pakistani origin, Syrian paramilitary formations, and local self-defence groups.

Any attempt to confront western forces militarily would have been pointless and counter productive. Most remarkably, there were also very few direct confrontations between the government's Syrian Arab Army and ISIS fighters. Usually, fighting the Islamic State or other strong Jihadi groups was done by the Kurds, by Hizballah and by Russian and western air strikes. If we take the sentence 'the enemy of my enemy is my friend' seriously, we must admit that the west itself has become an ally of Assad. This may be very disturbing insight. But looking at results and not at intentions, the western air campaign against ISIS 'counter DAESH' not only destroyed Islamist fighting formations and infrastructure on behalf of humanity and the safety of the Syrian population, it finally paved the way for Assad to restore order in Syria. And this without firing a single round at ISIS.[17]

Assad's strategic horizon is indefinite, whereas the western one is one legislature of four to five years at maximum. No western political leader wants to go into the next election campaign with an ongoing war of attrition. This creates high pressure for achieving quick gains on the side of political and military strategy designers. For Assad the opposite is the case. The longer the conflict lasts the higher the likelihood of survival, because sooner or later people want back peace and stability at all cost – this is true for western political leaders as well as for Assad's unhappy population. So, protracting has been (and will be) the key factor in successfully resisting the west. As soon as the critical early days of a conflict are overcome, the balance can shift to the supposedly weaker opponent of western military supremacy. For Damascus, lessons were learned from Iraq and Libya, where both rulers tried to brutally keep up public order with their own forces.[18] As the west was not aware which internal tensions would break out after the removal of the dictator, political leaders were tempted to topple Gaddafi and Saddam. One of the insights that came from the Iraqi and the Libyan interventions was that in case of an imminent western strike, it was necessary to rapidly escalate unrest to a full-grown civil war in order to assign intolerable collateral damage to a 'regime change'. This would discourage western decision makers before they could act. This brings us to the last factor in Assad's strategy. Protracting the conflict in Syria was not enough. Only when total escalation with major damages and losses could be achieved very quickly and in a way that would last for a long time, would regime change become unattractive to international decision makers and to dissatisfied parts of Assad's population. The way to manage a steep and lasting escalation curve was the release of militant Islamists and the retreat of core elements of the security and armed forces into safe strongholds without engaging the insurgents. The loss of control, population and territory was not a sign of the near end of the regime, as often misinterpreted by western observers, but of a strategy that created more problems for internal and external enemies than for the regime itself.

As stated in the introduction: the nation state is a western construct. In Syria the understanding of 'nation' is much narrower. So, the necessity of defending the 'nation' is much more focused on core groups and regions, turning great parts of territory and population into bargaining pieces.

The above outlined picture of the Syrian conflict looks disturbing to many readers, but its aim is to present a completely different worldview, leading into a completely different concept of winning. All strategic steps as taken by the Syrian regime were ready and their intended outcomes were relatively predictable already in 2011. This is simply because there are certain rules and obligations that each actor must obey and because there are individual weak points that are essential for the survival of each actor around the Syrian theatre. Knowing these obligations and weak points made the actions and reactions of each actor relatively predictable.

Additionally, the inner constitution of the Syrian society made it easy for Assad to absolutely rely on certain parts of the population and to sacrifice others.

The same applied to the constitution of Assad's international enemies and allies. Neither Russia nor Iran would ever switch sides, at least not while Putin and the Ayatollah regime are in power. On the other hand, the US, Europe, the Gulf states and Turkey were struggling with manifold obstacles in their strategy design and execution towards Syria. They were predictably unstable if the situation in Syria got out of control.

For sure, no strategy runs as smoothly as planned. Some steps do not work, others run better than anticipated. For Assad one part of the strategy was extremely successful, to a degree that was surely unexpected, which finally tipped the balance to his favour. Putting the entire region at strain with millions of refugees was a central element of Assad's survival strategy. If it went well it contained a potential domino effect of unrest and chaos in Lebanon, Jordan, Turkey and Iraq. The Syrian crisis would spill over on other fragile neighbours. That it went so well, that in 2015 a refugee wave of millions of migrants arrived in the middle of Europe, was 'too good to be true' for the regime in Damascus, because it entangled the survival of many European governments with stability and peace in Syria – as quickly as possible. Assad's comeback had begun.

## *Notes*

1 G. Shrimpton, 'Persian Strategy against Egypt and the Date of the Battle of Kitium', *Phoenix*, Vol. 45, No. 1 (1991), 1–20; A. B. Bosworth, 'The Government of Syria under Alexander the Great', *The Classical Quarterly*, New Series, Vol. 24, No. 1 (May 1974): 46–64; A. Hourani, *A History of the Arab Peoples* (London: Faber and Faber, 1991); W. Kaegi. *Byzantium and the Early Islamic Conquests* (Cambridge: Cambridge University Press, 1992), 1–25, 147–180; A. Cameron and L. I. Conrad, *The Byzantine and Early Near East* (Princeton: Darwin Press, 1992), 277–315; Kasaba, Fleet, Faroqhi, *The Cambridge History of Turkey, Vol. 2* (Cambridge: Cambridge University Press, 2006–2012), 87, 96, 139, 153–154.

2 T. E. Farah, *Political Behaviour in The Arab States* (Boulder: Westview Press, 1983), 113–133; Hourani. *A History of the Arab Peoples*; N. Richani. *Dilemmas of Democracy and Political Parties in Sectarian Societies: The Case of the Progressive Socialist Party of Lebanon 1946–1996* (New York: Macmillan, 1998), 15–51; R. Hrair Dekmejian. 'The Arab Republic of Syria', in T. Y. Ismael and J. S. Ismael, *Politics and Government in the Middle East and North Africa*, chapter 8 (Miami: Florida International University Press, 1991); T. Philipp and C. Schumann, *From Syrian Land to the State of Syria and Lebanon* (Beirut: Ergon Verlag, 2004), 1–26.

3 N. van Dam, *The Struggle for Power in Syria* (London: Croom Helm, 1981); S. Alianak, *Middle Eastern Leaders and Islam–A Precarious Equilibrium* (New York: Peter Land, 2007); P. Seale, *Asad: The Struggle for the Middle East* (Berkeley: University of California Press, 1990), 3–8, 9–11, 92–94.

4 M. M. Bar-Asher and A. Kofsky, *The Nusayri-Alawi Religion* (Leiden: Brill, 2002); A. K. Kazi and J. G. Flynn, *Muslim Sects and Divisions–The Section on Muslim Sects in Kitab al-Milal wa l-Nihal by Muhammad b. Abd al-Karim Shahrastani* (London: Kegan Paul, 1984).

5 To put it in the words of Antoine Saadeh, founder of the SSNP: 'Religion is for every single one of us, but the nation is for all of us.' J. Waardenburg, *Muslim Perceptions of Other Religions* (New York: Oxford University Press, 1999), 3–69, 85–101; J. Waardenburg, *Muslims and Others. Relations in Context* (Berlin: De Gruyter, 2003), 162–191, 192–196.

6 Farah, *Political Behaviour in the Arab States*, 113–133; N. S. Yared, *Secularism and the Arab World (1850–1939)* (London: Saqi, 2002), 10.

7 Waardenburg, *Muslim Perceptions of Other Religions*, 85–101; Waardenburg, *Muslims and Others. Relations in Context*; Hourani, *A History of the Arab Peoples;* D. D. Grafton, *The Christians of Lebanon–Political Rights in Islamic Law* (London: Bloomsbury Academic, 2003).

8 https://www.britannica.com/place/Tunisia/The-Jasmine-Revolution#ref341075, accessed 2 April 2020.

9 https://www.britannica.com/place/Egypt/The-Mubarak-regime, accessed 2 April 2020.

10 https://www.britannica.com/place/Iraq/Iraq-under-Saddam-Hussein, accessed 2 April 2020.

11 https://www.mei.edu/publications/lion-and-eagle-syrian-arab-armys-destruction-and-rebirth, accessed 2 April 2020.

12 https://www.bbc.com/news/world-europe-51733595, accessed 5 April 2020; https://carnegieendowment.org/2019/10/24/brief-guide-to-russia-s-return-to-middle-east-pub-80134, accessed 5 April 2020; https://jamestown.org/programs/rme/, accessed 6 April 2020.

13 https://www.mei.edu/publications/irans-unconventional-alliance-network-middle-east-and-beyond, accessed 5 April 2020; https://moderndiplomacy.eu/2020/01/12/irans-middle-east-strategy/, accessed 6 April 2020; https://www.iiss.org/publications/strategic-dossiers/iran-dossier/iran-19-03-ch-1-tehrans-strategic-intent, accessed 7 April 2020.

14 https://www.thenational.ae/world/assad-regime-abetted-extremists-to-subvert-peaceful-uprising-says-former-intelligence-official-1.319620, accessed 5 October 2020.

15 A. Dixit and B. Nalebuff, *Thinking Strategically–the Competitive Edge in Business, Politics and Everyday Life* (New York: W. W. Norton, 2008), 119–158.

16 https://www.britannica.com/place/Kurdistan, accessed 9 April 2020; https://www.britannica.com/place/Turkey/The-Kurdish-conflict, accessed 9 April 2020.

17 https://www.gov.uk/government/news/update-air-strikes-against-daesh, accessed 9 April 2020; https://www.britannica.com/topic/Islamic-State-in-Iraq-and-the-Levant, accessed 9 April 2020.

18 https://www.britannica.com/biography/Muammar-al-Qaddafi, accessed 10 April 2020; https://www.britannica.com/place/Libya/Qaddafi-toppled-and-killed, accessed 10 April 2020; https://www.britannica.com/event/Libya-Revolt-of-2011, accessed 10 April 2020.

CHAPTER 10

# The Impact of History, Politics and Religion: Three Contrasting Conceptions of 'Winning' in Iran since 1979

Ali Parchami

Modern Iran has three distinctive and often contrasting identities that shape and inform its conceptions of victory. The first of these revolves around the cultural heritage of a people who identify themselves as 'Iranians' by tracing their lineage to Aryan invaders of the 1st millennium BC and, in particular, to the establishment of the Achaemenid Empire of the Persians in 550 BC by Cyrus the Great. The second strand centres on the clerical regime that governs present-day Iran within the framework of the Islamic Republic. With its anti-western ideology, opaque structures, parallel institutions and factional politics, the Islamic Republic is a notoriously difficult topic even for the most seasoned of political analysts. Predicated on centrifugal forces and internal contradictions, the regime's behaviour and motives can often be paradoxical – with its conception of victory reflecting these schisms. The third strand in Iran's conception of victory should be sought in the country's Shia identity and within the political outlook of its Shia clergy. The latter originated in the late 15th century with the re-emergence of a unified Persia under rulers who deliberately employed Shiism as a religious–political tool to forge an empire to rival the Sunni Ottoman Turks. Organised around an intricate network of theological seminaries, the Shia clerical movement spread from Iran to other parts of the Middle East – and Asia – and soon became a community with its own transnational aspirations. While all three identities are constituent parts of modern Iran, we must recognise that there are significant distinctions between how 'winning' is perceived by the Iranian School, the rulers of the Islamic Republic regime, and within the broader context of Shia Islam and its clergy.

The Iranian School associates victory with national identity and patriotism. Drawing on the past glories of Persia, its outlook contains facets of cultural supremacy and imperialistic attitudes: most notably the belief that the Iranians are a culturally superior people and are the rightful overlords of the Persian Gulf area. Within this narrative, Iran is celebrated as a polity with over 2,500 years of history whereas neighbouring countries tend to be dismissed as 'upstarts', post-colonial creations, and more pejoratively as 'tin-pot runts'. These sentiments are tempered, however, by the experience of Iranians as a people who were conquered brutally and successively by Graeco-Macedonians, Arabs, Turkic nomads and Mongols before becoming the pawns of the British and the Russians during the 19th-century 'Great Game'. The retention of Persian identity is, therefore, central to the school's conception of winning. In contrast, the Islamic Republic's notion of victory is primarily shaped by an instinct for survival and the preservation of the regime's security against a myriad of internal and external adversaries. Its secondary ambitions are calibrated around revisionism and projecting power throughout the Middle East. In the pursuit of these ambitions, the Islamic Republic's ostensibly dogmatic ideology is quite deceptive as it is infused with a surprising degree of pragmatism. Similarly, its unconventional *modus operandi* often obscures its long-term strategy of eclipsing its regional rivals by creating an intricate network of states-within-states. The third strand – Shiism – is often preoccupied with the welfare of the Shia community at large. Accordingly, its conception of victory lends itself into an expansive ideology with little interest in Iranian national interests or its people and cultural heritage.

## Victory and the 'Iranian School'

This school's conception of victory centres on Iran, its national interests, culture and history. It has a broad constituency that prides itself on the ancient roots of the Iranian people and on a cultural heritage permeated with poetry and the visual arts. History play a prodigious role for this school as past events weigh heavily on the Iranian national psyche. Whereas in many cultures history might be treated as dates and occurrences from a bygone age, for Iranians it has an undeniable resonance in their daily lives. It is reflected in their everyday language and cultural allusions and in how they interact with one another as well as with the outside world. Western scholars and diplomats have noted that Iranians continue to 'live in their history' and sometimes even replay the struggles of their past.[1] This is not to suggest that the average Iranian has a nuanced appreciation of the country's long and colourful history. Most will only have a sketchy knowledge of its contours and an even hazier grasp of its many twists and turns. Yet awareness of key moments from Iran's past has made a strong and lasting impression on the collective memory.[2] Therefore, an explanation of how winning is interpreted by the Iranian School is contingent upon an understanding of how Iranians perceive themselves, their past and the world around them.

Iranians have a strong sense of identity: Iran has existed as a discernible polity for over two thousand years, albeit with notable periods of interruption. For centuries different incarnations of Iranian empires dominated their geopolitical vicinity.[3] This meant that Iran's frontiers were fluid and its population changed many times. As a result, Iranian identity is not fixated upon physical boundaries or ethnicity. This makes it quite distinctive from western notions of a nation-state, or from ideas that emphasise ethnicity and blood. Likewise, it is predicated on a different set of criteria to Iran's neighbours, many of whom emerged as independent entities in the 19th century or were post-colonial creations with a carefully manufactured top-down sense of nationhood. In contrast, 'Iranianness' is grounded in cultural identity. It is the ethereal bond that unites a multi-ethnic country with different languages, a mix of local dialects and traditions, and a varied geographical terrain. Once viewed from this prism, it becomes clear why Iranians regard modern Iran as the cultural core of a 'Greater Iran' that spans from the borders of India to the shores of the Mediterranean: from the Farsi-speaking territories of Tajikistan and Afghanistan in the east, to the Ossetian lands of the northern Caucasus, to small pockets of 'Iranianness' spread along the west in Turkey, Iraq, Syria and Lebanon.[4]

Accumulatively, these factors have produced a number of impulses that shape the Iranian conception of victory. First and foremost, is the Iranian notion of the 'long game': the idea that however difficult the challenge or however bad the circumstances, Iran will eventually prevail by biding its time. It stems in part from the experience of being conquered by four irrepressible waves of Graeco-Macedonians, Arabs, Mongols and Turkic nomads. In each instance, the Iranians succeeded in retaining not only their identity but also gradually engrossing their conquerors. Most notably, Iran was the only territory across the Middle East and North Africa to be subjugated by Muslim Arabs and yet did not become Arab.[5] Even Egypt and Syria – with their illustrious histories – became Arabic-speaking nations. Iran, however, retained the symbol of its heritage and identity – its language – despite a prohibition on the use of Persian. Nearly a century of concerted attempts by the Umayyad Caliphate to repress Iranian identity, sometimes by vicious means, failed. In the end it was with Iranian assistance that the Umayyad were overthrown and replaced by the Abbasids who, though maintaining the appearance of an Arabic Caliphate, were significantly influenced by Iranian culture, including many Persian courtiers and viziers.[6] The same fate befell the Ilkhanate – the Mongol descendants of Genghis Khan; and a succession of Turkic dynasties, including the powerful Ghaznavid and Seljuq Sultanates who each adopted the Persian language and Iranian traditions to such a degree as to become indistinguishable from Iranians. As we shall see later, playing the 'long game' is a practice that has also been adopted by Iran's current clerical regime, even if it is Islamist in its ideology and orientation.

A second impulse produced by 'Iranianness' is the sense of cultural superiority which, in turn, sustains an imperialistic mind-set that justifies interventionism and periodic expansion. It is true that since the 18th century, Iran has not been an expansionist power in the traditional sense of seeking to acquire territory by overtly attacking its neighbours. But intervention can assume many forms and

Iran's neighbours continue to be wary of Tehran's meddlesome behaviour and of its use of third-party proxies and surrogates to achieve its geopolitical objectives. As discussed above, 'Iranianness' rests on the idea of an expansive culture which explains why successive Iranian regimes have not felt constrained by Iran's internationally recognised borders or inhibited from interfering in regional affairs. Although the meddlesome proclivities of Iran's current clerical regime are rightly censured by the international community, we should recognise that this behaviour is merely a continuation of pre-revolutionary Iranian behaviour dating back to at least the early modern era: the Pahlavi monarchy in the 20th century, Iran's Qajar dynasty in the 18th to 20th centuries, and the Persian Empire of the Safavids in the 15th to early 18th centuries were all interventionists.[7]

This predisposition to dominate – be it militarily, culturally or, in more recent years, ideologically – is an impulse that is reinforced in the Iranian psyche in line with the country's past glories. Regardless of their social background, education or, indeed, actual ethnicity, Iranians learn from a young age that they are descendants of the warrior race of Indo-Europeans after whom Iran was named the 'Land of the Aryans'. They are taught that the original Semitic inhabitants of Iran were conquered by these Aryan invaders who established first the Kingdom of the Medes followed by the Persian Empire of the Achaemenids. While notable for its religious tolerance and humane treatment of conquered peoples, the latter was an expansionist imperial project. Described by modern scholars as the 'first world empire', at its peak it covered a territory spanning from the borders of modern Ethiopia in the west to India in the east. The empire established by Cyrus was conquered by Alexander the Great but, within a couple of generations, the second of the great Iranian polities was founded by the Parthians who vied with Rome in Mesopotamia and the Caucasus. Its successor state, the Neo-Persian Empire of the Sasanians, was even more aggressive and for centuries sparred with Rome and later the Byzantines in the Levant. It is telling that the rulers of all these dynasties styled themselves as Shahenshah or the 'King of Kings' – a title that deliberately signified their superiority and the dominance of Iranians over neighbouring monarchs. The Sasanian Shahs went even further by proclaiming themselves the *Shahs* of Iran and 'Aniran' – literally 'non-Aryan'.[8] It was a pejorative way of describing non-Iranians in the same way that the term 'barbarian' was applied by the Graeco-Romans. It is a supremacist worldview which still resonates in modern Iran.

These historical allusions are instructive because, while they emphasise Persia's past glories, they also serve as reminders that Iranian hegemony has been interspersed with periods of decline and failure. Acknowledging defeat, even humiliation, is integral to the Iranian narrative of victory. After all, it is an irrefutable part of the history of Iranians that they were subjected to repeated invasions and, on more than one occasion, were conquered and turned into vassals. The Arab conquest of the 7th century is a particularly bleak stain on the national psyche and is remembered as a cultural travesty for its partial destruction of pre-Islamic Iran's heritage. It is an event that serves as a watershed between pre-Islamic and post-Islamic Iran by separating the ancient warrior race who ruled the Middle East from Medieval Iranians who abandoned their martial proclivities but went on to thrive in other ways, most notably in the sciences and the arts.[9] Although the Arab conquest is held in great disdain by modern Iranians, it paled to the carnage inflicted by the Mongol invasion of the 13th century. The Mongols systematically destroyed most of the urban centres of the Iranian plateau by razing entire cities to the ground and massacring three-fourths of the Iranian population – an estimated figure of 10,000,000–15,00,000 inhabitants.[10] Few nations could have recovered from such destruction, especially as it was repeated on a smaller scale only a century later by the Mongol-Uzbek conqueror, Timur or Tamberlane.[11] The survival and subsequent successes of the Iranian people, in spite of these catastrophes, is arguably the greatest single enunciation of victory in the national conception of 'winning'.

Iranian art and literature capture the nation's experience of coming close to existential destruction by comparing Iranians to the mythical phoenix: even if they are put to the sword or conquered by foreign invaders, they will be re-borne and rise again by fire. It is a narrative that simultaneously expresses pride in the succession of Iranian empires while stressing Iranian resilience in the face of terrible catastrophes and national humiliation. Victory, in this narrative, is synonymous with the idea of overcoming occupiers by subverting them from within. Ancillary to it is the age-old Iranian practice of luring opponents into an ambush – often by feigning weakness. Perhaps the most widely recognised example of this in the west gave rise to the expression 'the Parthian Shot' – named after the Iranian ambush that destroyed the invading Roman army of Marcus Licinius Crassus in one of the most disastrous defeats in Roman history.[12] Relying on unconventional methods of warfare and, in particular, tempting the enemy into a trap, are age-old Iranian tactics. Though employed with mixed results during the protracted conflict of the 1980s with Saddam Hussein, it is a type of warfare that conforms to the idea of the 'long game'. Through a mixed combination of deterrence, ambush, proxy warfare and degradation of the enemy's resources, it aims for long-term strategic victory – often spanning decades – even if it entails short-term military setback and failure. It is a conception of winning that underlined how the ancient Parthians and Sasanians waged their wars against the Romans and the Byzantines, or the 16th-century Safavids against the Ottomans, and can be seen in the Islamic Republic's 21st-century style of 'hybrid' warfare in Lebanon, Iraq, Afghanistan, Syria and Yemen.

One final national characteristic has to be mentioned to help make sense of the Iranian conception of victory. Iranians are highly individualistic as a people and conceive of 'winning' as an outcome for the individual. They tend to be in their element in activities that require individual initiative and resourcefulness but are more liable to struggle as part of a team. This is exemplified in the sporting world where Iranians excel in fields such as wrestling but have seldom produced successful teams. Conversely, as individual traders and merchants, Iranians are world renowned for their bartering and negotiation skills and the ability to secure the best deals. A by-product of this proclivity for individualism is a willingness to put self-benefit before national interest. Among the most notorious episodes in Iranian history is that of courtiers and officials who, during the Anglo-Russian 'Great Game', prioritised their own status or financial gain by serving foreign powers to the detriment of Iranian interests.[13] Fervent individualism is an attribute that has served Iranians well by allowing them to adapt quickly to new circumstances. For instance, the large Iranian expatriate community in the west has, for the most part, assimilated seamlessly into their host nations and unlike many other ethnic and cultural groups seldom congregate to create a 'Little Italy' or 'China Town'. The reverse is also true in that individualism is often an impediment to Iranians securing victory politically or militarily because they disdain hierarchy and leadership. So as a people they have only excelled historically when strong leaders have imposed themselves and welded the unruly Iranians into a unified force. Yet as much of a handicap as it might appear, Iranian disunity and factionalism can sometimes be a source of strength. It was the component that prevented Iran's Muslim conquerors from turning Iran into an Arab State. Nevertheless, it can also lead to fragmentation and factionalism – attributes that define Iran's Islamic Republic regime.

## Victory and the Islamic Republic system

The second conception of Iranian victory must be seen through the eyes of the Islamic Republic – the clerical regime that was instituted by the Ayatollah Khomeini following the 1979 revolution. Scholars have long recognised that the Islamic Republic is a paradoxical system with embedded contradictions: while harbouring a democratic constitution, in practice it is a pseudo-theological autocracy that exercises social, political and economic repression. As a republic, it holds elections – albeit not fair and free – and derives some of its political legitimacy from popular will and token displays of accountability. As a Shia Islamist State, however, the Office of the Supreme Leader has progressively imbued itself with powers which are near absolute and claims to be beyond reproach. Since its inception, the highest priority of the regime has been self-preservation: both Khomeini and his successor, Ayatollah Sayed Ali Khamenei, have openly stated that the security of the regime overrides every other consideration.[14] It is not surprising then that the regime is built around parallel institutions that operate as a system of checks and balances with rival officeholders given overlapping responsibilities and charged with similar functions.

Consequently, for outsiders, it is often difficult to pinpoint where power lies in modern Iran. This also makes it difficult to establish the regime's objectives and, self-preservation notwithstanding, how it views 'winning'. The structural schisms of the Islamic Republic can produce bizarre circumstances where different organs of the state not only compete with one another, but on occasions may also set out to neutralise, even reverse, the successes of rival branches or agencies. A recent example is the conflict between the Ministry of Intelligence – which is headed by a cabinet minister appointed by the president of the country – and the intelligence branch of the Pasdaran – the Islamic Revolutionary Guard Corps (IRGC) – which is an organisation that answers only to the supreme leader.[15] Accordingly, the organs of the Islamic Republic are structurally predisposed to producing contradictory strategies, which harbour contrasting conceptions of winning, but are just as liable to deliberately undermine one another. Scoring points against competing state organs, at the cost of major policy setbacks at home and abroad, is a well-recognised feature of the paradox that is the Islamic Republic. Few modern states are riddled with such institutional contradictions; and whenever the supreme leader and the president – representing, respectively, the theocratic and elected branches of the regime – disagree on an issue or advocate different strategies, these tensions become more palpable.

The competing military structures of the Islamic Republic are riddled with these tensions: the regular Armed Forces are entrusted with protecting Iran's territorial integrity. The IRGC, however, are the guardians of the regime: an ideological movement created to protect the leadership of the Islamic Republic and to export its revolutionary zeal beyond Iran's borders. The IRGC have incrementally branched out and become a 'state-within-state'. Alongside forming their own Air Force and Navy, they have assumed control over Iran's non-conventional assets, including its nuclear and missile programmes; and the funding, training and arming of regional proxies. They have also developed non-military interests by setting up numerous commercial companies. The Revolutionary Guard now control a significant portion of the Iranian economy with vested interests in areas ranging from oil and gas extraction to manufacturing, industry and farming.[16] Critics claim that this growing empire is also involved in money laundering, contraband shipments, racketeering, and possibly narcotics as far afield as Latin America.[17]

As a burgeoning military and business conglomerate, the Guard can no longer be characterised as a monolithic force with the singular objective that was assigned to them at the outset of the 1979 revolution. Their business empire mandates activity across a range of spheres within and outside Iran. But the covert nature of its operations, and its illicit dealings, make it extremely difficult to gauge what it is doing and what type of interests it is pursuing. Other than self-interest and the preservation of the Islamic Republic regime, 'winning' for the Guard spans military security, commercial contracts and financial gain. The Guard can be likened to a large multi-national conglomerate with the key difference that it has, at its disposal, the resource and political protection of a regional power. As such, its extra-territorial activities are not necessarily commensurate with Iranian national interests or, indeed, the policies of the elected government. While broadly committed to the guidelines laid down by the supreme leader, the Guard has a significant degree of autonomy. It can generate money through its many commercial subsidiaries, thus making it less dependent on the official state budget. It has a fiscal self-sufficiency that is not normally seen in the regular military branch of a conventional state; and this makes it even less accountable to the Iranian executive or legislative oversight bodies.[18]

In addition, the Guard allegedly funds some of its extra-territorial activities through nefarious means, which gives it a shadiness associated with Transnational Criminal Organisations (TCO).[19] Considering the degree of discretion it exercises in its operational and tactical decision-making, one has to assume that its definition of 'winning' occasionally skirts the lines between what is in the interests of the regime and its own corporate interests or even the self-interests of individual Guard commanders. This is particularly relevant in any assessment of Iran's relationship with regional proxies and extra-territorial operations. Specifically, one should not assume that the financing, training and arming of non-state acolytes by the Guard is always undertaken with the full knowledge, much less approval, of all the key elements of the Iranian regime.

Nor is Iran's policy in Middle Eastern operational zones necessarily set and directed by the government. For instance, the expeditionary unit of the Revolutionary Guard – the Quds (Jerusalem) Force – has a track record of activities that have blindsided even the most senior officials of the Islamic Republic. While we might not be able to present tangible evidence of such behaviour operationally, we do know that President Assad of Syria arrived in Iran in February 2019 at the invitation of General Qassem Soleimani – the former shadowy commander of the Quds Force – without the knowledge of the Iranian Foreign Ministry. In a rare expression of public disgust, Iran's Foreign Minister immediately tended his resignation at the slight.[20] The episode made is manifestly clear that it was General Soleimani – and his Quds Force protégées – who were running Iran's policy in Syria. These peculiarities of the Islamic Republic are a source of great frustration for adversaries and political analysts alike because deciphering aims, including conceptions of winning, is invariably contingent upon determining which of its many competing State organs are active in a given situation and what is their strategy.

It is not just structural divisions that make the Islamic Republic difficult to read. The Iranian establishment is blighted by a fragmentation of another sort: deep-seated divisions within the clerical ruling elite – a fact that plays out in the form of factional politics in every sphere of public life. This factionalism has its own far-reaching implications for how 'winning' is perceived and pursued by the Islamic Republic. At any one time, there are at least two broad and competing currents that determine decision-making within the regime. One is represented by a pragmatic wing which recognises that success needs to be calibrated around the national interests of the Iranian State and its people. The other is more ideological with a narrower conception of victory predicated on an Islamist and anti-western worldview. Both wings are united in their desire to secure the welfare of the regime but otherwise they pursue strategies that can be contradictory. The Islamic Republic can concurrently send conciliatory signals while behaving pugnaciously; or it can enter into negotiations with adversaries but conduct itself in such a way as to negate the process. It is not uncommon for a senior official to make announcements only to be forced to retract or find themselves contradicted by other officials. For instance, in August 2019 President Rouhani hinted he would be willing to meet with the President of the United States in what was seen by the international community as a political opening. Internal critics, however, saw the offer as a contravention of the Supreme Leader's directive prohibiting negotiations with the US. Within hours Rouhani was forced to backtrack under intense pressure and dismiss the idea of meeting with US officials.

Although few have reason to question the reality of an Islamic Republic system that is plagued by parallel structures and factional politics, there is suspicion that these divisions are sometimes exaggerated by the leadership as a ploy to confuse internal and external critics while permitting the regime to pursue conflicting objectives. This leads us to two other sources of tension that make it difficult to fathom the Islamic Republic's concept of 'winning'. The first facet centres around what the average Iranian characterises as 'Mullah Culture'. Within Iran the clerics – popularly called 'mullahs' or 'akhounds' – form a well-known sub-culture. Educated from a young age at theological seminaries that inculcate them with a recognised and distinctive worldview, they have their own peculiar mores and way of speaking including idiosyncrasies that easily set them apart from the rest of the population even when they are not dressed in their customary clerical robes. Among Iranians the mullahs are popularly associated with cunningness and deception. They are known as skilful negotiators with an ability to twist things, misrepresent them and secure a good deal by offering very little in return; and for never giving ground.[21] Although these attributes stem largely from stereotypes, it nevertheless stands to reason that a regime that is run by mullahs should display some of these characteristics.

Certainly, from the point of view of 'winning', the Islamic Republic is known not to give ground unless it feels existentially in peril; and, when under intense pressure, it is more likely to attack than cave-in and surrender. Looking at the chronology of events over the past four decades, whenever the regime has been cornered, it has struck back – sometimes overtly but more often covertly using the levers of terrorism or by resorting to low intensity warfare. Even though conventionally weak compared to Israel, some of its Arab adversaries and the United States, the Islamic Republic has shown little inhibition when it comes to escalation or confrontation. While all authoritarian regimes eschew losing face lest they appear weak, the regime's Mullah culture amplifies this tendency with a leadership that is adamant about never giving ground nor buckling under pressure. As Saddam Hussein was to discover in the 1980s, the regime can be intransigent and dig in deep when weak, and relentless when it has the upper hand and is in a position of strength. This intransigence was instrumental in how Iran helped to turn around the Syrian Civil War. When all appeared lost for the Syrian regime between 2012–15 – with even Moscow contemplating the end of the Assad presidency – it was the Iranians alone who insisted on keeping the beleaguered Syrian president in power by not giving-in and instead launching a brutal offensive against the rebels. Of course intransigence, alone, is not always a recipe for victory. The mullahs are also known for their pragmatism. For all the Islamist rhetoric coming out of Tehran, the Islamic Republic is not an Islamist entity in the mould of Al Qaeda or Daesh. This is a regime that can make friends with Marxist Venezuela and Cuba, communist North Korea, or support Christian Armenia against Muslim Azerbaijan. Russia's brutal wars in Muslim Chechnya were conveniently ignored by a clerical elite with pretensions of being the 'protectors of all Muslims'; neither is much said in Iran about the systemic persecution and internment of Uyghur Muslim minorities by the government in Beijing.

A final facet that contributes to the paradox of the Islamic Republic is 'Iranianness'. Once it achieved power in 1979, the new clerical regime sought to do away with many vestiges of Iranian identity, including the country's pre-Islamic traditions and, at one point, even the Persian language. All these attempts proved self-defeating and, with the passage of time, it was the regime that became more Iranian in its composition and outlook. So, in spite of the affectation that the Islamic Republic reflects 'Mohammedan Islam in its purest form', the last two decades have betrayed signs of Iranian nationalism creeping into the regime's outlook and conduct. So much so that the government of President Mahmoud Ahmadinejad – known in the west for its inflammatory and intractable Islamist rhetoric – was itself assailed by hard-liners within Iran for giving a platform to the 'Iranian School'; and deviating from the correct Islamic path by pursuing 'Iranian Islam'.[22] The Islamic Republic, like any other regime, cannot operate in a vacuum and its character and orientation were always bound to be shaped by its cultural surroundings. The Iranian characteristics of 'playing the long game', interventionism, and seeking hegemony over neighbours are all recognisable hallmarks of the Islamic Republic. Even the title used by the regime for its leader – Vali-amr Muslimin or the 'Guardian of all Muslims' – has overtones that conjure-up memories of the 'King of Kings' designation preferred by Persian monarchs. In fact, Iranian cultural traditions have not only moulded much of the vernacular and behaviour of the Islamic Republic, but they are also responsible for creating yet another layer of schism within an already paradoxical system.

As we have seen, the Islamic Republic's conception of 'winning' is shaped by multiple variables which are at odds with one another. Yet, these complexities do not prevent us from outlining the strategic aims of the regime and how it perceives and pursues victory. Strategically, the Islamic Republic seeks to create a favourable environment for itself both domestically and within the Greater Middle East. As we have already noted, authoritarianism and repression – packaged and justified in the language of Islam – is how the clerical establishment keeps its internal critics at bay. Externally, its self-declared aim is to bring about a new Middle East which could guarantee the security of the regime and also reflect its worldview and ideology. To achieve this the Islamic Republic seeks to weaken and eventually replace the governments of neighbouring countries with more friendly-regimes, preferably Shia-leaning in orientation. This is an ambitious goal that requires a two-fold strategy of promoting radical forces who can challenge the conservative governments of neighbouring states while, at the same time, expunging the influence of western powers and patron protectors of Israel and hostile Arab regimes. Accordingly, Tehran has periodically encouraged regional instability and radicalisation and has relied on terrorism and proxy warfare to make it costly for the United States and its European allies to remain in the region. Undermining the status quo, and fostering change across the Middle East, is the natural and favoured outcome for an Iranian regime that sees itself as isolated and surrounded by countries that harbour ideological and cultural hostility towards it.

As part of this strategy, the Islamic Republic has been nurturing violent and armed non-state actors who either share its ideological worldview or can be co-opted in the short-term. Proxies and affiliates are often a relatively cheap but highly effective mechanism for Iran's extra-territorial operations. They allow the Islamic Republic to project a degree of power disproportionate to its size and economic and military capabilities. Indigenously-bred militias in neighbouring countries have the added advantage of providing the Islamic Republic with a measure of plausible deniability: many such groups are Arab-speakers and some, such as Hamas, are Sunni rather than scions of Iran's Shia community. These groups vary in size, strength and resources and have a variable relationship with the Islamic Republic. It would be misleading to describe the Lebanese Hezbollah as a mere 'proxy' of Iran as it is tantamount to an actual extension of the Islamic Republic's clerical regime. Conversely, the Mahdi Army of Muqtada al-Sadr in Iraq has always had a transactional relationship with Tehran with little convergence of interest between the two despite a shared Shia ideology.

The Islamic Republic may have only limited influence over the behaviour and activities of some of its so-called proxies and its relationship with them tends to be fluid and can quickly sour and become antagonistic. Yet they provide the Iranian regime with a formidable transnational lever that can be mobilised to alternatively intimidate or serve as an instrument of deterrence. Tehran prefers to have many different types of 'proxies' on its payroll because, whatever their disposition and however transitory the relationship, violent non-state actors will contribute to its overarching objectives by simply existing: first, by structurally hollowing out the authority of rival governments; and, second, by keeping powerful regional adversaries – such as Israel, Saudi Arabia or Turkey – busy and distracted. More broadly, they are a useful strategic tool against the order-building disposition of great powers, such as the United States and its European allies, and can be deployed as impediments against their efforts to promote political and social stability. The Islamic Republic has successfully fielded such proxies to stir up anti-western sentiments, sow divisions among rival political parties, and to galvanise support for itself in Lebanon, Iraq and Afghanistan.

## Victory and Shia Iran

For a more thorough understanding of Iran's regional behaviour and its strategy of securing victory, we must discuss Shia Islam. Iran is the leading Shia power in the Islamic world. For this reason, Shia Islam is sometimes disparagingly described by Sunni Muslims as 'Iranian Islam'. Historically, Shiism did not originate in Iran or with Iranians: The movement can be traced back to a 7th-century sectarian rift within a Muslim community that was still predominantly Arab. The Shia would almost certainly have remained a peripheral minority sect were it not for the re-emergence of the Persian Empire under the Safavid dynasty in the early 16th century.[23] In 1501 Ismail I, a young charismatic warrior, succeeded in reuniting a divided Iran and its population for the first time since the Arab conquest of ancient Persia. With the aim of creating a powerful centralised state, Ismail sought an ideology that could inspire his subjects and invigorate them to counter the Turks and perhaps one day to eclipse the Ottoman Empire. With the Ottomans styling themselves as a reconstituted Muslim Caliphate, Ismail and his successors made Shiism the official religion of the Safavid Empire and used the Shia brand to forge a new imperial identity. The Safavid dynasty proved so effective in this endeavour that within a couple of centuries the distinction between Iranianness and Shia Islam became blurred.

Historically regarded by Sunni Muslims as a deviation, Shia Islam is politically derided these days less for theological reasons and more because it is synonymous with Iran. Since the early days of decolonisation, Arab regimes have viewed their Shia communities with suspicion and as a potential Iranian 'Fifth Column'; or, at least, susceptible to Iranian influence. This explains why ambitious Arab nationalists have introduced harsh measures against their own Shia populations. In the Arab Gulf States – most notably in Bahrain and Saudi Arabia – Shia communities are treated as second class citizens and are subject to cyclical persecution even though ethnically and culturally they are Semite Arabs rather than Persian/Iranian. Latent misgivings about the Shia in Arab countries intensified exponentially after the 1979 Iranian Revolution that created the Islamic Republic. Iran's clerical Shia rulers made it known that they sought to 're-balance' the historical wrongs inflicted upon Shia minorities in neighbouring Arab countries; and, more egregiously, Ayatollah Khomeini's controversial 'Velayat-e-faghih' (Supreme Leader of the Islamic Republic) doctrine gave out that Shias everywhere were under the direct protection of his regime. The fact that many Shia communities outside of Iran abhor Khomeini's brand of

Islam has made little difference to the boastful claims of the Iranian leadership nor has it reduced the repressive measures of some Arab governments against their Shia constituencies. The net effect of these trends has been the start of a new Shia-Sunni sectarian conflict with many Shias – as well as Sunnis – becoming helpless pawns in the struggle between Iran and its numerous Sunni adversaries.

Shia identity is integral to the Iranian regime's conception of victory for two key reasons. First, the ideological wing of the Islamic Republic has always viewed Iran as merely the heartland of the Shia world and has been open about its aspirations to unite all Shia communities under the direct rule of the Vali-e-faghih. For these ideologues, the Iranian State and its people are no more than a repository for the resources necessary to 'liberate' repressed Shia brothers and sisters in other Middle Eastern countries.[24] The ideological wing of the Islamic Republic has always defined itself by its Shia identity rather than Iranian nationality. This propensity has caused considerable popular anger within Iran and in December 2017 translated into uprisings against the regime in eighty towns and villages. The protestors could be heard chanting against the generosity of the Islamic Republic in giving away billions of dollars to the Shia communities of Iraq, Syria, Lebanon and elsewhere while average Iranians were suffering from a shortage of basic goods and bearing the brunt of the US-led economic sanctions. Nevertheless, the regime has remained steadfast in pouring money and resources into Arab Shia communities not least because the Islamic Republic's clerical leadership is culturally more attune with Shia Arabs than with the Iranian population.

Ideologically, the regime has always sought legitimacy from the wider Shia community to sustain its pretensions as the theological and political bastion of Shia Islam. In return, it has nurtured these communities financially and politically in order to keep them on side. This relationship has progressively deepened over the forty-year lifespan of the Islamic Republic and now has taken an added urgency with the Iranian public growing manifestly tired of the regime's rhetoric while living standards continue to decline. Four decades of Islamism, conflict, sanctions and repression have made the clerical regime so unpopular with the average Iranian that its leadership feels it has little choice but to find popular support and recognition outside of Iran's borders. At the same time, the Iranian public is steadily becoming more secular or turning to religions other than Islam. Iran's pre-Islamic Zoroastrian faith has seen a revival in recent years and, according to statistics released by Christian organisations, Iran is also witnessing one of the fasting growing church movements of any country in the world.[25] As the Iranian masses become more alienated from Islamism, Shia radicalism is on the rise in parts of the Arab world. These developments are a factor in why the regime has become more interventionist in recent years and why it is unlikely to change course even under intense international censure.

Second, for a leadership that prioritises regime security but also recognises its international isolation and growing domestic unpopularity, there are political reasons that make it necessary to remain active across the region's Shia communities. Years of arms embargoes have taken their toll on the conventional capabilities of the Islamic Republic. So the regime has created power multipliers by organising thousands of Shia militias in neighbouring countries. This is a 'cascade strategy' that could prove highly effective in the event of a military confrontation with regional adversaries or a world power such as the United States. Tehran calculates that in this scenario the war cannot be contained and will not be limited to the regime or the Iranian heartland but can be spread like wildfire to vast areas of the Middle East with a potentially devastating impact on oil exports, financial markets and the environment. Such an outcome has been used as a trump card by the Islamic Republic to deter attacks against the vital interests of the regime and dissuade efforts at regime change by emphasising its huge costs and repercussions. In the event of an armed conflict, the Islamic Republic knows it cannot win but it is confident that nor will its adversaries. Denying the US and its allies victory is, in itself, the highest enunciation of winning for Iranian leaders. It is a spoiler's mind-set, but it has already proved highly effective in the conflicts in Iraq and Afghanistan and, more recently, in Syria and Yemen.

To this end, the Lebanese Hezbollah represents the most successful example of a strategy of implanting parasitical non-state actors in surrounding countries and then nurturing them financially and militarily to produce states-within-states. At the low-end, these parasitical actors provide Iran with access to the resources and the broader infrastructure of other countries and allow it to circumnavigate international sanctions augmenting the needs of the regime illicitly. At the high-end, they could create a haemorrhage inside a country such as Lebanon or Iraq or hold the governments of these states to ransom. Fear of the thousands of militias operating in Iraq, but with loyalty to the Iranian regime, has certainly been a variable in the political and regional calculations of successive Iraqi governments since the fall of Saddam. Even in ostensibly friendly Syria, Iran has created thousands of militia forces which are loyal to Tehran and not Damascus – lest the Assad regime one day fails to toe the line.

There is a more tangential consideration regarding Shia Islam and Iran's conceptions of victory. Ultimately, no Iranian regime – whatever their political or ideological disposition – will be willing to sit on the periphery of developments in the Islamic world. As Muhammad Iqbal, the renowned Pakistani Islamic philosopher, observed:

> [The] Muslim civilisation is a product of the cross-fertilisation of the Semitic and the Aryan ideas. It is a child who inherits the softness and refinement of his [Iranian] Aryan mother, and the sterling character of his [Arab] Semitic father. But for the conquest of Persia, the civilisation of Islam would have been one-sided. The conquest of Persia gave us what the conquest of Greece gave to the Romans.[26]

Iran regards itself as the quintessence of Islam just as much, if not more than the Arabs. As the cultural engine of the Islamic community, and not just the Shia sect, it will seek to dominate the Islamic world ideologically, politically and perhaps militarily for the foreseeable future. It is in this guise that Iran's current leadership defines its conception of victory.

## Conclusion

Iran's contrasting identities invariably produce divergent impulses in its policymaking, conduct and conceptions of winning. As we have seen, proponents of the Iranian School are at odds with Islamism while Iran's clerical rulers are fixated on sectarian concerns and geopolitical developments. These tensions are further accentuated by a regime that, though Islamist in its outlook, is also imbued with vestiges of Iranian nationalism and cultural thinking. Together, these contrasting impulses have produced a paradoxical entity that is at once a modern republican nation-state and an anachronistic Islamist polity; and a leadership that simultaneously pursues regime security, Islamist and sectarian geopolitical objectives, and a nationalist cause that sit uncomfortably with the other two. Within this perplexing framework, any understanding of modern Iran's conception of victory is contingent upon whether one is looking at it through the prism of Iranians as a people, the Islamic Republic system, or Shia Iran which is a cause in itself and an ideological oddity.

### *Notes*

1 See, for example, Jack Straw, *The English Job: Understanding Iran and Why It Distrusts Britain* (London: Birbeck Publishing, 2019).

2 See David Morgan, *Medieval Persia 1040–1797* (London: Routledge, 2016); and Sandra Mack, *The Iranians: Persia, Islam and the Soul of a Nation* (New York: Plum, 1998).

3 In antiquity, the empires of Achaemenid Persia, Parthia and the Neo-Persian Sassanids; in the Medieval era the great Ghaznavid and Seljuq Sultanates; and, in the early modern era, the Safavid Empire and the empire of Nader Shah – the last of the great Asiatic conquerors.

4 Historians sometimes refer to it as the Persianate. See Richard M. Eaton, *India in the Persianate Age: 1000–1767* (Berkley: University of California, 2019).

5 Iranians also re-interpreted and re-moulded Islam through the Shia movement or what Sunni Arabs sometimes describe scathingly as 'Iranian Islam'.

6 The later Abbasid caliphs actually became de facto puppets of emerging Iranian Sultanates, such as the Buyid and Samani dynasties. John McHugo, *A Concise History of Sunnis & Shi'is* (London: Saqi, 2017), 91.

7 See Kaveh Farrokh, *Iran At War 1500–1988* (Oxford: Osprey, 2011).

8 Touraj Daryaee, *Sassanian Persia: the Rise and Fall of an Empire* (London: IB Taurus, 2010).

9 Iranians produced a roster of great Medieval Islamic scholars. Morgan, *Medieval Persia*.

10 Steven R. Ward, *Immortal: A Military History of Iran and Its Armed Forces* (Washington DC: George University Press, 2009), 38–39.

11 Gerard Chaliand and Arnaud Blin, 'Manifestations of Terror Through the Ages', in Gerard Chaliand and Arnaud Blin (eds.), *The History of Terrorism from Antiquity to Al-Qaeda* (London: University of California Press, 2007), 87.

12 At the Battle of Carrhae in 53 BC.

13 Esmael Raeen, *On Britain's Payroll?* (Tehran: Elmi Publications, 2004).

14 See Ayatollah Khomeini, *Sahifeh-ye-Imam*, Volume 15 (Tehran: The Institute for Compilation and Publication of Imam Khomeini's Works, 2008): 365, https://farsi.rouhollah.ir/library/sahifeh-imam-khomeini/vol/15/page/365 and (http://farsi.khamenei.ir/newspart-index?tid=2994; and Ayatollah Khamenei, Speech Commemorating the First Anniversary of the Death of Imam Khomeini on 24 December 1991, https://hawzah.net/fa/Article/View/6176/یمالسا-ماظن-ظفح.

15 Maysam Behravesh, 'Iran's Spies Are at War With Each Other', *Foreign Policy* (9 August 2019).

16 Hesam Forozan and Afshin Shahi, 'The Military and the State in Iran: The Economic Rise of the Revolutionary Guard', *Middle East Journal*, Vol. 71, No. 1 (Winter 2017): 67–86.

17 Sami Kronenfeld and Yoel Guzansky, 'The Revolutionary Guard and the International Drug Trade', *Military and Strategic Affairs*, Vol. 5, No. 2 (September 2013): 105–120.

18 Emanuele Ottolenghi, *The Pasdaran: Inside Iran's Islamic Revolutionary Guard Corps* (Washington DC: FDD Press, 2011), 41–48.

19 Saeed Ghasseminejad, 'How Iran's Mafia-like Revolutionary Guard Rules the Country's Black Market', *Business Insider* (10 December 2015); Julian Borger and Robert Tait, 'The Financial Power of the Revolutionary Guards', *The Guardian* (15 February 2010); and Douglas A Livermore, 'Dismantling Iran's Illicit Networks', *Small Wars Journal* (August 2017) (https://smallwarsjournal.com/jrnl/art/dismantling-irans-illicit-networks).

20 Mahmoud Paragoo, 'How Assad's Tehran Trip Exposed Political Rifts in Iran', *The Interpreter* (1 March 2019).

21 While largely cultural and particular to Iranian Shia clerics, this form of cunningness also has a theological basis within Islam known as principle of *Taqiya* or precautionary dissimulation. See Moojan Momen, *An Introduction to Shi'i Islam* (New Haven: Yale University Press, 1985), 183.

22 See, for example, Robert Tait, 'Iranian President's "Religious-Nationalism" Alienates Hard-line Constituency', *Radio Free Europe* (19 August 2010).

23 See John McHugo, *A Concise History of Sunnis & Shi'is* (London: Saqi Books, 2017).

24 From the experience of the Iran-Iraq War, the leadership is confident that by appealing to Iranian nationalism, Iranians will always fight to defend their State irrespective of circumstances, including how much they may disdain the regime.

25 Duane Alexander Miller, 'Power, Personalities and Politics: The Growth of Iranian Christianity Since 1979', *Mission Studies*, Vol. 32, No. 1 (2015): 66–86.

26 Javid Iqbal (ed), *Stray Reflections: The Private Notebook of Muhammad Iqbal* (Lahore: Iqbal Academy Pakistan, 2006), 49.

CHAPTER 11

# The Chinese Concept of 'Winning'

Kerry Brown

In the Xi Jinping era, from 2012–13, Xi Jinping and the leaders around him grew increasingly fond of speaking of 'win–win' outcomes. Speaking about combatting climate change at the opening of the Paris Conference in 2015, Xi declared, 'Let us join hands and contribute to the establishment of an equitable and effective global mechanism on climate, global sustainable development at a higher level, and new international relations based on win–win co-operation.'[1] Win–win co-operation has become one of the buzz words of the time since then. But this arouses a number of questions about what, in fact, Chinese officials, and other actors from China, mean when they use a phrase like this.

The usual rejoinder is that in fact what China is saying is that it intends to win – twice! Win–win as a phrase could be interpreted as China playing at a different game to the rest of the world, and therefore winning the game it is playing on its own terms, while everyone else succeeds at their game – the two occurring in parallel. It could be that what is being referred to is more akin for a draw – the game in question being so complex that all those involved in it can come away feeling that they are can all validly claim victory. Or maybe China is saying that it is not the winning of a game that matters, and in fact in the issue they are talking about there are no real parties that come out on top. In the end, everyone is a winner. It just depends on how they feel.

One thing is clear. These days, particularly under Xi, the notion of the People's Republic of China (PRC) winning is a tangible one. The context for this is historical. China as a victim, as a place which was done down, defeated, and unfairly picked upon, is something that comes across strongly in the narratives of history promoted since the emergence of patriotic education campaigns in the 1990s. It is not hard to find evidence in official discourse of this kind of storyline. This is from the preamble to the country's current constitution, passed in 1982 and revised several times since then:

> After waging protracted and arduous struggles, armed and otherwise, along a zigzag course, the Chinese people of all nationalities led by the Communist Party of China with Chairman

> Mao Zedong as its leader ultimately, in 1949, overthrew the rule of imperialism, feudalism and bureaucrat-capitalism, won a great victory in the New-Democratic Revolution and founded the People's Republic of China. Since then the Chinese people have taken control of state power and become masters of the country.[2]

The text continues to document the era from 1949 in which China was on the path as a new entity to victory and renaissance. Elite military and political leader in the era under Mao Zedong, Lin Biao, in a famous speech in 1965, 'Long Live the Victory of People's War!' stated that:

> Today, the U.S. imperialists are repeating on a world-wide scale the past actions of the Japanese imperialists in China and other parts of Asia. It has become an urgent necessity for the people in many countries to master and use people's war as a weapon against U.S. imperialism and its lackeys. In every conceivable way U.S. imperialism and its lackeys are trying to extinguish the revolutionary flames of people's war. The Khrushchev revisionists, fearing people's war like the plague, are heaping abuse on it. The two are colluding to prevent and sabotage people's war. In these circumstances, it is of vital practical importance to review the historical experience of the great victory of the people's war in China and to recapitulate Comrade Mao Tse-tung's theory of people's war.[3]

It should be stressed here that at least since the late 1970s, when we talk about China as a winner, this is not about its victories on the physical field of battle. The aggressive and bellicose tone of the revolutionary rhetoric of the Maoist era up to 1976 has largely gone, along with the more abrasive actions like the border war with India in 1962 and the small clash with the USSR in 1969. In the contemporary context, China no longer being a loser or a victim, but an actor with agency, power and potency, does not relate to military fights so much as battles in other areas. Indeed, one of the anomalies of contemporary Chinese power is the extent to which it has avoided actual physical conflict despite growing more powerful. Its armies have not been deployed beyond its borders to any significant extent since the clash with Vietnam in 1979. In many ways, the PRC's last major international conflict for its armed forces was the 1950–53 Korean War. Despite this, under Xi, the reality of China as a power has to be taken seriously, but one where its field of conflict and the space where it is winning is the economic and the geopolitical, not the military. In this context, therefore, what does the concept of winning in contemporary China actually look like, and what kinds of cultural and intellectual foundations has it been built on?

## The story, the game and the endgame

China is a winner. That is the message of the Xi Jinping era. But what exactly is it winning at, and how? As the philosopher Wittgenstein famously pointed out, there are many different games. Some of these you play against yourself, some as teams, some against time, some in order to fulfil certain targets regardless of time. There are games that involve boards, fields, courts, tracks. What is the common element between them? Wittgenstein wrote:

> Consider for example the proceedings that we call games. I mean board games, card-games, Olympic games, and so on. What is common to them all? Don't say, 'There *must* be something common, or they would not be called "games"', – but *look and see* whether there is anything common to all. For if you look at them you will not see something that is common to *all*, but similarities, relationships and a whole series of them at that. I can think of no better expression to characterize these similarities than 'family resemblances'; for the various resemblances between members of a family.[4]

This prompted him to explore the role of rules in games, and to extend this to the notion that language itself was a kind of game. In complex ways, games, like language, followed rules – and the two (rules or games and language) were simply integral elements of one whole. Even rules themselves could take many different forms. 'What we call a rule of a language-game may have very different roles in the game.'[5] If the concept of games was so complex, then so, too, should the associated one of winning. Winning is clearly not straightforward!

This is why locating the contemporary notion of 'victory' and 'winning' for the People's Republic in the right interpretative context is important. Part of that means understanding the historical, and political, narrative alluded to above. One of the games China is playing is that of modernisation. The rules for this game were, at least till recent times, largely devised in the outside world. China's complaints about multilateral forum like the International Monetary Fund, and the whole post-World War II Bretton Woods order, particularly since the 2008 global financial crisis, partly arise from this sense that China has, without much choice, needed to play by rules it hardly had any say in when they were devised. Nor, however, does it have any real choice about having to adopt these rules – up to a point. This is because since reforms started in 1978, China's aspirations and its needs for growth have involved the outside world, and in particular partners like the US and the European Union, for whom these rules are important. They would not have played the game of 'economic engagement' with China had China not accepted these ground rules.

Nor was it just China that was playing a game – that of using foreign markets, technology, and involvement in order to help itself modernise and become economically stronger and more powerful. The developed world, led by America, was practicing its own game back. For those countries, engagement, underpinned by what were meant to be the rules of modernisation theory, implied that through economic change supported by the west, China would undergo a political transformation and end up with a system similar to most of the rest of the democratic world. This strategy after all had worked well in Eastern Europe, in the late 1980s and into the 1990s, when communist one-party regimes collapsed and multi-party ones replaced them. It even happened in the former Soviet Union. In the 1990s in particular, China looked vulnerable, the last major country with a communist system. The west had won the cold war, and were simply waiting for China to follow what President Bill Clinton called 'the right side of history.'

Coming to the end of the second decade of the 21st century, we can now appreciate that if these were the games that people were playing, towards the kind of victories they were expecting, in the latter decades of the last century, then the outcomes have not been straightforward. The Communist Party of China has, so far, won the fight to maintain its monopoly on power. The PRC seems to be winning the fight to become the most important single economy in the world. The west seems to have lost in its desire to see engagement via economic means lead to political change. Even this bald description however shows that we are talking about a number of different games, and different actors, and different kinds of outcomes.

Nor can we remotely talk about a China, in this context, that seems to have won in terms of changing the world around it to become more like it. Its political system remains isolated, and levels of distrust towards it according to surveys remains high. [6] China is not winning in terms of military combat too, for the simple reason, alluded to above, that since the 1980s, despite having an increasingly large army, navy and air force, it has not had any combat experience. This means it has fought no actual wars, and therefore has won none. It might be that China is winning fights in the virtual world, but these, being mostly invisible, are hard to assess in terms of success or failure. China seems to be winning certain metaphorical battles around producing world class technology. Because of Xinjiang and Hong Kong and clamp downs and unrest in these places respectively over 2018 into 2020, China does not seem to be winning any major soft power battles. This has only worsened in 2020 with the impact of Covid-19 and the widespread international criticisms of China's originally handling of the pandemic when news came to light of it in late 2019. For all these reasons, China's reputation in 2020 is a contentious one.

All of these are various different areas and contexts where the PRC under Xi might be said to be either winning or losing. The question is, however, whether China itself sees things in this way. Does it have some different attitude about what it is to win, or lose? Long term thinking, and a different attitude towards the composition of the world order, is something that people inside and outside China have attributed to it, marking it out as exceptionally exceptional. In this context, China does not want to be a winner according to the template that has existed till now. It does not see a world where any one party dominates, but one where there are many different, balancing players, none of them practicing hegemony. It says that it aims for a global order where multipolarity replaces unipolarity. In this interpretation, China is not playing the same game as others – particularly the US. It has opted out. It does not subscribe to the desire of the individualist-obsessed west to always be competing, trying to overtake or defeat others. Its vision of victory is one of harmony and balance. All it asks, it says, is that it is allowed to play this game, in its own way, in its own particular space, while everyone else is free to carry on playing theirs. There are no conflicts and incompatibilities here. Everyone has the space to do what they feel best for themselves, without impeding on others. That, at least, is the ideal China seems to subscribe to.

## The idea of efficacy

The work of Francois Jullien, a French philosopher who has worked between China and Europe, may help to think through this idea of what might validly be called a Chinese notion of winning. Jullien writes about the issue of strategy and action, and how plans get implemented. He is interested in the mindset behind those that attempt to influence reality and bring about outcomes, one of which might be something described as 'victorious' such as successful campaigns in a war, or attempts to implement policy that end up fulfilling their criterion of success. After a discussion of Clausewitz's classic work, *On War*, Jullien, in the *Treatise of Efficacy* describes where he locates a fundamental difference between the ideas in this book, and those found dispersed in the great corpus of Chinese philosophy from ancient times to the present day. 'Chinese thought,' he states, 'is way of out of our [the west's] rut, for it never constructed a world of ideal forms, archetypes, or pure essences that are separate from reality but inform it.' [7] This lack of interest in questions of ontology and metaphysics is something that historian of ancient China Roel Sterckx also recognises. Such questions as 'how the human mind works, or whether there exists a world or reality beyond this one' did not figure high on the agenda of Chinese thinkers.'[8] Notions of Platonic forms or Aristotelian categories do not figure in the almost contemporaneous thinking of Confucius and Mencius 2,500 years ago. They were concerned not so much with *what* humans were, but *how* they were. That frames Jullien's description of the Clausewitzian idea that the aim of strategy is to 'frame an ideal plan that they will then have to be incorporated into factual reality.'[9] The aim is to change, modify, transform reality.

In Chinese thinking, 'instead of setting up a goal for our actions, we could allow ourselves to be carried away by the propensity of things. In short, instead of imposing our [human] plan upon the world, we could rely on the potential inherent in the situation.'[10] Strategy is to read the potential within things, rather than to shape them to external ideals – to understand their constituent order, and work with this. In one classic text Jullien refers to, he states that 'in order to have the whip hand over another and make use of him as you please, the only way forward, after assessing him carefully, is to adapt yourself to him.'[11] There is an agenda for control in all of this: 'by constantly adapting yourself to another person and be never alienating him, and so never causing him to doubt or resist you, you gradually increase your power over him and so can manipulate him as you will.' For us, and here Jullien is referring to those he categorises in the western tradition, 'the means are always relatively artificial, a deliberate construction designed to exert pressure on things so as to bring about the desired end.' For them (those in the Chinese tradition) 'once the situation's potential has developed, it becomes a situation of strength.'[12]

For all the suggestiveness and stimulation of Jullien's description of differences between western and eastern views of the nature of action, and intervention by humans in the real world, it is prudent to be sceptical that the differences are as deep as his account suggests. The classical approach to warfare typified by Clausewitz may be informed by a deep cultural and philosophical background influenced by notions of Platonic Ideal forms and abstractions. But there is plenty of acknowledgement elsewhere of the interiority of things in themselves (witness the work of the great Kant with his distinction between Phenomena and Noumena) and the need to accept the fluid structure of reality, rather than imposing and intervening. The same goes for the rather passive, almost manipulative attitude Jullien attributes to Chinese ideas. There are plenty of examples of Sun Zi and other strategic thinkers of the need to prosecute plans of action and shape the direction of reality. These impacted Mao Zedong's own strategic stance, where there were always clear aims, and an idea of an ultimate outcome to be driven towards that needed to shape, rather than be shaped, by what was.

What Jullien does highlight, however, is this notion, at least within the Chinese intellectual and philosophical tradition, right until today of some kind of difference between it and other thought systems, and of plenty of people inside and outside this tradition thinking that it is markedly different. The very fact that this conviction exists is striking. That so many do believe their view of the world and how humans operate in it is marked by special characteristics, is itself something that distinguishes and marks these people out. Everyone, everywhere, is of course unique and can claim their own kind of exceptionality. But for the Chinese view, a very large number of people at the same time make the same kind of claim. That is worth trying to understand more deeply.

It is also clear that in the era in which China is a more significant global actor, the many claims that it, as a country and a political actor, has some kind of view of its own influence, and is trying to change the behaviour of others to suit this, is also important to reflect on. China, with its unique political model of sinified Marxism, does not seem to be like the Soviet Union during some phases of its existence when it was zealously promoting its system in the outside world and trying to achieve an act of conversion. But nor is China wholly passive, happy with its own mandated space and working in that, with no impact on the wider world. Instead, China seems to operate in some specific areas as an activist power – intellectual property theft, in the recent past for instance, and now facing claims that its agents are trying to gain influence in places like Australia, Europe, or the US. China has been accused of being involved in the dissemination of fake news, and of trying, particularly in Taiwan, of influencing the outcomes of elections in ways which are favourable to it. The perception of this influence, rather than its reality, is the important think to acknowledge. China is being attributed with an intention, and a mode of operation, that implies it does want to have influence, and dominate in some areas. It is not just happily enjoying passivity while events unfold, seeing what their intrinsic properties are going to reveal. Instead, it is the same business as others – actively trying to shape and control events to achieve outcomes that suit its understanding of its strategic aims. These are broadly to become a strong, powerful, wealthy country, with a central place in the world. China is into winning as much as anyone else. It is just the way it plays the game, rather than the game itself, which is the issue. And many, on this line of argument at least, find this unfair. Win–win is the mantra not of a generous competitor, on this reckoning, but the slogan of a cheat!

## China's unfairness

The passive China of Jullien's description, with its abstinence from overt action, its refusal to get involved in conflict, its reading with deep intuition of situations so that it fits in with the movement of the world, rather than making the world fit into it, for critics at least can be seen as a façade. What it masks is not a peace loving, harmony creating China, but one that in its practices and mindsets is wedded to unfairness. This is a China that wants to win, but only on its terms, and according to its rules. The China, in effect, of a free loader.

There are many areas to see examples of this kind of behaviour. It has been spelt out, with particularly starkness, by the Trump administration. China is a country that has enjoyed the benefits of a rules-based order, grown wealthy and powerful on this, but while abiding usually by the letter of the law, has definitely not taken heed of its spirit.[13] The main claims are that China has violated intellectual property conventions, entering into arrangements with foreign companies where it has demanded technology transfer, after which it has simply walked away, failing to observe its commitments when it had gained what it was looking for. This is a China that has been allowed generous access to US, EU and other markets and failed to reciprocate, protecting many of its most important sectors, like finance and services. This is a China, the citizens of which can, if they fulfil certain criterion, own land and property in places like Australia or the UK, but whose laws only allow time-limited leases on any property in China for foreigners. It is a country where social media, news websites and other sources of information are blocked, or controlled, when they irritate the government, but where the government behind these laws can, because of freedom of speech conventions abroad, can run information campaigns without hindrance in the outside world.[14]

Unfair China has become a particular bone of contention as, since 1978, it has also become successful China. The link is an obvious one to make. A powerful, rich strong China, its critics say, has achieved this through observing its own interpretation of the rules, and gaming the system. It has engaged with the international rules based system as a parasite, gaining from the predictability it provides, but not holding to the underlying values of this system – respect for individual human rights, for the rule of law, freedom of speech, and liberal values.

It is clear that those who accept the narrative that China is unfair in its attitude and practices, often also believe that behind what they see as these self-centred practices there is an endgame, one where China wants to 'win' in achieving this. Despite all of the protestations of Chinese leaders, therefore, and their agents, this endgame is not for harmony, peace, balance, and win–win. It is about Chinese dominance, control and hegemony, a world which is run in ways that benefit China, and only China. The Belt and Road Initiative (BRI) in particular has been a lightning rod for this kind of self-centred, opportunistic, ruthless behaviour, the international space where China is able to use its economic and other means, because of lack of military opportunity, to get what it wants – access to new markets, diplomatic support for things that matter to it, and an area where it can export its new technologies like high speed trains. This is also a place where it can export over-production and create debt traps. These are the standard criticisms of Chinese power in 2019.

Does China really have such a clear endgame, however? Can its language about wanting balance and harmony be wholly discounted? Is it truly guilty of the long list of crimes and misdemeanours imputed to it? Perhaps, for looking at this notion of what China means when it talks of winning, it would be sensible to not focus so much on the high level, grand geopolitical areas, but come to examples that are more prosaic – that of games. And in particular, the Olympic Games and the World Cup. What do these tell us about China as a power that wants to win?

## The winning streak – China and the Olympics

Few countries can have invested as much time, effort, and funding, as the PRC into the Olympic games. China as the Republic of China was involved in the games from as early as 1924. The People's Republic, due to the issue of Taiwan, did not formally take part in modern games till UN recognition of it, and conferral of US formal diplomatic relations in 1971 and 1979 respectably. The PRC's Olympic experience therefore is almost exactly the same age as that which marks the reform era, starting from 1978.

For the PRC, much like the quest for Nobel Prize recognition,[15] hosting an Olympic Games was regarded as a major objective, along with winning as many medals as possible. It is one for which China has been accused of ruthlessness and lack of principle. That is probably testament to the fact that in the almost four decades after the first games that a PRC delegation attended in 1984, the trajectory of the county has been that of a winner. It applied to host a games in 1993. Only four years after the international condemnation heaped on the country as a result of the clampdown on protestors and the massacre of a significant number of them during the 4 June 1989 events, the bid failed by two votes to that of Sydney. Many at the time speculated that this rebuttal would result in Beijing withdrawal from all future attempts. This proved wrong. By 2000, the cache of holding such a symbolically important event was too strong. Beijing's success in winning the right to host the 2008 games was regarded as one of the most important of the achievements of the Jiang Zemin era, marking a rehabilitation of the country to the international community, and a major moment to showcase its achievements to a global audience.

The Beijing 2008 Olympics can be viewed now as a moment to appreciate all the upsides and downsides of China as a winner. The opening ceremony of the event, on August 8th, was of unprecedented scale and grandeur. The architecture for the various games constructed across the capital, and in other venues, was also at a cost that had never been seen before. The Hu Jintao administration seemed to be showcasing the merits of a state led, planned, autocratic system, something particularly potent at a time when the global financial crisis was just starting to rear its head, afflicting the US, Europe and others with profound challenges. The attention that came China's way however highlighted some of the real issues about the country's international image. Earlier in the year, unrest in Tibet, claims that Chinese investment was being used by regimes with massive human rights issues in Africa, and intense scrutiny over the ways in which property and people were impacted on by the redesign of Beijing all helped to construct a powerful counter narrative to the one promoted by Chinese officials. This was of a country which was threatening, increasingly heedless of fundamental rights, not remotely likely to implement proper political reforms, and assertive in its regional and global role. The Beijing Olympics was arguably one of the key moments in cementing a notion of China as a threat, not as the peaceful harmonious power being promised domestically.

On one measure, however, the games were an unqualified success – that of gaining medals. In the period between the 1992 games in Barcelona, and the 2008 ones held in Beijing, China rose from fourth in the medal winners table to first. In this period, it effectively doubled its haul of gold, silver and bronze medals, rising from 54 to 100. Gold medals in particular shot up three times from 16 in Barcelona to 48 in Beijing.

Even this achievement was subject to intense scrutiny. It was noted that the majority of the country's success was in gymnastics, or sports which involved intense, regimented training. The PRC was accused of adopting the kind of inhumane approach which had been practiced during the Soviet Era, with talented children put into semi-slave like situations from an early age as they specialised in particular sports. This frequently, the argument went, involved decisions being made against their will, high levels of injury, and use of questionable to clearly illegal methods. One particularly famous example of this had, in fact, received coverage before the 2008 games – the methods adopted by coach Ma Junren, who had achieved notable success for a group of runners in previous Olympics, despite being accused of using doping and other methods. Some of the runners trained by Ma were still active in 2008.

The 2008 Olympics, for all its controversy and challenges, has clearly not put China off hosting further events like this. It is still due to hold the 2022 winter games, subject to Covid-19 restrictions no longer being in place. What it probably most effectively achieved was this notion of a China that could, in areas where it put the effort in, be number one, and where it had to be accepted as such. And while in subsequent Olympics, China's ranking slipped to second and then third in the overall medals table, it symbolised a winning habit. Within two years of 2008, China rose from the world's third to second largest economy, in terms of gross size. It has the number one slot in sight sometime in the next decade, something that has most probably been accelerated due to the impact of the global pandemic on US and European economies. The propaganda value of 2008 for the Communist Party was to see a China that not only said it desire to be number one in some areas, but actually achieved this.

## China the loser – the sad story of football

A far less happy tale is that of China's attempts to succeed at football. Current president Xi Jinping is an avowed football fan, someone who took time out during this state visit to the UK in 2015 to visit Manchester City football ground. He has been pictured kicking footballs around, and reportedly when seeing visitors shows genuine passion about the subject if and when it comes up. A campaign around 2016 started extolling the virtues of China improving its football prowess, with slogans appearing across the country. There was even ambitious talk of China not only hosting a FIFA World Cup event by 2050 but winning one.

Football has massive public following in China. World Cups get some of their highest audiences there. Premier League clubs like Liverpool, Manchester United, Chelsea and Arsenal enjoy enthusiastic audiences, with the television rights payments amongst the largest contributors to the league's coffers. Clubs like Manchester United have high recognition, and have, over the years, opened shops to sell their merchandise in the PRC. Their teams have sometimes gone there for exhibition games, and players as varied as Britain's Paul Gascoyne and David Beckham have played in the country.

Despite all of this, the PRC's football record, domestically and internationally, has been lamentable. As Rowan Simon described, in *Bamboo Goalposts,* provision of decent pitches for schools kids and others to play on is minimal, particular in Chinese cities.[16] The game has been blighted in recent decades by claims about corrupt refereeing and very poor professional standards. There is a lack of a properly and effectively regulated national league, with very few local teams garnering anything like the affection and support of those from Europe or elsewhere. Very few Chinese players have registered on the international stage. This is reflected in the tale of Chinese involvement in the international game. The one time in which the country was able to take part in a FIFA World Cup competition, in 2002, it did so after being one of the four countries that qualified in the Asian Football Confederation area. Its performance as part of Group C, playing against Brazil, Turkey and Costa Rica, ended up with zero goals scored, nine put in against it, and bottom ranking and elimination. The humiliation was particularly sharp because of the relative success of Japan that year, one of the hosts alongside South Korea. Since then, China has failed to qualify again.

In view of the significant resources, and the political attention given to the game, many have wondered why it is that at the world's sport, China has been so unsuccessful. One answer this question has been simply that while the Chinese mindset might suit more organised, regulated games like gymnastics, or table tennis, the more entrepreneurial nature of football means it fits with more democratic political cultures.[17] There has been extensive research on this issue. As ever, the conclusions are a little more complex, with wealth, cultural and other factors needing to be taken account of. In any case, were politics to be so decisive just in terms of the game itself, this would need to account for why women footballers in China have been far, far more successful than their male counterparts, qualifying for every World Cup except one in the last two decades, being runners up in one, and reaching the quarter finals twice.

Despite claims made by some of the more nationalistic within China that the game was, in fact, invented there, perhaps one explanation for football's poor showing in the male game, at international level, is that this is not an area that the country has ever really devoted as much attention to, nor really taken to heart, despite the public interest in the game. It is an entertainment, not as politically important as something like the Olympics, and in any case, it arguably fits in more with the values of the capitalist west, and is something that China can indulge itself with, rather than become completely devoted and married to. Some football competitions in China are all too often loaded with a nationalist sentiment which the government probably regards with trepidation. Even a low level friendly between China and Japan in the mid-2000s ended up with protests and riots in Beijing. This sort of conflict duplicated in a world cup final where China ran the risk of being pushed into runner up status by its most bitter foe in modern history would be hard to countenance. For football success, it seems, the Chinese state is happy to let this exist as a remote aspiration, a nice dream. This is the most likely place it will stay as the country tries to conquer other areas it can say it is a winner in.

## China the winner

The comparison of the case of football and Olympics shows that the notion of winning and losing is not an alien one for China. There are areas in which the country knows precisely what it means to succeed and fail, despite the cultural and philosophical issues talked about earlier in this paper. A more accurate description of the country's attitude, at least as far as its political elite go, is that engaging in any kind of competition, in any area, means handling the risks of failing, or losing – and that in some of these can be borne, in others they are not worth even the endeavour of starting off in the first place. The Communist Party of China, after all, is a risk management body. This is a fundamental part of its function. And it accepts the concept of winning and losing, and the distinction between them, on pretty similar terms to the rest of the world.

It is as ever in areas where one looks at the detail, that important contextualising distinctions become clear. The victim or underdog mindset which arises from narratives of modern Chinese history promoted particularly heavily since the 1990s has embedded and internalised to many in China this idea that the global rules were devised before they could have a voice in them, and therefore have an intrinsic bias against the country. In the era in which the PRC was indeed a poor, weak and undeveloped place, this sense of fighting in a game which had been predetermined against it even before it started helped inculcate a belief that adapting, bending or ignoring rules was justified – because they were intrinsically unfair. The issue since Xi Jinping came to power however is that China is no longer a weak underdog, but a huge, and ostensibly powerful, player – and one therefore who can be justifiably accused of not playing to the rules that others have to, and failing in its duty to set an example. This has been one of the powerful motivations behind the criticisms of the country's trade practices by the Trump administration. China can continue to say it can be a winner, but not against rules that are biased against it. The US and others can reply however that this is sharp practice, and something unacceptable from such a major entity now. All of these claims have only been sharpened and intensified by Covid-19. It can no longer legitimately claim any handicap, because it is an experienced participant, not a novice. We are in an era in which this fundamental conflict about what status China really merits is still ongoing. Does China win in its claim to continuing unique status, or does the US and those allied to it force it to accept the same terms as everyone else in the same spirit? The outcome of this particular battle will matter profoundly both to China and the rest of the globe.

While this essay has discussed the notion of the sort of games and actions China can be said to be engaged in which might have outcomes that can be described as 'winning' and victorious, it is worth concluding on some discussion of what this notion of a final victory might be. In the South and East China Sea for instance, from the late 2000s, there has been a clear set of objectives which China clearly does regard as important for it to achieve, and thereby claim 'victory'. One is the acceptance of the Nine Dash Line, scoping out vast parts of the region right down to the coast of Indonesia. In order to achieve this, the PRC has constructed artificial islands, and increased its naval presence in the area. It has also responded aggressively to any attempts to contest its rights to this area in the diplomatic arena. It has closed down the international space for Taiwan, particularly since the election of President Tsai Ing-wen of the more independence leaning Democratic Progressive Party (DPP) in 2016, and undertaking often highly muscular diplomatic work abroad, trying to pick off the few remaining formal diplomatic allies of the island. Even more importantly, it has attempted to use economic leverage (over 60 per cent of Taiwan's trade is with the Mainland) to coerce it. This has led to very mixed responses, with opinion now shifting much more supportively behind Tsai amongst Taiwanese. The main thing to note here is that the PRC clearly does have a vision of what winning on this issue is – reunification, and control of the maritime region around Taiwan. This is not a passive objective, even if it is one that Beijing needs to pursue within often constrained international space because of the role of the US and its close links to Taiwan.

This is important, because if two, three or more partners are involved in a competition where there is the concept of a winner, there has to be some criterion of what merits the conferring of a winning title. In this context, we do have a history of knowing what major powers like the US look like as winners. But we don't have any modern experience of what this means for China, in terms of geopolitical role and status, beyond these more local issues mentioned above. China as the PRC has been involved since 1949 in a long process of perpetual limbering up, getting ready for the final stages of a competition to prove what it is really capable of. It states that involves being a strong, powerful country. This is clearly the game it most wants to win at. But to win, it does need others to recognise this status. Being rich, strong and powerful in its own self-conceptualisation begs the question of how it can do this when there is no one else it is being strong to, or powerful against. Like it or not, the world outside is to some extent involved in this game and endgame that China has set itself.

That aside, because China as a national winner of this kind of grand geopolitical game never occurred before in modern history, not only the rest of the world has no idea what victory and winning in this context might mean – nor has China. Despite being the principle agent in all of this, it is as new to this game as everyone else. We as outsiders don't know what China as a global winner looks like. Nor does China itself.

The very final thing to say is that the subjective feeling of being a winner is very different for different players, even when they are in the same game. For Brazil, winning a FIFA world cup is amazing, but something they have done many times before. For the UK, with its one win in 1966, the mission for a second victory has taken on an almost unbearably burdensome spiritual importance. It might be the same game, and the same rules, but the winning quality depends very much not just on who wins but how they feel about this. For China, its conception of its history and the status of being a victim and underdog does mean that it probably sees winning not just as a novelty, but something which is freighted with far more emotional importance than perhaps for others. China the winner means a very different thing inside and outside the country. And while one can dispute whether the PRC in 2019 is a winner, or indeed will ever be one, one cannot ignore the fact that it often speaks like it does want to win – and that in itself is something we need as commentators, scholars and observers to take seriously.

### Notes

1 Xi Jinping, *The Governance of China, Volume 2* (Beijing: Foreign Languages Press, 2017), 581.
2 'Constitution of the People's Republic of China', revised 2004, available from website of the National People's Congress, www.npc.gov.cn/zgrdw/englishnpc/Constitution/2007-11/15/content_1372962.htm, accessed 6 December 2019.
3 Lin Biao, 'Long Live the Victory of People's War', at https://www.marxists.org/reference/archive/lin-biao/1965/09/peoples_war/introduction.htm, accessed 6 December 2019.
4 Ludwig Wittgenstein, *Philosophical Investigations* (Oxford: Basil Blackwell, 1958), 31e-32e.
5 Wittgenstein, *Philosophical Investigations*, 26e.
6 Laura Silver, Kat Devlin and Christine Huang, 'Attitudes Towards China', Pew Research Centre, 5 December 2019, available at https://www.pewresearch.org/global/2019/12/05/attitudes-toward-china-2019/.
7 Francois Jullien, *A Treatise on Efficacy: Between Western and Chinese Thinking*, translated by Janet Lloyd (Honolulu: University of Hawaii Press, 2004), 14.
8 Roel Sterckx, *Chinese Thought from Confucius to Cook Ding* (London: Pelican Books, 2019), xi.
9 Jullien, *A Treatise on Efficacy*, 3.
10 Ibid, 16.
11 Ibid, 25.
12 Ibid, 39.
13 On this distinction, see the work of St Augustine, 'On the Spirit and Letter', in *A Select Library of Nicene and Post Nicene Fathers of the Christian Church, St Augustine*, Vol. 5 (Grand Rapids, MI: W. B. Eerdmans Publishing Company, 1971), 80–116.
14 This and a long list of other grievances are most strikingly laid out in the speech on China by Vice President Mike Pence at the Hudson Institute in October 2018. See https://www.hudson.org/events/1610-vice-president-mike-pence-s-remarks-on-the-administration-s-policy-towards-china102018.
15 See Julia Lovell, *The Politics of Cultural Capital; China's Quest for a Nobel Prize in Literature* (Honolulu: University of Hawaii Press, 2006).
16 Rowan Simons, *Bamboo Goalposts: One Man's Quest to Teach the People's Republic of China to Love Football* (London: Pan, 2009).
17 'What Makes a Country Good at Football', *The Economist*, 9 June 2018, https://www.economist.com/international/2018/06/09/what-makes-a-country-good-at-football.

CHAPTER 12

# A Decisive British Victory? The Confrontation with Indonesia, 1963–1966

Christopher Tuck

This chapter explores the undeclared war fought against Indonesia by Malaysia and its British, New Zealand and Australian allies from 1963 to 1966. Known as 'Confrontation' (from the Indonesian *Konfrontasi*), this complex conflict provides a fascinating insight into the problems of determining the outcomes of wars. Ostensibly, the final political settlement of August 1966 represented a decisive defeat for the Indonesians, and an equally significant success for their adversaries, especially the British. Yet, as this chapter will show, on closer inspection the outcomes were much more ambiguous. In many respects, Indonesians were both the biggest losers *and* the biggest winners from this conflict. In making this argument, this chapter will provide a brief overview of Confrontation, laying out the basis upon which it has traditionally been categorised as a decisive British victory. The discussion will then reconsider the outcomes of the conflict from both an Indonesian and a British perspective. Ultimately, Confrontation demonstrates many of the critical problems and issues involved in determining the outcomes of armed conflict: the unreliability of political end states as a metric of success; the question of whose viewpoint provides the most legitimate assessment of the outcome of a war; the impact of the passage of time on questions of success and failure in conflicts; and, especially, the role that perception rather than fact plays in determining popular beliefs on the outcomes of wars.[1]

## Confrontation

The ostensible cause of Confrontation was the creation in September 1963 of the federation of Malaysia.[2] This new entity brought together the independent state of Malaya, the self-governing colony of Singapore, and the crown colonies of Sarawak and British North Borneo (the latter re-named Sabah on its accession to the federation). Britain was a key agent in the creation of Malaysia. British policymakers painted British objectives in altruistic terms, arguing publicly that the new federation reflected 'the wishes of the people concerned, the advantages for those peoples, both internally and externally, and the advantage for South-East Asia (and so ultimately the free world).'[3] In particular, the government argued that the welfare of the populations of Sarawak and Sabah was a paramount concern, the new federation enabling the populations of these two states to be placed 'in a constitutional framework which will give them the opportunity of the greatest measure of political stability, collective security and economic expansion when they become independent.'[4]

In early 1963, Indonesia, which had previously seemed indifferent to the Malaysia project, suddenly declared its opposition and commenced a strategy of *Konfrontasi* in order to undermine it. The Indonesian president, Sukarno, argued that the new federation was simply a reflection of British neo-imperialism and that Malaysia had crushed the legitimate aspirations for independence of the peoples of Sarawak and Sabah. Keen to avoid escalating the conflict to the level of a conventional war, which Indonesia would probably lose, Sukarno instead engaged in an undeclared sub-threshold war. Sukarno sought to destabilise the new federation using a panoply of measures that were just sufficiently limited that military escalation by Britain would not seem proportionate or legitimate. *Konfrontasi* involved a combination of diplomacy, subversion, economic warfare, terrorism, propaganda, and low-level military action designed to destabilise and fragment Malaysia. Britain was drawn into the conflict through activation of the 1963 Anglo-Malaysian Defence Agreement, an extension of defence agreements signed with Malaya in 1957. Over time, both Australia and New Zealand were also induced to provide forces for the campaign. Though technically only a supporting ally to Malaysia, the size of the British military contribution to Confrontation dominated the response to Indonesian aggression, especially in the military sphere, and the costs of the conflict over time weighed increasingly heavily on British foreign and defence policy.

The campaign was not an easy one for Britain and its allies. Concerned about sustaining the legitimacy of Britain's position, it was viewed as important that the conflict should not be seen in terms of a regional state, Indonesia, being bullied by a white, colonial power. Political restraint, therefore, was the order of the day. Military operations, for example, had to appear to be defensive, with a strong emphasis on discrimination and proportionality.[5] The key military front line in Confrontation was the nearly thousand-mile frontier on the island of Borneo between East Malaysia (Sarawak and Sabah) and Indonesian Borneo (Kalimantan). The operating conditions for the forces involved were extremely challenging. The Borneo frontier area was predominantly sparsely populated jungle, punctuated by ridges, swamps, rivers and streams. There were few roads. Tactical mobility and logistic support relied upon the use of rivers, where navigable and, especially, helicopters.[6] The nature of the security problem was also complex. One challenge was posed by enemy border incursions. Here Indonesian-supported irregulars and, increasingly over time, regular members of the Indonesian army, the TNI (*Tentara Nasional Indonesia*), sought to cross from Kalimantan into East Malaysia with the intention of conducting attacks on the security forces and linking up with anti-Malaysia elements.[7] A second threat was internal in nature and comprised mainly dissident elements within the substantial Chinese communities in Sarawak and Sabah. These communities, forming between a quarter and a third of the population, were heavily penetrated by a communist subversive movement known by the authorities as the Clandestine Communist Organisation (CCO).[8] By late 1963, for example, it was projected that there might be around 4,500 infiltrators based in Kalimantan, and a further thousand Clandestine Communist Organisation irregulars (CCO) in Sarawak.[9] But there was also a threat against West Malaysia. The Indonesians created infiltration bases on the island of Sumatra, and on the Riau islands off Singapore. From these bases, small groups of infiltrators attempted sea-borne incursions. Those that made the crossing successfully then set about acts of subversion and terrorism.[10]

Because of the strategically defensive posture adopted by British forces, the campaign became a protracted struggle which sucked in, in some shape or form, all of the British military assets in the Far East. British and Commonwealth forces were generally successful in defeating Indonesian incursions and in controlling the threat posed by the CCO. But the response of the Indonesians to these tactical failures was to increase the numbers and quality of its infiltration forces. Initially, these forces comprised mainly of dissident Chinese and Malays, many of the latter drawn from the remnants of the TNKU (*Tentara Nasional Kalimantan Utara*, or North Borneo National Army) which had launched an abortive rebellion in Brunei in December 1962.[11] The failure of these irregulars led to the deployment of more Indonesian regular troops including marines and special forces. At the same, the Indonesians shifted the kind of operations that they engaged in, moving from attacks on security force positions, which were well defended, to shallower incursions that would allow them to withdraw quickly back across the border. Early on in Confrontation, the British chiefs of staff sounded a note of concern that Britain faced the potential position in which 'more and more resources may have to be deployed in Sarawak and Sabah to meet any expanding scale of Indonesian-inspired guerrilla activity … this could lead us into severe financial, manpower, and equipment difficulties at the expense of our other commitments.'[12]

The political aspects of the campaign also were challenging for Britain. In the context both of the Cold War and decolonisation, many states were sympathetic to the Indonesian argument that Confrontation was a struggle against the manifestations of neo-colonialism. Organisations like the UN, therefore, were viewed as suspect tools in the fight to sustain Malaysian legitimacy.[13] Moreover, key British allies were also equivocal in their support for British policy, especially at the beginning of the conflict. In the US, for example, the administration of John F. Kennedy took the view that Indonesia mattered more to the west than Malaysia, and that Sukarno could be managed through diplomacy.[14] The political problems associated with Confrontation grew over time. Indeed, in August 1965 tensions within the federation led Singapore to leave, a circumstance that seemed to support Indonesian arguments that the federation was an artificial creation.[15]

But despite the political, military, and economic difficulties associated with prosecuting the campaign, it has long been seen as a decisive British triumph. General Sir Walter Walker, the joint military commander in the Borneo theatre until March 1965, argued that the conflict was for Britain 'a complete success'.[16] Denis Healey, the British Minister of Defence from October 1964, argued that the campaign was 'one of the most efficient uses of military force in the history of the world.[17] After the event, other commentators also framed the campaign as a major British success, one labelling it: 'The most successful use of armed forces in the twentieth century'.[18] The basis on which Confrontation is assessed to be a victory relates to two themes: political outcomes and the quality of strategy-making.

In relation to the attainment of political goals, Confrontation could be assessed as a decisive British victory because Britain seemed to have attained all of its objectives, whereas the Indonesians appeared to have achieved none of theirs. In defending Malaysia against Indonesia, British policymakers articulated two key goals. First, in the short-term Indonesia had to be induced to end *Konfrontasi*: to cease its opposition to the new federation, and to restore normal diplomatic relations. In the long-term, however, it was clear to Britain that Indonesia was a critical player in South-East Asian security. As a non-aligned power, Indonesia could play a pivotal regional role in the west's struggle to contain communism. Britain's second goal, therefore, was to ensure friendly relations with the Indonesian government. As one British review of options noted: 'If we are to defeat confrontation successfully it must be done in such a way that Malaysia is able to establish reasonable relations with her neighbour Indonesia'.[19] The problems in reconciling these short- and long-term objectives were obvious – how could Britain defeat Indonesia, and yet do so in ways that would foster a positive relationship afterwards? Yet, that is exactly what Britain seemed to have achieved.

A political agreement to end the conflict was reached in August 1966. In this, Indonesia agreed to cease Confrontation and to recognise Malaysia. No substantive concessions were made by Malaysia. As one commentator has noted: 'From the standpoint of the issues ostensibly involved in the Malaysia dispute, the final settlement was in fact a rather complete capitulation on Indonesia's part.'[20] Moreover, an attempted coup in Indonesia at the end of September 1965 had set in motion a train of events that led to Sukarno's effective removal from power. The new Indonesian government, led by General Suharto, proved to be pro-west and anti-communist, a significant result in the context of the wider Cold War. Thus, Britain won because it had achieved its political objectives, despite their apparent incompatibility; and Indonesia lost because it didn't achieve any of its over-arching goals.

The scale of British success was multiplied by the manner in which it was achieved. It seemed to be an 'intentional' victory: a victory brought about by the deliberate planning and execution by Britain of a campaign that linked tactical and operational activity to the desired strategic end-states.[21] This careful relating of means to ends was apparently nowhere more evident than in the military campaign in Borneo. The overall British strategy was one of protraction. As the Defence and Overseas Policy Committee put it: 'our policy should be to work for a negotiated settlement while showing the Indonesians that we were prepared to maintain an effective military defence of Malaysian territory, since it is only by a combination of economic and military pressure that the Indonesians might be induced to have second thoughts.'[22] By mid-1964, however, Britain's defensive stance risked handing the tactical initiative to the Indonesians. General Walker argued for the need to take the fight tactically to the Indonesians. Walker, though, also recognised that military action had to be subordinated to political imperatives. As he noted, '[t]he political factor … must be paramount and the services must accept this.'[23]

On this basis, Walker instituted Operation *Claret*, a secret campaign of small-unit Commonwealth attacks across the Borneo frontier. These operations were heavily constrained – they had to be sanctioned, for example, at the highest level; and the pace and type of operations was modulated to meet changes in the political situation.[24] *Claret* seemed to transform the war. Commonwealth troops inflicted disproportionate losses on the Indonesians and induced them to withdraw from the frontier the bases that supported their operations into East Malaysia. Walker was clear that *Claret* 'changed the fortunes of the war for both the Indonesians and their Commonwealth opponents'.[25] The skill with which Britain nested tactical military activity within the broader context of political constraints was something noted by the troops involved. One British officer argued that in Borneo: 'Instead of progressive escalation, often so inevitable, military force intelligently deployed produced conditions which helped and perhaps even promoted sensible negotiations.'[26] *Claret*, argued Walker, 'brought the Indonesians to the conference table in 1966.'[27] Over time, Sukarno's failure to end Confrontation successfully led to growing tensions between two key mutually antagonistic domestic players – the Indonesian army and the Indonesian communist party, or PKI (*Partai Komunis Indonesia*). As Sukarno's frustrations grew, he moved closer to the PKI, creating growing concerns within the army. The confused and increasingly polarised political situation in Indonesia led to the attempted coup on 30 September 1965, an event used by the army as an excuse to move against the communists. The army's success in this also led gradually to Sukarno's removal from power and his replacement by General Suharto.

Thus British success seemed comprehensive. Tactically, the military outcomes were weighted heavily in the Commonwealth favour, with 1600 Indonesian troops killed, wounded or captured, as against only 300 Commonwealth.[28] More importantly, however, Britain won politically: the conflict ended on terms overwhelmingly favourable to Britain and its allies.

The idea that winning in war should be conceptualised in terms of the relative attainment of political goals at least acknowledges that military successes do not automatically constitute overall victory.[29] But the concept of victory is itself a contested and problematic concept. As the introduction to this book has illustrated, the literature on success in war highlights a range of difficult challenges intrinsic to any attempt to assess the outcomes of a conflict. The impact of these factors can be understood by exploring the two propositions that 'Indonesia lost' Confrontation and that 'Britain won'.

## An Indonesian defeat?

In exploring the assertion that Indonesia lost Confrontation, a number of questions arise. One clear reason to argue that Indonesia lost would be that Indonesia achieved none of its stated goals. Confrontation didn't work for Indonesia in that it did not deliver the desired political goals. But this judgement is rendered problematic for at least three reasons. First, it isn't wholly clear what Indonesia's real objectives were.

As Rex Mortimer has noted 'Sukarno's ultimate objectives in the confrontation of Malaysia were never clearly defined and have been the subject of controversy'.[30] In negotiations in 1963, President Sukarno argued that peace was contingent on Malaysia agreeing to a plebiscite in Sarawak and Sabah to determine whether they wished genuinely to be part of the federation.[31] But later pronouncements from Sukarno were much more bellicose, threatening to *ganyang* ('crush' or 'mouth') Malaysia, and to destroy it by 'cock crow on 1 January 1965'.[32] The latter objectives certainly weren't achieved. The August 1966 agreement did include a secret codicil which identified prospective Malaysian elections in 1967 as the means to ascertain whether the populations of North Borneo wanted to remain part of the federation. But these elections returned strong results for the Malaysian government: unsurprising given the government's dominance of the election process.

But other Indonesian goals also seemed to exist. One was anti-colonial sentiment and the associated concern with British regional intentions. Sukarno argued, for example that Malaysia represented 'a new concentration of colonial forces on the very frontiers of Indonesia' and 'an unacceptable colonial intrigue'.[33] In dealing with this threat of British encirclement, Confrontation therefore sought also to drive Britain from its bases in Singapore. Confrontation did not achieve this directly, and the growing economic pressures on the British government were probably the prime determinant of the government's increasingly equivocal view on the utility of the Singapore bases.[34] However, Confrontation did contribute to accelerating this process, illustrating the unsupportable costs for Britain of waging extended regional conflicts. Confrontation sharpened these debates and provided incontrovertible evidence of the growing gap between Britain's resources and its commitments.[35] Confrontation did, then, act materially to encourage Britain's physical exit from South-East Asia. In this sense, John Subritzky notes that through Confrontation Indonesia 'had succeeded beyond expectation'.[36]

Indonesia also pursued other, more amorphous, objectives including those relating to national honour, credibility and prestige.[37] Here Indonesia could also claim a measure of success. *Konfrontasi* demonstrated the costs of having Indonesia as an enemy and so in many respects reinforced for Indonesia's adversaries the importance of sustaining a positive and cooperative relationship with Indonesia, especially in the context of the Cold War. As a result, both Britain and Malaysia were keen to avoid presenting the August 1966 peace settlement as a defeat for Indonesia. When Suharto declared publicly that peace was for Indonesia 'not capitulation but a compromise in the interests of both parties and peace in South-East Asia',[38] his recent enemies were keen to support this line. The Malaysian prime minister, for example, argued that the settlement was, indeed, 'good for all'.[39]

A second reason to assert that Indonesia lost Confrontation would be to point to the outcome for President Sukarno himself. At the apex of a chaotic and highly personalised foreign policy-making process, the president was the key architect of Indonesian *Konfrontasi*. Whatever the objectives that he pursued, his own political demise was unlikely to have been one of them. Yet, that is exactly what Confrontation delivered. His Guided Democracy regime was replaced by General Suharto's New Order, and Suharto emasculated Sukarno politically, stripping him of his presidency and isolating him until his death in 1970.

But this metric of Indonesian failure can again be questioned. Here, it is important to draw in the issue of the referent object. In relation to the referent object, who are we talking about when we use the term 'Indonesia'? Sukarno, certainly, was a major loser in the outcomes of Confrontation; but there were other Indonesian actors involved, each with their own agendas. *Konfrontasi* performed important domestic functions in Indonesia, and we need to consider who supported Sukarno's policies and why. As Michael Leifer has noted, Confrontation was in many ways 'the external expression of the very nature of Guided Democracy serving as a common denominator for political elements joined in adverse partnership'.[40]

This 'adverse partnership' consisted principally of the Indonesian army on the one hand, and the Indonesian communist party on the other. Both of these constituencies supported Confrontation at its inception. But for both, the primary motive for doing so was essentially domestic. In other words, whether or not Malaysia was actually defeated was of secondary importance to them. Instead, in their mutual struggle against one another, the real metric of success for both the TNI and the PKI was domestic political power at the expense of the other.[41] For the army, for example, the febrile atmosphere created by Confrontation staved off cuts to military budgets, and participation in Confrontation allowed it to bolster its nationalist credentials. For the PKI, Confrontation offered the opportunity to continue to build its influence and, through increased conflict with the west, encourage Indonesia in its move leftwards.[42] But the ultimate goal for each was ascendancy over the other. The army, for example, deliberately interfered with Indonesian military Confrontation operations in order to make them less effective, fearing that, if the conflict escalated into conventional war, the army would suffer heavily and that this would be exploited by the PKI.[43]

Taking these points into account, assessing the outcomes of Confrontation for 'Indonesia' is too crude because it implies a unity of goals that did not exist. Sukarno certainly lost Confrontation. In this he was also certainly joined by the PKI. The army portrayed the September 30 coup as attempted communist coup and mobilised its supporters in a violent campaign to destroy the organisation.[44] In this campaign at least 500,000 people were killed (some argue up to a million), and a million and half imprisoned.[45] For these reasons, it is clear that the Indonesian communist party was also a clear loser with respect to Confrontation. The PKI was eliminated as an organisation: the party was banned, and its membership suppressed. But this, of course, actually represented an enormous victory for the Indonesian army. By March 1966, General Suharto had effectively side-lined Sukarno and destroyed the army's key domestic adversary, the PKI. This provided the foundation for the creation of Suharto's authoritarian New Order regime. This regime cemented the influence of the army in Indonesian politics a condition that lasted until the Asian Financial Crisis of 1997. Overall, for General Suharto personally, and the TNI generally, the outcomes of Confrontation were hugely beneficial.

Third, and reinforcing the success of the TNI, we can also examine the outcomes of Confrontation through the lenses of perception and the passage of time. In many respects the success of the army increased over time because of their strategy, in Orwellian fashion, of manufacturing a new history of Confrontation and so moulding popular Indonesian perceptions of the outcome. This new history reframed the popular understanding of Confrontation and the crisis associated with Sukarno's period in office. Whereas Sukarno had focused Indonesian attentions externally, on the neo-colonial threat, the new narrative focused on the communist threat posed by the PKI, and the way in which the PKI and Sukarno had taken Indonesia to the brink of destruction.[46] Confrontation, then, was portrayed as one discrete element in a broader crisis brought about by communism. Of course, in this new narrative it was the army that had saved Indonesia by successfully warding off external attacks during Confrontation and by dealing once and for all with the most fundamental threat to Indonesia – the communists.[47]

The ability of the army successfully to create this narrative lay first in the fact that, given the propagandising already conducted under Sukarno, the truth was already in short supply. The new government then not only withheld the facts, by such expedients as limiting access to official documents, it then went on to create a new history of Confrontation. Leveraging the government's control over the education system, the army produced new school and university textbooks advancing a highly partisan view of the Sukarno period. It also engaged in a concerted propaganda campaign through such media as books, films, the creation of memorials and museums, and by building into the fabric of Indonesian life an inescapable programme of commemoration. This programme included compulsory school visits to anti-communist monuments and the creation of such national holidays as 'Armed Forces Day' and 'National Awakening Day'.[48] The success of this new narrative is indicated by the fact that it survived the end of the New Order regime itself. The regime's anti-communist narrative remains deeply embedded in Indonesian society. Even today, the army has been able to use the spurious spectre of a recrudescence of communism to legitimate actions designed to protect their interests.[49]

Thus, the question of the extent to which Indonesia lost Confrontation is open to more ambiguous answers than might be supposed. Indonesia, in aggregate, did achieve some of its goals. More specifically, though, whilst some Indonesian actors did indeed suffer calamitous outcomes, not least the PKI, for the Indonesian army circumstances ended up being much more favourable. The TNI succeeded in Confrontation not by defeating their Malaysian and Commonwealth adversaries but instead by defeating their domestic adversaries and shaping perceptions of the outcomes in the minds of their key constituency: the Indonesian people.

## A British victory?

Similar ambiguities apply when we assess the extent to which Britain won Confrontation. Here, it is worth returning to the problems associated with using political goals as the key metric for victory in war.

One key difficulty with using this metric is that Britain had private as well as public objectives. The public objectives articulated were goals largely designed to massage perceptions of the legitimacy of the federation of Malaysia policy – to support the idea that it was essentially a local initiative and not an exercise in neo-colonialism. However, the private objectives reflected the actual objectives that Britain sought to achieve, and these were much less altruistic. The key objective sought by Britain was the survival of the new Malaysian federation. But the purpose of this federation was to allow Britain continued access to its bases in Singapore at reduced political and economic costs. These reduced costs would come about because the federation, and not Britain, would be responsible for providing internal security in Singapore, and because giving Singapore full independence within the context of Malaysia would deflect criticism of a continued British colonial presence.[50] Clearly, however, the outcome delivered in August 1966 was problematic in regards to this objective. First, the very need to fight Confrontation raised significantly the costs of remaining in Singapore. Indeed, as has already been noted, the costs of Confrontation contributed materially to the decision by the government of Harold Wilson to end the British presence in Singapore. Second, in August 1965 Singapore left the federation anyway, largely as a result of tensions between the respective ruling parties and their leaders, Lee Kuan Yew and Tunku Abdul Rahman.[51] Singaporean secession was a major blow to British interests. As the Minister of State for Commonwealth Relations, Cledwyn Hughes, argued: 'we are faced with a radically new situation in which the whole basis of our position in regard both to confrontation and to the tenure of the bases has been drastically undermined'.[52] As parallel failures in relation to federations in South Arabia and central Africa demonstrated, British policymakers fatally overestimated the viability of such heterogenous political constructs.

To compound the challenge with using British political goals as a metric for success, these goals changed over time, as goals often do in war if a conflict does not end very quickly. By the middle of 1965, for example, wider British foreign and defence policy was undergoing change as part of the Wilson government's defence review process. The outcome of this process was a focus on cutting defence spending, especially in the Far East where it was judged British economic interests were relatively small.[53] As part of this, it was assessed that a British presence in Singapore could not be sustained in the long-term. However, Britain could not make these necessary defence cuts until Confrontation had ended. Thus, wider changes in context turned Confrontation on its head. Instead of being the means to achieve British objectives, in particular sustaining a British presence in Singapore, Confrontation had become an obstacle to making what the government considered to be essential cuts to defence spending in the Far East. For this reason, the British private objective had become by mid-1965 to end Confrontation on any terms acceptable to Malaysia.[54] With further economic pressures intensifying the British imperative to end Confrontation, by the end of 1965 British objectives had shifted again and had become to seek a negotiated solution with Indonesia irrespective of Malaysian wishes.[55] A British review of its position argued that: 'We must seek a negotiated end to confrontation. It would be best if this arose from the initiative of someone else, but if this does not occur soon, we shall have to seek to promote such an initiative indirectly and covertly ourselves'.[56] By this stage of the conflict, then, Britain wanted to jettison its short term goal of securing Malaysia in order to secure its long-term goal of creating better relations with Indonesia. The problem for Britain was that the US position towards Indonesia had changed. As Confrontation progressed, Sukarno's policies became increasingly left-leaning and anti-American. In conjunction with the US' growing commitment to Vietnam, Lyndon B. Johnson's administration argued that Britain could not afford to be seen to be weak – this would only embolden the west's regional adversaries.[57] The US threatened economic and political consequences for Britain if it tried to end Confrontation. For this reason, and despite the growing feeling that Confrontation could not be won, British decision-makers felt compelled to continue with the conflict. By the end of Confrontation, therefore, the most important British objective pursued through Confrontation was to sustain Anglo-American relations. Victory over Indonesia had become increasingly irrelevant.

Judging British outcomes in relation to political goals is therefore complex. Confrontation did indeed end; and it did so in a way that did not undermine British prestige nor result in a fundamental break in Anglo-US or Anglo-Malaysian relations. But the goals being serviced by Confrontation in 1966 were very different from those originally intended in 1963. We can build further on these ambiguities by drawing in the issues of referent objects, intentions, and the passage of time. In particular, rather than simply assessing outcomes for 'Britain' it is illuminating to examine the views of an important sub-actor: the British military. Through this, we can qualify the extent to which Confrontation was an intentional British success.

Victory, one would hope, would be a product of deliberate intention – the conscious relating of means and ways to produce the desired ends. As was noted earlier in this chapter, this is indeed the way in which many commentators have viewed the outcomes of Confrontation for Britain. Interestingly, however, this was not necessarily the view taken by the British military during Confrontation itself. Rather, the private consensus by 1965 was that Britain was losing the conflict and this conclusion was one factor that reinforced the British search for an end to Confrontation.

It is certainly the case that aspects of the British campaign were extremely successful. As has been noted, the tactical military handling of the conflict in Borneo was extremely good. Operation *Claret* did have a salutary effect on Indonesian forces deployed in the border area of Kalimantan. But the fact that the tactical military aspects of the campaign generally went well didn't mean that the strategic political effects were equally positive. During Confrontation itself, key military commanders including (despite his comments after the war) General Walker, thought that Indonesia was winning the campaign. The view taken was that the political constraints placed on Commonwealth military action were so severe that the tactical military successes won by Commonwealth forces simply could not impose sufficient costs on the Indonesians to change their commitment to prosecute Confrontation.[58] Instead, the view taken was that the costs of Confrontation weighed more heavily on Britain and that British military success was simply not sufficient to change Sukarno's political goals. Sir John Grandy, commander in chief, Far East, argued in mid-1965 that Confrontation 'cost him [Sukarno] nothing and suited him very well'.[59] With the benefit of hindsight, of course, this was not the case: but at the time, Confrontation seemed to be being won by Indonesia. Increasingly, therefore, the military began to advocate for escalation as the only way to win. In this vein, Operation *Claret*, which after the war Walker argued was central to British success, at the time was viewed by him as only a stopgap. Small-unit deniable operations comprised only the first step towards what he really wanted, which were undeniable brigade-scale operations into Indonesia, supported by air power.[60] Perhaps the apogee of this military pessimism came in October 1965 when the Commander-in-Chief, Far East, arguing that the British campaign had thus far had 'no strategic effect' on the Indonesians, proposed implementing a staged process of military escalation up to, and including, conventional war in order to force an end to the conflict.[61]

The degree of military pessimism at this stage in the campaign is important because it illustrates that the 30 September coup, which in hindsight was so pivotal to ending the conflict, at the time was viewed by British decision-makers in a much more equivocal way. The coup surprised those outside of Indonesia, including Britain and the US. Even today, its origins are not clear. At the time, those involved in the coup argued that they were trying to pre-empt a CIA-backed effort to topple Sukarno. It may well have involved the communists and even have been conducted with the knowledge of Sukarno himself. General Suharto argued that it was indeed an attempted communist coup and used it as the pretext to crush the PKI and seize power. But at the time, its impact on Confrontation was very uncertain. It simply was not clear, until at least March 1966, that the coup would produce an Indonesian government any less committed to Confrontation.[62] Indeed, the fear was that the military might be just as committed to undermining Malaysia but might also actually be more effective than Sukarno at doing it. As one British review of the potential implications of the coup noted: 'It is still not clear how confrontation will be affected by those major events [i.e. the coup] [but] there is considerable evidence that the Indonesians intend to pursue it with greater vigour, especially across the frontiers of Sarawak where greatly increased activity has already been noted'.[63]

Thus, asserting that Britain won Confrontation needs to take into account that for most of 1965 and into 1966, the British military thought that it was losing. The key transformative event, the September 30 coup, was neither at the time viewed necessarily as transformative and nor was it an intended consequence of any deliberate British strategy. This pessimism continued even with the signing of the August 1966 peace settlement. Walker's successor in Borneo, General George Lea, believed that the peace deal did not reflect a change in Indonesian intentions, only their methods. As late as September 1966, he believed that the Indonesians had signed the peace deal in order to get rid of British troops and that, once they had gone, they would simply start Confrontation again.[64]

On the other hand, however, did any of this really matter to the British government of the time? These problems were hidden, exposed only in the private debates between British policymakers and consultations with close British allies. In many respects the constituency that mattered most to the government in August 1966 was the British public. During Confrontation itself there was substantial media coverage of the campaign. As part of this, some of the problems facing the British government, such as the cost of the campaign and the political restrictions that limited British options, were exposed.[65] But in general, the public was unaware of the changes in British objectives or of the profound problems that afflicted British strategy that led to the strong measure of pessimism that pervaded British thinking by the end of 1965. The public mainly saw British tactical successes, in the context of a commitment designed merely to honour alliance commitments, won against a rogue Mussolini-like regional dictator. The perception of the British public, therefore, was that Confrontation was a signal success and one in which the costs of the conflict needed to be set against what the government argued was 'the tragedy that could have fallen on a whole corner of a continent if we had not been able to hold the situation and bring it to a successful termination.'[66]

## Conclusion

In theory, then, winning in war could be construed as the deliberate attainment of the desired political objectives. Confrontation has been assessed by many as just that sort of victory: the patient, highly disciplined application of military power in the service of carefully considered policy goals. As this chapter has explored, however, the reality is often very different. The concept of victory, despite being so important to strategy, in practice is highly ambiguous. Political goals provide a deeply flawed basis upon which to judge the success of a campaign, not least because they are often vague, or undeclared, or they change over time. Other factors also complicate the calculations: the question, for example, of whose perspective we take; or the timescale over which to choose to judge outcomes; or the importance, not of facts, but of the perceptions of key audiences on the outcomes of a conflict.

With these issues in mind, it is possible to question the scope of British victory against Indonesia in Confrontation. Actually, Indonesia did achieve some of its apparent goals. More importantly, some constituencies within Indonesia, especially the army, benefitted hugely from the conflict. At the same time, success in August 1966 for Britain clearly was a contingent outcome. British decision-makers during the conflict became increasingly pessimistic about the likely outcome of the conflict, and changing British objectives robbed Confrontation of much of its original utility.

In the final analysis, though, Confrontation illustrates that victory is not necessarily a zero-sum condition. Both the British and Indonesian governments in the end were able to convince their key domestic constituencies that the outcomes of Confrontation had been favourable to them. Whatever the actual complexities, these perceptions trumped reality. In many respects, therefore, victory can be construed as an imagined condition.[67]

### *Notes*

1 For an introduction to the problems of assessing victory and defeat in war, see: Col. E. A. de Landmeter, 'What constitutes victory in modern war?', *Militaire Spectator*, 20 March 2018. https://www.militairespectator.nl/thema/strategie/artikel/what-constitutes-victory-modern-war-

2 For a discussion of the origins of Malaysia, see Matthew Jones, 'Creating Malaysia: Singapore Security, the Borneo Territories, and the Contours of British Policy 1961–63', *The Journal of Imperial and Commonwealth History*, Vol. 28, No. 2 (2000): 85–109.

3 'Guidance for Ministers: Malaysia', 12 July 1963 (The National Archive (TNA), CAB 21/5520).

4 House of Lords Debate, 6 February 1963, *Hansards*, Vol. 246, cc. 593–692.

5 The best analyses of Confrontation are: David Easter, *Britain and the Confrontation with Indonesia, 1960–66* (London: Taurus Academic Studies, 2004); Matthew Jones, *Conflict and Confrontation in Southeast Asia, 1961–65: Britain, the United States, Indonesia and the Creation of Malaysia* (Cambridge: Cambridge University Press, 2002); John Subritzky, *Confronting Sukarno: British, American, Australian and New Zealand Diplomacy in the Malaysian-Indonesian Confrontation, 1961–65* (Basingstoke: Macmillan, 2000).

6 For first-hand accounts illustrating the physical challenges of the campaign, see, for example, Christopher Bullock, *Journeys Hazardous: Gurkha Clandestine Operations, Borneo 1965* (Upton Upon Severn: Square One, 2000) and N. C. Smith, *Nothing Short of War: With the Australian Army in Borneo 1962–66* (Victoria: Mostly Unsung Military History, 1999).

7 See, for example: General Sir Walter Walker, 'Borneo', *The British Army Review*, No. 32 (August 1969): 7–15; Nick Van der Bijl's, *Confrontation: The War With Indonesia, 1962–66* (Barnsley: Pen and Sword, 2007); Harold James and Denis Sheil-Small, *The Undeclared War* (London: Leo Cooper, 1971); Tom Pocock, *Fighting General: The Public and Private Campaigns of General Sir Walter Walker* (Bungay: Chaucer Press, 1973); J. A. C. Mackie, *Konfrontasi: The Indonesia-Malaysia Dispute, 1963–66* (London: Oxford University Press, 1974); Major Peter J. Kramers, 'Konfrontasi in Borneo 1962–1966', *Military Review* (November 1990): 64–74.

8 'The Joint Report on the Borneo Campaign', 27 January 1967 (TNA: DEFE 5/172).

9 COS 329/63, 'Requirements to Counter Increasing Indonesian Action Short of Overt Hostilities Against Sarawak and Sabah', September 1963 (TNA: DEFE 5/143).

10 Lt. Col. J. P. Craw, 'Indonesian Military Incursions into West Malaysia and Singapore Between August 1964 and 30th September', *RUSI Journal*, Vol. 111, No. 643 (August 1966), 208–219.

11 Harun Abdul Majid, *Rebellion in Brunei: The 1962 Revolt, Imperialism, Confrontation and Oil* (London: IB Tauris, 2003).

12 COS 329/63, 'Requirements to Counter Increasing Indonesian Action Short of Overt Hostilities Against Sarawak and Sabah', 27 September 1963 (TNA: DEFE 5/143).

13 Secretary of State to Foreign Office, 27 September 1963 (TNA: PREM 11/4870).

14 Ibid.; Washington to New Zealand High Commissioner in London, 'Indonesia/Malaysia', 10 October 1963 (TNA: FO 371/169908).

15 Djakarta to Foreign Office, No. 1732, 10 August 1965 (TNA: FO 371/181454).

16 General Sir Walter Walker, *Fighting On* (London: New Millenium, 1997), 151.

17 House of Commons Debate, 27 November 1967, *Hansards*, Vol. 75, cc. 52–53.

18 Roger Annet, *Drop Zone Borneo: The RAF Campaign 1963–65* (Barnsley: Pen and Sword, 2006).

19 COS 73/65, 'Graduated Military Options in Kalimantan', 8 April 1965 (TNA: DEFE 5/158).

20 Franklin B. Weinstein, *Indonesia Abandons Confrontation: An Inquiry into the Functions of Indonesian Foreign Policy* (Jakarta: Equinox, 2009), 91.

21 See, for example: Emile Simpson, *War From the Ground Up: Twenty-First-Century Combat as Politics* (Oxford: Oxford University Press, 2013); Paul Rosenthall, '"Confrontation": Countering Indonesian Insurgency, 1963–66' in Gregory Fremont Barnes (ed), *A History of Counterinsurgency*, Vol. 2 (Santa Barbara, CA: Prager, 2015), 94–123.

22 Trend to PM, 'Indonesia', 17 October 1963 (TNA: PREM 11/4870).

23 Gen. Sir Walter Walker, 'How Borneo Was Won', *The Round Table* (January 1969): 19.

24 For a discussion of CLARET operations see Raffi Gregorian, 'CLARET Operations and Confrontation, 1964–1966', *Conflict Quarterly* (Winter 1991): 46–72.

25 Walker, *Fighting On*, 167.
26 A Battalion Commander, 'Reflections on Borneo', *The Infantryman*, No. 83 (November 1967): 20.
27 Walker, *Fighting On*, 207.
28 General Sir Walter Walker, 'Brunei and Borneo, 1962–1966: An Efficient Use of Military Force' in Julian Thompson (ed.), *The Imperial War Museum Book of Modern Warfare: British and Commonwealth Forces at War, 1945–2000* (Basingstoke: Pan, 2002), 219; Van der Bijl, *Confrontation*, 241–242.
29 See Brian Bond, *The Pursuit of Victory: From Napoleon to Saddam Hussein* (Oxford: Oxford University Press, 1998).
30 Rex Mortimer, *Indonesian Communism Under Sukarno: Ideology and Politics, 1959–1965* (London: Cornell University Press, 1974), 213.
31 Michael Leifer, *Indonesian Foreign Policy* (London: George Allen and Unwin, 1983), 86, 102.
32 J. D. Legge, *Sukarno: A Political Biography* (London: Allen Lane, 1972), 369.
33 Leifer, *Indonesian Foreign Policy*, 77.
34 See, for example, Malcolm H. Murfett, 'What's in it for us? Rethinking the British Defence Commitments to Singapore and Malaysia from Macmillan to Wilson' in Malcolm Murfett (ed), *Cold War Southeast Asia* (Singapore: Marshall Cavendish, 2012), 251–294.
35 P. L. Pham, *Ending 'East of Suez': The British Decision to Withdraw from Malaysia and Singapore, 1964–68* (Oxford: Oxford University Press, 2010), 6–7.
36 Subritzky, 'Britain, Konfrontasi, and the End of Empire', 209.
37 Leifer, *Indonesian Foreign Policy*, 80–89.
38 Ibid.
39 'The Picture that Made 1966', *Straits Times*, 1 January 1967.
40 Leifer, *Indonesian Foreign Policy*, 80.
41 Mortimer, *Indonesian Communism*, 205.
42 Harold Crouch, *The Army and Politics in Indonesia* (Ithaca: Cornell University Press, 1988), 63–67.
43 See, for example, Anton Aliabas. 'Joint Warfighting: Lessons Learned from the *Dwikora* Operation Experience During the *Konfrontasi*, 1963–66', *Indonesian Quarterly*, 46/4 (2018), 344–362.
44 Mortimer, *Indonesian Communism*, 387–399.
45 For a discussion of the killings, see John Roosa, *Pretext for Mass Murder: The September 30th Movement and Suharto's Coup d'Etat in Indonesia* (Madison, WI: University of Wisconsin, 2006).
46 Katherine E. McGregor, *History in Uniform: Military Ideology and the Construction of Indonesia's Past* (Honolulu: University of Hawaii Press, 2007), 62.
47 'Suharto Disbands Alert Theatre set-up', *Straits Times*, 4 June 1967; McGregor, *History in Uniform*, xviii.
48 Grace Leksana, 'Reconciliation Through History Education: Reconstructing the Social Memory of the 1965–66 Violence in Indonesia' in Birgit Baruchler (ed), *Reconciling Indonesia: Grassroots Agency for Peace* (London: Routledge, 2011), 180–181.
49 See, for example, 'Defence Minister Claims Leftist Book Raids to Pre-empt 'Communist Revenge'', 25 January 2019. https://www.indoleft.org/news/2019-01-25/defense-minister-claims-leftist-book-raids-to-preempt-communist-revenge.html, accessed 16/3/2020>
50 'Greater Malaysia: Report by the Chairman of the Official Committee', (D (61) 62), 24 October 1961 (TNA: CAB 131/26); JP(62)155 (Final), 'Long Term Strategy', 28 January 1963, (TNA: DEFE 7/2155).
51 See Albert Lau, *A Moment of Anguish: Singapore in Malaysia and the Politics of Disengagement* (Singapore: Times Academic Press, 1998).
52 Hughes to Wilson, 13 August 1965 (TNA: PREM 13/431).
53 Pham, *Ending 'East of Suez'*, 21.

54 OPD(O)(65)32, 'South-East Asia and the Far East', 5 May 1965 (TNA: CAB 148/43).
55 Subritzky, *Confronting Sukarno*, 163.
56 Rogers to Wilson, 'Singapore: (OPD (65)123)', 27 August 1965 (TNA: PREM 13/431).
57 Healey to Wilson, 13 August 1965 (TNA: PREM 13/431). See also: Bradley R. Simpson, *Economists with Guns: Authoritarian Development and U.S.-Indonesian Relations, 1960–1968* (Stanford, CA: Stanford University Press, 2008); Matthew Jones, 'U.S. Relations with Indonesia, the Kennedy-Johnson Transition, and the Vietnam Connection, 1963–1965,' *Diplomatic History*, Vol. 26 (Spring 2002): 249–281; H. W. Brands, *The Wages of Globalism: Lyndon Johnson and the Limits of American Power* (Oxford: Oxford University Press, 1995), 155–182.
58 Annex to COS 73/64, 'Indonesian Intentions and Prospects up to the End of 1966' Particularly in Relation to the Malaysian Area', 28 February 1964 (TNA, DEFE 4/149).
59 CINCFE to CDS, SEACOS 201, 2 October 1965 (TNA: DEFE 25/170).
60 See Christopher Tuck, '"Cut the bonds which bind our hands": Deniable Operations during the Confrontation with Indonesia, 1963–1966', *The Journal of Military History*, Vol. 77, No. 2 (April 2013): 599–623.
61 Annex A to COS 176/65, 'Measures to Counter Indonesian Confrontation', 21 October 1965 (TNA: DEFE 13/475).
62 Ibid.
63 Annex A to COS 30/66 'ANZAM Defence Committee Meeting, March 1966–Review of British Defence Policy', 9 March 1966 (TNA: DEFE 5/166).
64 'The Director of Borneo Operations Weekly Assessment of the Threat to East Malaysia and Brunei… 3 Sep–9 Sep 1966', 10 September 1966 (TNA: WO 305/3327).
65 For example: 'British in Borneo Take Lesson from Mao', *The Guardian*, 30 March 1965; 'The Reason Why ….', *The Guardian*, 2 June 1965.
66 House of Commons Debate, 27 November 1967, *Hansards*, Vol. 755, cc. 52–53.
67 Dominic D. P. Johnson and Dominic Tierney, *Failing to Win: Perceptions of Victory and Defeat in International Politics* (Cambridge, MA: Harvard University Press, 2006), 6.

CHAPTER 13

# The Ambiguity of Victory: The Spectrum of 'Winning' in African History

Richard Reid

## Introduction

What is 'winning'? In the context of African history, as elsewhere, our assessment of what constitutes victory depends on two important considerations. The first relates to the actual *objectives* of the respective combatants, and an analysis of whether or not these were achieved. This is (arguably) the more straightforward of the two. The second concerns the timeframe within which we are positioning our analysis. In other words, what often looks like a victory (or a defeat) *in the moment* – that is, in the immediate context of the armed encounter itself – may not actually prove to be such over a more extended timeframe. If we adopt a longer *durée* approach, how much does winning actually look like winning? The longer-term the perspective, of course, the less anything looks very permanent at all; rather, only perennial *issues* remain at the core of the human condition, and solutions come and go, applied with variable but ultimately only temporary success. But this is a somewhat nihilist take on human history. What is proposed in this necessarily concise overview is that in African history, again as elsewhere in the world, two broad forms can be discerned. First, we can identify individual wars, campaigns or battles in which one side achieves a more or less decisive win over another, in conventional terms, and in which that victory is owed to superior armed force, willpower, morale, and occasionally a dash of good fortune. The outcome is a fundamental alteration of the political as well as the military landscape. Second, however – and probably more common in the grand sweep of Africa's past – 'winning' is in fact a complicated and frequently ambiguous business; advantageous positioning was the aim of those involved in wars of attrition, and putative wins were compromised even at the point of victory by a complex weave of circumstances and dynamics enfolding the win itself. In many ways this is as true of the more recent past as in Africa's deeper history.[1]

Throughout, explicit or implicit reference is made to the three core concepts related to winning underpinning this volume, namely *ownership*, *effect* and *ideas*. The first, ownership, concerns what we can essentially define as territorial struggle – whether in the form of civil war, invasion or annexation, and empire building. There are plenty of examples of civil conflict and imperial aggression in African history, although it is worth noting at this juncture that instances of conventional interstate invasion are relatively rare in the postcolonial era, though not unheard of. In terms of effect, the practice of short-term intervention with a view to extending influence, changing regimes and altering local behaviours has a long history in Africa, and continues to the present day in some areas, perhaps most dramatically in the Horn of Africa. Finally, struggles over ideas are often difficult to assess in the deeper past, owing in large part to a paucity of source material, although there are some very visible examples of religious contest involving Islam and, in the Ethiopian context, Christianity. But in the 20th and early 21st centuries, battles over ideas have been very much in evidence, from broadly-defined anticolonial struggle, to national liberation and renewal, to religious terrorism.

## Assessing 'victory' in the deeper past

There are many examples from the precolonial African past of apparently unambiguous wins – the kinds of decisive battles which form the backbone of military history in the global north and which were once learnt by rote at school, if not quite as famous as those found in the illustrious rollcall from Metaurus and Agincourt to Stalingrad and The Bulge. Two cases from the 13th century might suffice in the first instance. The momentous clash between the rebel Sunjata and the Susu king Sumanguru in *circa* 1235 led to the establishment of Mali, one of West Africa's mightiest empires. The Mande epic of Sunjata – the lyrical narratives developed around the life and exploits of the great warrior-king – relates the importance of this critical battle.[2] Later that century, on the other side of the continent, a series of individual engagements between a Shoan army led by Yekuno Amlak and the forces of the ruling Zagwe dynasty between 1268 and 1270 led to the creation of Solomonic Ethiopia – another of Africa's great empires.[3] Both Mali and Ethiopia represent recurrent themes in Africa's military history, namely the righteous insurgency and the glorious restoration of legitimate government through *coup d'etat*. Nonetheless, these also represent something other than a clear, winner-takes-all pivotal moment. In the case of Sunjata, it is just as likely that Mali was forged as the result of a much longer-term war of attrition across the region in the first half of the 13th century; more importantly, for our purposes, Mali struggled – as did all of the West African savannah empires – to impose itself convincingly on local communities, although the durability (to the present day) of the *Epic of Sunjata* undoubtedly reflects a rather stunning victory in itself. And that is an important theme, resonant with the Ethiopian case, too: like Mali, the Solomonic state would experience dramatic changes in fortune in the centuries following Yekuno Amlak's seizure of power, until its overthrow (including the ouster of the last Solomonic emperor, Haile Selassie) in 1974. But the idea of the great triumph of Christian legitimacy in 1270 owed much to the burnishing of historical memory, in this case a text in the form of the *Kebre Negast* ('Glory of the Kings') compiled in the late 13th and early 14th centuries, in which highland Ethiopians awarded themselves a covenant with God and which undergirded national identity (and a marked sense of exceptionalism) for centuries to come.[4]

The battle of Tondibi in 1591 offers another, slightly different, opportunity to question the relevance of thinking in terms of winners and losers. Tondibi, fought in March 1591 on the north-east bend of the Niger River, was long regarded as a pivotal moment in the political and military history of West Africa. Fought between a Moroccan force of cavalrymen and musketeers and the horsemen and infantry of the Songhay empire, then the major power of the western savannah, the clash supposedly brought about a significant shift in the balance of power across the region: Songhay's defeat meant the end of the era of the great savannah empires, dominant for close to a thousand years. After 1584, a series of Moroccan diplomatic missions to Songhay involved the gathering of intelligence, and in 1591, a Moroccan invasion force of some 5,000 – mostly Iberian mercenaries and Muslim converts – set off across the Sahara under the command of Judar Pasha. Around half the army perished on the hazardous journey, but the force that reached the banks of the Niger several weeks later had time to recuperate while Songhay mobilised. When the armies met, on 12 March 1591, the Moroccans were vastly outnumbered. Around 2,500–3,000 Moroccan musketeers confronted much a larger Songhay force, although estimates for the latter vary in the sources from between 12,500 and 18,000 horsemen and between 9,700 and 30,000 infantry. The Songhay army advanced with its infantry concentrated in the centre and with cavalry positioned on the wings, while the Moroccan cavalry attacked from both sides. The Songhay drove a thousand cattle into the centre of the Moroccan force, while the infantry advanced behind it; however, this backfired when the Moroccan guns caused the cattle to stampede back into the Songhay force. The Moroccan infantry now attacked, although they encountered a stubborn arrow assault by elite Songhay archers who had fastened their ankles to their calves in order to render flight impossible. The Moroccans continued to advance, however, and repulsed a further attack by Songhay cavalry. The Songhay force withdrew, and the Moroccans carried the field. The Moroccans devastated much of the surrounding area, inflicted heavy damage on Gao, the Songhay capital, and occupied the cultural and spiritual centre of Timbuktu.

Again, an older interpretation of Tondibi and its aftermath held that the battle represented a total victory for the Moroccans and singularly spelled the end of the Songhay empire, and that the Moroccan deployment of firearms spelled the beginning of the end of cavalry in the savannah. Nonetheless, recent revisionism has suggested that the military impact of the Moroccan invasion – in particular, the impact of the introduction of firearms – has been exaggerated, and that the Moroccan victory was far from total. In fact, the invasion initiated a prolonged war of attrition which endured for the next two decades. The Moroccan commander had expected the occupation to be temporary, and awaited orders to return; the *askiya* or ruler of Songhay offered to acknowledge Moroccan suzerainty and to pay an indemnity, in return for which the Moroccan forces would leave the country. However, the Sultan rejected the offer and reopened hostilities against Songhay. In the course of the 1590s, a further 23,000 men were sent to the region. Overcoming their initial shock at the new weaponry, Songhay horsemen in fact adapted and increasingly held their own against marauding Moroccans, who were also weakened by disease and the obstacles presented by difficult terrain, including mountainous districts and swampy land. It is nonetheless the case that Songhay's power was broken, and the empire was torn apart by the resumption of internecine conflicts, as well as prolonged war with the Moroccans; a series of smaller successor states would emerge in the course of the 17th century. As for the Moroccans, the grand plan to seize control of the gold trade ultimately failed. The invading force, in the face of continued local resistance and disease, was unable to impose itself beyond the Niger bend and in any case gradually lost contact with Marrakesh. While the Sultan attempted to establish a military government over the occupied region, in the years that followed local commanders built up their own power bases and Morocco quickly had little direct control. Abandoning allegiance to Marrakesh, these soldiers and their descendants settled around Timbuktu and constituted an unruly and disruptive new community in the 17th century. Morocco itself, meanwhile, entered a period of decline. It has been argued that Tondibi ultimately destroyed both Morocco and Songhay, leaving the region vulnerable over the longer term to European commercial and, eventually, military dominance in West Africa.

Meanwhile, in terms of weaponry, it is true that firearms swung the battle itself decisively in the Moroccans' favour (although their value probably lay as much in their noise as in their lethal impact); certainly Tondibi represents one of the earliest recorded uses of firearms in precolonial West Africa, and the Songhay army had no effective response to the new technology. However, guns had a limited impact in the savannah environment, and cavalry remained the dominant mode of warfare north of the forest until well into the 19th century – witness the recovery of Songhay horsemen in the years after the battle.[5]

Terrain and distance, and the resultant logistical challenges, meant that the 'win' was not in fact what it first appeared to be, and thus Tondibi prompts us to consider the more common forms of tactic, strategy and outcome in precolonial African warfare. There are, of course, plenty of examples of singular engagements in which antagonists sought a decisive and enduring victory, with varying degrees of success.[6] But 'winning' usually took more nuanced forms, and a long-term theme in African history has been the emergence of armed groups – whether at state- or sub-state level

– which have had little or no chance of achieving outright wins against opponents, in conventional military terms, or even (in some cases) a desire to do so. The aim, rather, was to create what we can broadly call 'unsettlement', and to pose persistent if episodic challenges to the status quo. Raiding war – akin to the 'skulking way of war' characteristic of early modern North America[7] – involved war of attrition, the gradual erosion of the enemy's position, and the struggle to gain advantageous positions, altering the balance of power, in a given area over the long term. In some respects, this reflects the fact that in a historically underpopulated continent, war in Africa was fundamentally concerned with the control of people rather than land *per se*, and thus normative decisive engagements involving the annexation of territory often made little sense when land was plentiful but people were relatively scarce. Control of people, moreover, is less easy to achieve through singular-event 'wins', and is more a question of long-term jostling for comparative advantage, one of the long-term strategies of which being the capture and absorption of resources, including people. Thus limited-objective raiding campaigns were common, and were aimed at the seizure of livestock and slaves, including (and often especially) women and children. Slavery as a practice was in large part sustained by 'winning' campaigns over a sustained period.[8] The slave trade – involving the export of people captured for the purpose – likewise involved strategies of predation targeting neighbouring communities over the long term.[9]

The evidence for the 19th century in Africa suggests that a revolution in military affairs involved a more 'total' kind of warfare, including the professionalisation of military organisation, greater firepower, heightened levels of violence, and an increase in the scale of war.[10] This may have led, in some areas, to more immediately decisive encounters and thus more immediately discernible 'wins'. The emergence and consolidation of the Zulu state exemplifies the pattern.[11] Another example comes from West Africa. In 1804, at the battle of Tabkin Kwotto in present day northern Nigeria, the Jihadist army of Uthman dan Fodio inflicted a decisive defeat on the cavalry and infantry of the ruling Hausa. The victory owed much to archers' fire from an apparently impromptu square, a tactic which was subsequently repeated as what became known as the Sokoto Caliphate consolidated and established itself as the dominant Islamic state in the region.[12] Both the Zulu and the Sokoto states demonstrate the growing capacity for new polities to deploy armed force in increasingly effective ways and achieve battlefield wins which were consolidated by the hegemony of ideas – the rejuvenation of the Islamic faith, in the case of Sokoto; ethnic identity incorporating martial ideology, in the case of the Zulu. In East Africa, the kingdom of Buganda was able to use both land armies and a fleet of war canoes on Lake Victoria to impose itself on an ever-wider area over several decades, though the evidence for decisive individual armed engagements is less forthcoming.[13] Nonetheless, despite – or, perhaps, because of – the intensification of war and the efficacy of armed force across the continent, the military revolution was attended by political turbulence and instability; fortified centres struggled to impose themselves on centrifugal frontier zones which were violent and creative in equal measure, and where dissident groups often successfully denied state-builders outright 'wins'.

## Illusions of victory? The imperial age

One of the factors driving Africa's 19th-century military transformation was a sharp increase in the import of foreign technology, namely firearms. But in the late 19th and early 20th centuries, imperial partition apparently pitted much better organised and better equipped armed forces – those of European colonial armies – against Africans. The 'win' seems unambiguous at the macro level. European or European-led armies crushed African resistance with a supposed finality which guaranteed the stability of the colonial order, and the win was apparently the product of the decisive utilisation of superior military technology, or military technology which was more efficiently deployed.[14] And yet, in almost all cases, anticolonial rebellions in the 1890s and 1900s – while defeated in military terms – were sufficient to prompt some degree of reform within colonial administrations, whose agents granted a range of concessions to formerly insurgent districts and who learnt to tread a little more carefully among 'conquered' subjects. Moreover, later nationalist movements laid claim to those early insurgencies as their progenitors, and though defeated in the short term, those rebels were retroactively positioned as the pioneers of new protest identities, laying the foundations for incipient nationhood.[15] Who had 'won', after all? Imperialism contained the seeds of its own destruction, and so the wins – while their effects endured for a couple of generations – were rather more superficial than they first appeared: the wonderful blocs of colour on imperial maps belied the fragility of colonial territories. While colonial officials and strategists might laud the *pax colonia* and the success of the civilising mission,[16] European empire-builders in Africa were confronted with many of the same problems as indigenous political elites: the struggle to impose themselves on distant peripheries and to transform the societies into which they had intruded. The *pax colonia*, indeed, was something of a myth in many places; overt armed resistance persisted for a number of years in Angola, Somalia, Libya, and Algeria, for example. It was not the last time in the modern era that unambiguous wins were not guaranteed by superior firepower alone.

At the same time, an interesting juxtaposition is offered by Ethiopia – the sole case of an African army successfully resisting European colonial invasion. At the Battle of Adwa in 1896, the Ethiopian army of Emperor Menelik II overwhelmed the Italians, becoming the only African army to defeat an invading European army in the age of the 'Scramble'. Ethiopian independence was secured, and in the years that followed, the 'win' seemed incontestable.[17] Only briefly was Ethiopia occupied by a foreign power – between 1936 and 1941, when the Italians returned, this time with a much clearer military advantage along with (illegal) poison gas – and even then, swathes of Ethiopia remained outside direct Italian administration.[18] And yet, tensions soon arose within Ethiopian political and intellectual culture, and a paradox, too. On the one hand, for some, Ethiopia had merely maintained a semblance of territorial sovereignty, for most of its modernising projects were in fact in the hands of foreigners on whose support Emperor Haile Selassie in particular relied. On the other hand, there were those who wondered whether Adwa had not in fact been the worst of all battles to 'win'. Ethiopia's battlefield victory had led to its chronic underdevelopment, mired in unchanging traditions of political absolutism and ecclesiastical tyranny, and reformist intellectuals gazed enviously across their borders at what they thought were the great projects of economic and political modernity in British colonial Africa.[19] Perhaps more pertinently for our immediate purposes, 20th-century Ethiopia follows the mould of earlier, precolonial African empire-states in struggling to impose itself on its outer-lying provinces – in this case, despite a landmark military victory over an external foe.

Be that as it may, arguably the great struggle of Africa's 20th century was that to decolonise, a cause in which independent Ethiopia was a shining symbol to several generations of African nationalists. The ending of colonial rule proved to be Africans' most dramatic 'win', evidence of the eventual triumph of the principle of national self-determination, one of the most important ideas of the modern age. The story, of course, is a little more complex than it might appear at first glance. While the nationalist narrative emphasises the victory of the African masses led by Great Men, the fact is that only a handful of African states actually became independent through violence, with military wins being a relative rarity. Most decolonisation took place as the result of constitutional process and political compromise between nationalist leaders with neither need for nor means of violent insurgency and metropolitan administrations with changing economic and political priorities.[20] It is also the case that even in those territories in which armed struggle was most ferocious, it was not military might alone – or even at all – which achieved the ultimate goal of independence. The case of Algeria is illustrative. The battle of Algiers (1956–57) awarded the French control of the urban centre in their increasingly brutal war against the insurgency of the *Front Liberation Nationale* (FLN), and indeed the FLN could never match French military power; but the political battle was already being lost, and just five years later de Gaulle withdrew from Algeria.[21] A supposed win – control of the urban environment and the destruction of FLN operations there – was not quite what it first appeared. Something similar could be said for Kenya in the same period, even if Kenya was a slightly more complicated scenario. The British had defeated the Mau Mau insurgency more or less definitively by 1956; yet the forces for change had been unleashed, and the Mau Mau rebels' effort and sacrifice led directly or indirectly (depending on one's viewpoint) to an independent Kenya in 1963 – albeit one which was governed in large part by a moderate group which had

had little to do with the rebellion itself.[22] In both Algeria and Kenya, moral and political victory swiftly emerged out of military defeat or at least extreme setback.

In Rhodesia/Zimbabwe, the combined liberation forces of the Patriotic Front, led by Robert Mugabe and Joshua Nkomo, had made significant advances against the Rhodesian security forces by the late 1970s; but it was ultimately a military stalemate which necessitated a political settlement, facilitated in large part by external parties, Britain included. The 'win' – as defined by the Patriotic Front's avowed objectives, namely the dismantling of the white minority regime and the transfer to majority rule – had been achieved; but it was not solely the result of military action, rather a combination of internal military and external political pressure. As Mugabe himself acknowledged, 'We did not win a military victory … We reached a political settlement … a compromise.'[23] At least Mugabe could consolidate his state (in increasingly brutal fashion) around that 'compromise' in the years to come; elsewhere, however, no 'win' was possible amidst the violence of anticolonial struggle. The withdrawal of the Portuguese from Angola and Mozambique in 1974–75, for example, did not signify a military 'win' in either territory, but rather the beginning of a period of civil war in both.[24]

## The elusiveness of 'winning' in contemporary Africa

Angola and Mozambique were not unusual in the postcolonial context, for civil war has been all too common.[25] But relatively rarely have these been resolved by one side clearly and definitively achieving a 'win' over the other(s). In part, this was because of the fractured nature of civil wars themselves, which often involved multiple actors rooted in particular ethno-regional constituencies. The chaotic fragmentation which attended the liberation of Angola from Portuguese rule, for example – a supposed 'win' for the forces of revolutionary anticolonialism – is captured in a poignant contemporary account, albeit one by a foreign observer.[26] But the larger issue, again, is that in these struggles over ownership of territory, straightforward 'wins' were almost impossible to achieve, owing to a combination of factors: lack of decisive capacity in one group or another; logistical and geographical constraints; the internally fractured nature of territories themselves. In two cases from the 1960s – Nigeria and Congo – internal revolts (Biafra and Katanga respectively) were more or less effectively crushed through a combination of military superiority and external pressure.[27] In both cases, if in slightly different contexts, the 'win' represented the triumph of the idea of the postcolonial nation-state, and the assertion (rendered constitutionally legible in the Charter of the Organisation of African Unity) that colonial-era boundaries were sacrosanct and not to be tampered with. And yet in both cases, internal division continued to blight the nation-building project. Wins, in fact, were defined not by complete military and subsequently bureaucratic control over defined territories in the Weberian sense, but rather by the dominance of the state and the associated benefits by one particular group and the exclusion of all others.[28] Control of the security apparatus was an extension of such dominance. More generally, as in the precolonial era, African states struggled to 'broadcast' power over the territory in its entirety.[29] These could be situations which stopped short of overt 'civil war', though often insurgency on the edges of the territory kept the state in an unsettled state. In this they shared much in common with their precolonial predecessors. Turbulent frontiers on the outer edges of polities were the cause of much instability, although these were also at the root of much that was creative and dynamic about African political culture.[30]

The weakness of the state is demonstrated, albeit *in extremis*, in Sierra Leone and Liberia, where civil wars in the 1980s and 1990s swiftly exposed the incapacity of incumbent regimes.[31] In those cases, the conflicts were brought to an end through a combination of military stalemate, amnesties and compromises, material incentives, and to some extent external intervention – most dramatically in Sierra Leone, where British armed forces intervened directly on the ground. External pressure was key, too, in Sudan, where Africa's longest-running civil war ended with the Comprehensive Peace Agreement in 2005 – not because of any outright military win, which had by then proved impossible for both sides, but because of a recognition that compromises, concessions and incentives were required. South Sudan's independence in 2011 certainly looked like a win for the Sudan People's Liberation Movement (SPLM); but two years later, SPLM rule collapsed into civil war, and longstanding tensions within the nationalist movement came to the fore.[32] No such resolution seems to be in sight for those other internal wars in which no side seems capable – or even, arguably, desirous – of an outright win, notably Somalia, the eastern region of the Democratic Republic of Congo or Libya.

Military wins are more obvious when achieved by those guerrilla organisations which might best be described as 'states in waiting', and which in the late 20th century succeeded in challenging incumbent regimes through a combination of firepower and agendas for social revolution. A series of such insurgencies captured power between the mid-1980s and the mid-1990s, and capital cities rang to the sound of rebel ordnance and political organisations winning control against seemingly insuperable odds:[33] Kampala in 1986 (National Resistance Army), Addis Ababa in 1991 (Ethiopian People's Revolutionary Democratic Front), Asmara in 1991 (Eritrean People's Liberation Front), Kigali in 1994 (Rwandan Patriotic Front). The African National Congress's victory in South Africa in 1994 was not an explicitly 'military' win but nonetheless belongs very much in this category, representing the remarkable triumph of political will.[34] In each of these cases, again, the 'win' seemed unambiguous. The anticolonial struggle had morphed into a second wave of 'national liberation' which had renewed moral power, based on ideas about social reform, behind it. But in each case, too, seeming 'wins' – and their aftermath – were more nuanced than might first appear. It was clear that military victory was not automatically (or at all) attended by lasting political and social settlement. The RPF in Rwanda has arguably been the most successful in this respect, consolidating the military victory which ended the genocide in 1994 through the development of an authoritarian political system in which dissent was essentially deemed illegitimate, indeed potentially treasonous.[35] Eritrea under the EPLF has witnessed something similar.[36] But in all cases, former liberation movements have not been able to establish the total control to which they have aspired. In Ethiopia and Uganda especially, internal violence and various forms of armed dissent have persisted.[37] Military wins, in that respect, have been compromised and ultimately weakened by the absence of a lasting and comprehensive political settlement.

Still, there is more than one way to 'win' – or at least to achieve a period of ascendancy. At this point it is worth introducing the issue of postcolonial interstate war – another example in modern Africa of the attempt to win through the assertion of 'ownership' or the achievement of 'effect'. Conventional interstate war has in fact been relatively rare,[38] and rarely decisive. The Tanzanian invasion of Uganda – involving the Tanzanian armed forces in conjunction with Ugandan armed opposition – in 1979 was a clear 'win' in that it led precisely to the objective as defined: the overthrow of Idi Amin.[39] Mission accomplished, the Tanzanians withdrew a few months later, although for Ugandans a period of brutality and political disorder ensued. And that was only ended by the NRM's 'win' in 1986 – at least according to the Movement's own narrative – thus supposedly ushering in the era of 'peace and development'.[40] The Ethiopia–Somalia wars in 1963–64, and again 1977–78, were examples of conventional national armies clashing.[41] Somalia lost both: it was ill-prepared, outgunned, and diplomatically isolated in 1963; and in 1977, the Soviet Union switched its military support to Ethiopia at the eleventh hour. The net gain for the latter was undoubtedly ascendancy in the region. Yet that hegemony was tested in 1998 when Ethiopia went to war with newly independent Eritrea over a stretch of contested territory around Badme in what looked like a reprise of decades of Eritrea's earlier independence war. The two sides fought a conventional ground and air war, and the result was a stalemate by mid-2000 – though the Ethiopian position was always that they agreed to a ceasefire only for political reasons, and could in fact have gone on to destroy Eritrea militarily.[42] The war reminded Eritrea of the fragility of their earlier great 'win' – independence in 1991 – and in fact the government built a national ethos around the need for seemingly eternal vigilance against a host of enemies. But arguably the 1998–2000 war was ultimately 'won' by Ethiopia, which drew on much larger reserves of international diplomatic influence and global status, whereas Eritrea ended up isolated and weakened, its famed militarism undermined by sagging morale and an illicit exodus of the young seeking to escape seemingly indefinite national service. War alone, it seemed, was not sufficient to consolidate the 'win' – much to the EPLF's annoyance, no doubt; other things mattered, such as diplomatic clout and economic growth.[43] Indeed, the eventual signing of a peace agreement in 2018 formally ending the conflict was among the considerations behind the awarding of the Nobel Peace Prize to Ethiopian PM Abiy Ahmed the following year. Ultimately, in other words, victory for the EPLF in 1991 did not necessarily mean some kind of enduring triumph; rather, in recent years, Ethiopia appears to have won this particular phase in the contest. As for the EPLF's 'win' – the achievement through armed struggle of a sovereign Eritrea – was clear, but as the EPLF discovered, the durability and significance of such wins is entirely contingent upon subsequent developments and, more specifically, policy choices at home and abroad. At the same time, these cases from Eastern Africa remind us of patterns of intervention in African history in order to have some effect or influence in order to 'win'. This has long been common in African history. It is in evidence for the precolonial era, when kings and princes intervened in neighbouring states to achieve favourable regimes and win influence. It was sometimes successful, though such success was usually time limited. In the postcolonial era, proxy war and intervention in the affairs of regional neighbours has been common. The Horn of Africa and the Great Lakes exemplify the pattern.[44]

The question of sustained 'wins' is more complex and in some ways more interesting. There are examples of winning interventions, at least in a limited way; Ethiopia's intervention in Somalia (2006–09) might be considered one such.[45] For others the actual long-term success is more questionable. Eritrean attempts to involve themselves in Somalia, or indeed in Darfur, for example, were met with little success.[46] But then if we consider the notion of unsettlement as one of the patterns of 'winning' in African history over the longer term, then these kinds of interventions make more sense. Even a short-term, limited-range destabilisation might be considered a win in context. Contemporary Africa exhibits a multitude of armed insurgencies, challenging or defying the state in its various manifestations:[47] these often have had no hope of 'winning' in the conventional battlefield sense, but rather the goal has been unsettlement and destabilisation, much as has been the case in the deeper precolonial past. In other words, unsettlement and destabilisation have not simply been the *consequence* of armed action; they are the *primary objective*, because if one can disturb, disrupt, unsettle, then one has 'won', according to a certain set of criteria. Often the aim of unsettlement has been effect or influence; at other times the spread of ideas, and the permanent influencing of belief or behaviour and the planting of ideologies.

Terrorism is an interesting aspect of this, and there can be little doubt that terrorism has been on the increase in Africa (as elsewhere) over the past twenty years or so.[48] Terrorists may not even envisage, realistically, some kind of total 'win'; but again the aim of unsettlement and the ideological and (more frequently) religious mission are key to understanding what might, in the eyes of the terrorists themselves, constitute a 'win'. For some, of course, terrorism is a distinctly modern phenomenon, and indeed is the product of modernity. It is ostensibly fought, often, by people who *cannot* win, at least by other means. But in many ways, in Africa, it is only the latest manifestation of much older tactics of destabilisation and unsettlement, and thus the 'winning' element must be understood in that context. Terrorism – profound unsettlement, political, emotional and psychological – is both means and end.

## Conclusion

What general principles and broad patterns can we discern in assessing the meaning of 'winning' in African history? It is, of course, very difficult to generalise: it will be obvious enough that Africa is vast, its history of military engagement diverse, and its internal regional histories equally distinct. At the same time, one model – decisive battlefield wins – does not preclude the other, namely ongoing wars of attrition and unsettlement which do not involve clear, long-term 'wins'. Often, the two have existed in parallel.

Perhaps the first thing to say is that winning is historically defined by the actors themselves, often refashioned to suit political purposes, and often sometime after the event. Winning involves historical memory, in other words, and public celebration. Of course, the same is true of 'losing': defeat can be harnessed and be converted into a 'win' for political causes; burning resentments harnessed; spirit of national or ethnic rejuvenation can fuel causes for many years.[49] And the cause itself is a form of 'winning' – regardless of the actual outcome. The 'winning' bit is the 'taking part', as it were; cultures of war and militarism which exist for their own sake. Indeed, it is worth considering the notion of 'losing' more broadly – for if we accept that winning is often ambiguous and incomplete, then logically so is 'losing'. The French departure from Africa by the early 1960s did not mark the end of French influence in the continent, for example; the predations of *Françafrique* only increased in the years to come. Sarkozy could still make a speech in 2007 declaring that Africa remained intrinsically backward and outside history.[50] British commentator-turned-politician Boris Johnson could likewise assert that the problem was not that the British once ruled swaths of Africa; the problem was they had left.[51] So winning on the ground did not put paid to deep-rooted, racialised perceptions of the continent. Likewise, more broadly, the victory of the forces of progressive, anticolonial liberation across the continent between the 1960s and the 1980s – and the supposed 'loss' experienced by colonial empires – did not mean the end of external influence, for neo-colonialism, in various guises, persisted.[52] At the same time, military 'liberation' did not represent an unalloyed 'win' for all citizens.

In modern African history it is relatively rare that a purely military, battlefield win brings about enduring political victory and settlement. Militaries, no more than the states which underpin them, rarely have had the capacity to impose themselves decisively. This may be because African history supposedly lacks a Westphalian moment, or its equivalent; there has been no '1918' or '1945'. In Africa, winning is not often what it seems – it involves compromise, negotiation, and a measure, perhaps, of exhaustion, and then particular parties can claim to have 'won'. Much depends, too, on how long a *durée* we adopt. Alternately short- and long-term perspectives can produce very different interpretations of 'winning' – in other words, what looks like a win in the immediate context can look rather less so a few years (or decades) down the track. The aftermath of the fall of Muammar Qaddafi in Libya in 2011 illustrates the point. The history of armed contest in Africa is characterised by grey areas and zones of transition between polities, and these have made it logistically difficult to 'win' in outright or decisive terms. States and empires alike have long faced the challenge of 'broadcasting' their power across territorial spaces.

On the one hand, this has rendered African political and military culture dynamic and creative – and this has often been misunderstood by outside observers who have long viewed the continent through a racialised lens. On the other hand, by the same token it makes 'winning' look very different from what we might term conventional military history. Control of people versus control of land, and consequently unsettlement and destabilisation have long been the pre-eminent form of 'winning'. As a result, 'winning' is nuanced and shaded. It happens by degrees. Compromise and settlement have often been key. Dissent is perennial, and inevitable, and potential if not always kinetic. Winning is usually achieved over the long term, and is the gradual outcome of a combination of military and non-military elements.

## *Notes*

1 The history of African war is an expanding field but remains, in relative terms, minority sport. A sample of key introductory literature would include: J. Lamphear (ed), *African Military History* (Aldershot: Ashgate Publishing Company, 2007); B. A. Ogot (ed), *War and Society in Africa* (London: F. Cass, 1972); R. J. Reid, *Warfare in African History* (New York: Cambridge University Press, 2012); R. J. Reid, *War in Pre-Colonial Eastern Africa: the patterns and meanings of state-level conflict in the nineteenth century* (Oxford: British Institute in Eastern Africa, 2007); W. Reno, *Warfare in Independent Africa* (New York: Cambridge University Press, 2011); R. S. Smith, *Warfare and Diplomacy in Pre-Colonial West Africa* (London: Currey, 2nd ed., 1989); J. K. Thornton, *Warfare in Atlantic Africa, 1500–1800* (London: Psychology Press, 1999); B. Vandervort, *Wars of Imperial Conquest in Africa, 1830–1914* (Bloomington IN: Routledge, 1998); P. D. Williams, *War and Conflict in Africa* (Cambridge: John Wiley and Sons, 2nd ed., 2016).

2 For example, Bamba Suso and Banna Kanute, *Sunjata: Gambian Versions of the Mande Epic* (London: Penguin, 1999).

3 Sergew Habte Selassie, *Ancient and Medieval Ethiopian History to 1270* (Addis Ababa: United Printers, 1972); Taddesse Tamrat, 'Ethiopia, the Red Sea and the Horn', in Roland Oliver (ed), *The Cambridge History of Africa. Vol 3, from c.1050 to c.1600* (Cambridge: Cambridge University Press, 1977).

4 M. Brooks (ed & tr), *A Modern Translation of the* Kebre Negast *(The Glory of the Kings)* (Lawrenceville NJ: Red Sea Press, 1995); see also E. A. Wallis Budge, *A History of Ethiopia, Nubia, and Abyssinia* (London: Methuen, 1928), Vol 1.

5 N. Levtzion, 'North-West Africa: from the Maghrib to the fringes of the forest', in R. Gray (ed), *The Cambridge History of Africa. Vol 4: from c.1600 to c.1790* (Cambridge: Cambridge University

Press, 1975); L. Kaba, 'Archers, musketeers, and mosquitoes: the Moroccan invasion of the Sudan and the Songhay resistance (1591–1612)', *Journal of African History*, Vol. 22, No. 4 (1981): 457–475.

6 Smith, *Warfare and Diplomacy*, is useful on specific battles and sieges in the West African context, for example.

7 Patrick M. Malone, *The Skulking Way of War: technology and tactics among the New England Indians* (Lanham: Madison Books, 1991).

8 Reid, *Warfare in African History*, ch. 1; John Lamphear, 'Sub-Saharan African warfare', in Jeremy Black (ed), *War in the Modern World since 1815* (London & New York: Psychology Press, 2003).

9 Paul Lovejoy, *Transformations in Slavery* (Cambridge: Cambridge University Press, 2011).

10 Richard Reid, 'The Fragile Revolution: rethinking war and development in Africa's violent nineteenth century', in E. Akeampong, R. H. Bates, N. Nunn and J. Robinson (eds), *Africa's Development in Historical Perspective* (Cambridge: Cambridge University Press, 2014).

11 Donald R. Morris, *The Washing of the Spears: the rise and fall of the Zulu nation* (London: Jonathan Cape, 1965); Jeff Guy, *The Destruction of the Zulu Kingdom* (London: Longman, 1979); and for a recent overview, John Wright, 'Turbulent Times: political transformations in the north and east, 1760s–1830s', in C. Hamilton, B. K. Mbenga, and R. Ross (eds), *The Cambridge History of South Africa, Vol 1: from early times to 1885* (Cambridge: Cambridge University Press, 2009).

12 J. P. Smaldone, *Warfare in the Sokoto Caliphate: historical and sociological perspectives* (Cambridge: Cambridge University Press, 1977).

13 Richard J. Reid, *Political Power in Pre-Colonial Buganda: economy, society and warfare in the nineteenth century* (Oxford: James Currey, 2002), chs. 10 and 11.

14 Vandervort, *Wars of Imperial Conquest*; and a robust assertion of the argument, Daniel Headrick, *Tools of Empire: technology and European imperialism in the nineteenth century* (Oxford: Oxford University Press, 1981).

15 T. O. Ranger, 'Connexions between primary resistance movements and modern mass nationalism, I & II', *Journal of African History*, Vol 9, Nos. 3 and 4 (1968): 437–453.

16 Frederick Lugard, *The Dual Mandate in British Tropical Africa* (Edinburgh & London: Blackwood and Sons, 1922).

17 See the contemporary assessment in G. F. -H. Berkeley, *The Campaign of Adowa and the Rise of Menelik* (London: A. Constable, 1902); and Paulos Milkias & Getachew Metaferia (eds), *The Battle of Adwa: reflections on Ethiopia's historic victory against European colonialism* (New York: Algora Publishing, 2005).

18 See for example Anthony Mockler, *Haile Selassie's War: the Italian-Ethiopian campaign, 1935–1941* (New York: Random House, 1984).

19 Sara Marzagora, 'History in twentieth-century Ethiopia: the "Great Tradition" and the counter-histories of national failure', *Journal of African History*, 58:3 (2017).

20 For useful overviews, see J. D. Hargreaves, *Decolonization in Africa* (London: Longman, 1996); W. R. Louis and R. Robinson, 'The Imperialism of Decolonization', *Journal of Imperial and Commonwealth History*, Vol. 22 (1994): 462–511; A. G. Hopkins, 'Rethinking Decolonization', *Past and Present*, Vol. 200, No. 1 (2008): 211–247.

21 The classic account in English is Alistair Horne, *A Savage War of Peace: Algeria 1954–1962* (London: MacMillan, 1977); and see also the classic film, *The Battle of Algiers* (1966), directed by Gillo Pontecorvo.

22 D. Branch, *Defeating Mau Mau, Creating Kenya: Counterinsurgency, Civil War, and Decolonisation* (Cambridge: Cambridge University Press, 2009); Huw Bennett, *Fighting the Mau Mau: the British Army and Counter-Insurgency in the Kenya Emergency* (Cambridge: Cambridge University Press, 2013).

23 Interview with Robert Mugabe, *New York Times*, 28 April 1980. More broadly, see D. Martin and P. Johnson, *The Struggle for Zimbabwe: the Chimurenga war* (London: Monthly Review Press, 1981); T. O. Ranger, *Peasant Consciousness and Guerrilla War in Zimbabwe* (London: James Currey, 1985).

24 Reno, *Warfare in Independent Africa*, chap. 2.

25 There is a growing body of work on civil war in Africa in the contemporary era. See for example T.M. Ali & R.O. Matthews (eds.), *Civil Wars in Africa: roots and resolution* (Montreal & Kingston: McGill-Queen's University Press, 1999); and two volumes under the stewardship of A. Nhema and P. Tiyembe Zeleza (eds), *The Roots of African Conflicts: the causes and costs* (Oxford: OSSREA and James Currey, 2008) & *The Resolution of African Conflicts: the management of conflict resolution and post-conflict reconstruction* (Oxford: James Currey, 2008).

26 Ryzsard Kapuscinski, *Another Day of Life* (London: Penguin, 1987; 1st pub. in Polish, 1976).

27 Michael Gould, *The Struggle for Modern Nigeria: the Biafran War, 1967–1970* (London & New York: Bloomsbury, 2012); Georges Nzongola-Ntalaja, *The Congo from Leopold to Kabila: a people's history* (London & New York: Zed Books, 2002), chs. 2 and 3.

28 Frederick Cooper, *Africa Since 1940: the past of the present* (Cambridge: Cambridge University Press, 2002); Jean-Francois Bayart, *The State in Africa: the politics of the belly* (Cambridge: Polity Press, 2009).

29 Jeffrey Herbst, *States and Power in Africa: comparative lessons in authority and control* (Princeton, NJ: Princeton University Press, 2000).

30 See for example Igor Kopytoff (ed), *The African Frontier: the reproduction of traditional African societies* (Bloomington & Indianapolis: Indiana University Press, 1987).

31 Paul Richards, *Fighting for the Rain Forest: war, youth, and resources in Sierra Leone* (Oxford: International African Institute, 1996); Stephen Ellis, *The Mask of Anarchy: the destruction of Liberia and the religious dimension of an African civil war* (London: C. Hurst and Co., 1999).

32 Edward Thomas, *South Sudan: a slow liberation* (London: Zed Books Ltd., 2015); Douglas H. Johnson, *The Root Causes of Sudan's Civil Wars: old wars and new wars* (Woodbridge: Boydell & Brewer, 2016).

33 See for example Christopher Clapham (ed.), *African Guerrillas* (Oxford: James Currey, 1998).

34 The ANC's military wing, *Umkhonto we Sizwe* (MK) was not regarded as a particularly decisive force in the toppling of apartheid. See Nelson Mandela's own account of MK in *Long Walk to Freedom* (London: Abacus, 1995).

35 See 'Rwanda under the RPF: assessing twenty years of post-conflict governance', special issue of the *Journal of Eastern African Studies*, 8:2 (2014).

36 Martin Plaut, *Understanding Eritrea* (London: C Hurst and Co Publishers Ltd., 2016).

37 Gérard Prunier & Éloi Ficquet (eds), *Understanding Contemporary Ethiopia: monarchy, revolution, and the legacy of Meles Zenawi* (London: Oxford University Press, 2015); Aili Mari Tripp, *Museveni's Uganda: paradoxes of power in a hybrid regime* (Boulder, CO: Lynne Rienner Publishers, 2010).

38 Armies in Africa have historically been inward-looking and concerned more with internal control rather than external adventures: J. M. Lee, *African Armies and Civil Order* (London: Frederick A. Praeger, 1969); Samuel Decalo, *Coups and Army Rule in Africa: motivations and constraints* (New Haven & London: Yale University Press, 1990).

39 T. Avirgan & M. Honey, *War in Uganda: the legacy of Idi Amin* (London & Westport: L. Hill, 1982).

40 The spirit of the moment is captured in H. B. Hansen and M. Twaddle (eds), *Uganda Now: between decay and development* (London: J. Currey, 1988).

41 Gebru Tareke, 'The Ethiopia-Somalia war of 1977 revisited', *International Journal of African Historical Studies*, Vol. 33, No. 3 (2000): 635–667.

42 Kjetil Tronvoll and Tekeste Negash, *Brothers at War: making sense of the Eritrean-Ethiopian war* (Oxford: J. Currey, 2000); Dominique Jacquin-Berdal and Martin Plaut (eds), *Unfinished Business: Ethiopia and Eritrea at War* (Lawrenceville, NJ: Red Sea Press, 2005).

43 See my own account of this in *Shallow Graves: a memoir of the Ethiopia-Eritrea war* (London: Hurst, 2020).

44 Gerard Prunier, 'Rebel movements and proxy warfare: Uganda, Sudan, and the Congo (1986–1999)', *African Affairs*, Vol. 103, No. 412 (2004); Gerard Prunier, *Africa's World War: Congo, the Rwandan genocide, and the making of a continental catastrophe* (Oxford: Oxford University Press, 2009); F. Reyntjens, *The Great African War: Congo and regional geopolitics, 1996-2006* (Cambridge: Cambridge University Press, 2009); Richard Reid (ed), *Eritrea's External Relations: understanding its regional role and foreign policy* (London: Chatham House, 2009).

45 Ken Menkhaus, 'The crisis in Somalia: tragedy in five acts', *African Affairs*, Vol. 106 (April 2007): 357–290.

46 International Crisis Group Africa Report No. 163, *Eritrea: The Siege State*, Nairobi & Brussels, 21 September 2010.

47 Morten Boas and Kevin C. Dunn (eds), *African Guerrillas: raging against the machine* (Boulder, CO: Lynne Rienner Publishers, 2007).

48 Christopher Clapham, 'Terrorism in Africa: problems of definition, history and development', *South African Journal of International Affairs*, Vol. 10, No. 2 (2003): 13–28; Princeton N. Lyman and F. Stephen Morrison, 'The terrorist threat in Africa', *Foreign Affairs*, Vol. 83, No. 1 (2004); Alex de Waal (ed), *Islamism and its Enemies in the Horn of Africa* (London, 2004); Robert I. Rotberg (ed), *Battling Terrorism in the Horn of Africa* (Washington DC: Brookings Institution Press, 2005); Richard Reid, 'Terrorism in African History', in R. English (ed), *The Cambridge History of Terrorism* (Cambridge: Cambridge University Press, forthcoming).

49 See for example Wolfgang Schivelbusch, *The Culture of Defeat: on national traumas, mourning and recovery* (London: Granta, 2003).

50 'Africans still seething over Sarkozy speech', Reuters, 5 September 2007 (https://uk.reuters.com/article/uk-africa-sarkozy/africans-still-seething-over-sarkozy-speech-idUKL0513034620070905)

51 Boris Johnson, 'Cancel the guilt trip', *The Spectator*, 2 February 2002.

52 Kwame Nkrumah, *Neo-Colonialism: the last stage of imperialism* (London: International Publishers, 1965).

CHAPTER 14

# The Provisional IRA and the Elusive Concept of 'Winning' since 1969[1]

Aaron Edwards

'A military stalemate had developed where each side had, not necessarily brought the other to the point of exhaustion – and that's a relative term – but had explored almost all means of conscionable confrontation.' – Author interview with Danny Morrison, former Director of Publicity for Sinn Féin, 23 November 2010

'The IRA was beaten. I've absolutely no doubt about that.' – Author interview with Anthony McIntyre, former member of the Provisional IRA, 20 October 2019

## Introduction

The two extracts headlining this chapter are from interviews undertaken almost a decade apart. Danny Morrison and Anthony McIntyre were prominent members of the Provisional Irish Republican Movement from its formation in 1969. Both were well-acquainted with one another during the Provisional IRA's (hereafter the IRA) 'armed struggle' and both served prison sentences for the republican cause. Once the IRA announced its first prolonged ceasefire in 1994, both men went their separate ways in terms of their analysis of the group's so-called 'peace strategy'.[2] Even today they remain fundamentally at odds with the interpretation of why the IRA called a halt to its violent campaign, which eventually ended in 2005. They also gave divergent answers to the question I asked them during interview: Did the IRA win or lose? Nowadays it is fashionable for commentators to seek refuge in the postmodernist mindset that winning and losing are meaningless concepts. Nevertheless, in Strategic Studies at least, they retain considerable currency. As Prussian general Carl von Clausewitz reminds us in his magisterial *On War*, war is 'an act of force to compel our enemy to do our will'.[3] In war, therefore, the aim is to disarm the enemy, which Clausewitz likened to a contest between a pair of wrestlers who each try, 'through physical force to compel the other to do his will.'[4] The competitor who 'throws' their opponent is generally regarded as the winner.

Significantly, for Clausewitz, it was the engagement between the belligerents that remained 'the only means of destroying an enemy's forces', which would hasten their defeat. For Clausewitz that outcome was essential, though it could be brought about directly – in the form of total destruction of the enemy – or indirectly, such as the capture of territory or the outmanoeuvring of one's enemy.[5] As Clausewitz argued:

> Not every war need be fought until one side collapses. When the motives and tensions of war are slight we can imagine that the very faintest prospect of defeat might be enough to cause one side to yield. If from the very start the other side feels that this is probable, it will obviously concentrate on bringing about this probability rather than take the long way round and totally defeat the enemy.[6]

The important point to note is that one's enemy can only be said to be truly defeated if their will to fight has been completely eroded. The outcome of the engagement, therefore, plays a far-reaching role in creating the kind of peace that follows war. Drawing on primary source material – including recently released archival material in the UK and Ireland – and on first-hand interviews with former members of the British security forces and their opponents in the IRA, this chapter applies Clausewitz's concept of winning to the Northern Ireland case-study.

## The concept of winning in small wars

It was fashionable during the recent, long years of western military intervention in Afghanistan and Iraq to wheel out the case of Northern Ireland to demonstrate how Britain had been a market leader in the dark arts of counterinsurgency. Observing the Northern Irish case against the backdrop of other post-war campaigns like Malaya, Kenya, Cyprus and Aden, British doctrine writers noted a paradox: 'It is curious to record that despite the extensive experience gained by the British Army in counterinsurgency during the 20th century, relatively little has been recorded as official doctrine in military publications'.[7] Had they more closely scrutinised the cases, however, they would also have noted a clear distinction. While these colonial hotspots were 'predicated on the presumption of Britain's military defeat. Britain was getting out. Therefore, to an extent, the other side had won'.[8] If we accept Hew Strachan's elementary, yet fundamentally important, point we can, perhaps, understand a certain reticence amongst contemporaries for wishing to close this final chapter on imperial rule. In Northern Ireland the political circumstances were different. Britain was not so much 'getting out' as 'getting in' and staying in until violence subsided. In this regard, Northern Ireland offered the perfect antidote to how failure in counterinsurgency campaigns might be avoided.

Nevertheless, in returning to the Northern Ireland case, doctrine writers putting together the new Army Field Manual on Counterinsurgency reached some surprising conclusions:

> The only real indication that the experience gained in Northern Ireland had been reflected in army units, was by the training of those units before they deployed to Northern Ireland for duty. The training was up to date, effective and extremely flexible. Hostile tactics were studied, changes in training promulgated quickly, and troops informed immediately. None of this seems to have been reflected in any official doctrine, although a large number of Aide Memoires were produced by the General Staff over the years.[9]

This view was also reflected in the thinking of the serving Chief of the General Staff (2006–9), General Sir Richard Dannatt.[10] While it is true that lessons were not generally captured in official doctrine publications, they were disseminated in training provided by the highly capable Northern Ireland Training and Advisory Team (NITAT) and also in the folk wisdom carried forward by the regimental system.[11] As the British Army discovered in relation to its interventions in Iraq and Afghanistan in the first decade of the 21st century, even if doctrine had been available, there are serious questions over whether it would have been read by the officer corps or left to gather dust in military libraries.[12] However, the perceived lack of doctrine capturing Tactics, Techniques and Procedures (TTPs) obscured more than it revealed for no attempt was made to address broader strategic questions, including the one that remained on the minds of the belligerents living in Northern Ireland long after the withdrawal of British troops: who exactly won the war?

Although it is tempting to ask a different question of a war that has long been marked by its irregular character, we must nonetheless persevere. As strategic thinker Colin S. Gray has observed, 'Irregular warfare is an old, old story, and so are the methods applied to wage it, on both sides ... Irregular warfare ... is about political power: who gets it, and as a rather secondary matter, what to do with it. That may seem a banal point, but really it is not'.[13] Gray makes clear that we should resist the temptation to draw a distinction between regular and irregular war in favour of calling them what they are: wars. What is important to note, however, is that the shape of war is dictated by shifting strategic circumstances. Distinctions matter and clearly it is not without utility that concepts such as asymmetry creep into our analysis of the character of war.[14] The point is that, while there is 'much more to war than warfare', warfare is warfare and the core competency of soldiers is 'skill in inflicting pain, killing people, and breaking things'.[15] Regardless of whether we prefer to see irregular warfare as possessing a 'sharply distinctive character', it is still war.[16] Michael Howard concurs with Gray's view, adding that war remains a useful concept, which should be expanded to 'embrace all armed conflict between political entities, whether or not these are or claim to be recognised as sovereign states'.[17] However, Howard adds one significant qualification: For military force to be decisive in such circumstances, two conditions must be satisfied:

> First, the defeated people must accept the fact of defeat and realise that there is no chance of reversing the verdict in the foreseeable future, whether by military revival, skilful diplomacy or international propaganda. Second, they must become reconciled to their defeat by being treated,

> sooner or later, as partners in operating the new international order. Honour must be satisfied, unless the defeated peoples are to be massacred or reduced to perpetual slavery.[18]

There is certainly evidence that the IRA were preparing to escalate their violent campaign in the mid to late 1980s. However, it was through a mixture of covert action and secret intelligence that the British state was able to deter and coerce the terrorists while reminding them of the 'attritional consequences of their violence'.[19]

Thus, against this unfavourable strategic backdrop, the idea of an 'honourable peace' was first articulated by senior IRA military commander Martin McGuinness in 1991.[20] McGuinness recognised, for the first time, the difficulty facing his organisation in achieving its strategic objectives by escalating its campaign.[21] Instead the IRA opted to call a prolonged ceasefire. In the wake of their 1994 ceasefire, the IRA's political associates in Sinn Féin began to sell a narrative to their supporters that a 'military stalemate' remained extant.[22] The British government continued to insist on its strategic goal of disarming the IRA for the next decade, something which eventually happened in September 2005. In response, the Army ran down its operations, with one official 'doctrine note' confusingly observing shortly afterwards how security forces 'do not 'win' insurgency campaigns militarily; at best they can contain or suppress the level of violence and achieve a successful end-state'.[23] The doctrine note then insisted in the next breath that the 'insurgent form of PIRA was defeated'.[24] Some scholars have seized on this doctrine note, like Sinn Féin did at the time it was released, to justify its narrative of 'military stalemate'.[25] However, by doing so, they decoupled the British government's political objectives throughout the war – the defeat of terrorism and the return of Northern Ireland to 'normality' – from the peaceful outcome which followed. Paul Dixon has observed how this was the result of 'pragmatic realism and even deception in driving the peace process forward which involved uncomfortable compromises with paramilitaries'.[26] However, a more persuasive analysis would resist the urge to obfuscate Britain's clear strategic goals with the IRA's decision to abandon its own. As Alex S. Wilner has argued in his work, 'Success in counter-terrorism requires two things: diminishing a group's capability to organise acts of violence and undermining a group's motivation to employ violence'.[27] As the remainder of this chapter will suggest, the Northern Ireland case should be seen as a counter-terrorism success.

## The Provisional IRA: from armed struggle to ceasefire

At its heart, the Northern Ireland conflict is about ethnicity and national identity, fought mostly between a majority Protestant community who wish to maintain the union with Great Britain and a minority Catholic community who wish to break that link. Known colloquially as the 'Troubles', the major armed hostilities which broke out in the late 1960s and persisted until the 1990s was one of the longest-running ethnic conflicts in modern Europe. Almost 4,000 people lost their lives and over 40,000 were injured in a population of approximately 1.5 million people. The re-emergence of the conflict in the 1960s came after the rise of a popular protest movement that took to the streets to demand a redistribution of political, social and economic rights it claimed were denied to the minority Catholic community. Violence soon followed in August 1969, overwhelming the local police force, the Royal Ulster Constabulary (RUC), and precipitating a request by the local Unionist government to the British government in London to deploy the Army onto the streets.[28] Within a few months, troops and police came under sustained attack from the more militant fringes of both communities. The most significant armed challenge to emerge was the IRA, which demanded the immediate withdrawal of the Army and the removal of the Unionist government.

Although in existence in one form or another since the Easter Rising of 1916, the IRA was proscribed by the southern Irish government in the 1920s and remained a subversive organisation for the next half century. The IRA carried out several notable armed campaigns, from bombings in England in 1939 through to its so-called border offensive, 'Operation Harvest', in 1956–62. The outbreak of intercommunal antagonism in 1969 saw the IRA reborn. However, the organisation remained divided on the question of taking a more offensive posture and eventually split over the issue in December 1969 between Official and Provisional wings.[29] The Provisionals were to become a highly sophisticated guerrilla organisation within a few months. One of its most active members was Tommy Gorman. He explained how the grouping was structured in the city in the early 1970s:

> It was based upon a British Army thing. You had companies, battalions and brigades. Companies were based on geographical areas … When Seán Mac Stíofáin [the Provisional IRA's first Chief of Staff], God love him, paid us a visit in 1972, everybody in the First Battalion got together in a school hall to meet the commander in chief. And I looked round and thought 'I don't know half of these people.' And that's par for the course. You have 300 people in a company and maybe five would be active, you know. You call upon the others to watch out and do something, keep their eyes open for targets, you know. But it was a minority of people who were active. The other people were … a lot of deadwood.[30]

The Provisional Army Council (PAC), the IRA's overall leadership, provided strategic direction for the organisation while its General Headquarters Staff (GHQ) provided operational guidance, though neither micro-managed the campaign.[31] It was not until the formation of the IRA's 'Northern Command' that the organisation stepped up its offensive. As Danny Morrison, recalled:

> I suppose it came to a head in 1976 when the then leadership made some concessions and agreed to the formation of a Northern Command in late '76, which meant that a lot of autonomy was given away from GHQ towards the North, which tended to be people based in the North. And they set out to reorganise the IRA. And that was a story for a very long, long time[32]

Northern Command soon 'penetrated every crevice of the IRA',[33] with its key operational units coalescing around prominent figureheads in Derry City, West Belfast and South Armagh. The latter area soon acquired a reputation as the most formidable component of the IRA.

Led by half a dozen men on a 'Brigade Staff', South Armagh PIRA was a close-knit and highly experienced group of terrorists with almost 200 suspected members, many of whom were concentrated in the townlands of Crossmaglen, Camlough and Drumintee, situated adjacent to the Irish border. The operational effectiveness of South Armagh PIRA was enabled by its organisational culture, which promoted and implemented innovation.[34] As the biggest IRA unit outside Belfast, South Armagh PIRA presented the single greatest security challenge to the security forces during the Troubles. Lieutenant Colonel Ian Lowis was Commanding Officer of 1st Battalion of the King's Own Scottish Borderers on their emergency tour of South Armagh in 1985. In his Post Operational Tour Report filed in July 1985, he noted how:

> In many ways it would be surprising if we had any positive success to claim. The South Armagh terrorist has 15 years' experience behind him. He operates on a different timescale to a Roulement Battalion and measures his success over a much longer period. The Roulement Battalion's aims tend to be short-term and incompatible with those of the enemy. Because of this it is difficult for roulement units to operate in a way that has not been tried before. Unfortunately, short-term measures to counter the South Armagh terrorists do not work. The dearth of intelligence precludes predictive operations. Speculative operations are unlikely to be successful against the PIRA, yet this remains the only method of operation when no useful intelligence is available. Sadly, this situation is unlikely to improve. Until a political solution is found which prevents the terrorist from using the Republic of Ireland as a sanctuary, all that can be done is to attempt to contain the level of the PIRA's activity at a tempo acceptable to the Government.[35]

This was to be an incredibly prescient appreciation of the ongoing difficulty of meeting the IRA challenge in the absence of a decisive victory by the British state.

By the end of the decade, Lieutenant Colonel John Sutherell, the Commanding Officer of the 1st Battalion of the Royal Anglian Regiment, reported no significant change in the threat posed by South Armagh PIRA. As 'the most dangerous and effective in the Provisional IRA … They are experienced, determined, clever and well-armed; an enemy to be respected', he told his subordinates at the end of their operational tour in 1990.[36] Apart from a culture of innovation, South Armagh PIRA's success was attributable to the continuity of its leadership and the IRA's gestapo-like internal policing, which one academic has called the 'less public face of community control and discipline' of the organisation.[37] This made South Armagh PIRA almost impenetrable to British Intelligence until the late 1990s.[38] As one former British agent who did successfully infiltrate the fringes of the group admitted, 'These guys were tight and ruthless. Two ingredients which made it nearly impossible to compromise them.'[39] Even former members of the RUC Special Branch have acknowledged the IRA's organisational prowess as a terrorist group, with one referring to the group as 'an intelligent organisation'.[40]

Despite these internal organisational advantages, the IRA was still constrained on a day-to-day basis by the British state. By 1990, leading members of the group were 'aware that the armed struggle was ineffective' and had authorised some members to discuss the possibility of a ceasefire.[41] We now know that Sinn Féin leader Gerry Adams had been in close liaison with Father Alec Reid, a Redemptionist priest in West Belfast, who carried communiques across the ethnic divide at this time.[42] A similar process was set in motion under Martin McGuinness in Derry, which saw a de-escalation in British Army and IRA activity in the city.[43] In addition to these on-the-ground talks with their opponents in loyalist paramilitary organisations and the British Army, Adams and McGuinness also entered into a high-level dialogue with the British government via a 'back channel' of communication serviced by Derry businessman Brendan Duddy. Duddy 'was at the heart of intensive contact between the two parties in the early to mid-1970s and acted as intermediary again in negotiations aimed at ending the 1980 and 1981 Republican hunger strikes in which ten IRA and INLA prisoners died'.[44] As his influence grew, Duddy became something of an unofficial advisor on strategy for the IRA, even going as far as to facilitate meetings between Martin McGuinness and the third most senior MI6 officer at the time, Michael Oatley.[45] The outcome of this dialogue was soon played out in public discourse choreographed by Duddy. In November 1990, Secretary of State Peter Brooke gave a speech in which he stated that the British government had 'no selfish or strategic or economic interest in Northern Ireland: our role is to help, enable and encourage. Britain's purpose ... is not to occupy, oppress or exploit but to ensure democratic debate and free democratic choice. Partition is an acknowledgement of reality, not an assertion of national self-interest'.[46] Martin McGuinness responded by telling the annual Sinn Féin *Ard Fheis* (party conference) a few months later in February 1991:

> Republicans want peace. We want *an honourable peace*, no papering over the cracks or brushing under the carpet the humiliations, degradation and injustices inflicted upon us by a foreign power.[47]

Eamonn O'Kane has argued that it would be wrong to see these exchanges as a sign that the British 'decided to embrace inclusiveness and abandon exclusiveness'.[48] London remained cautious in its approach to dealing with the IRA in public and in private at this time.

News of the secret, behind-the-scenes dialogue finally leaked in November 1993. The Secretary of State for Northern Ireland, Sir Patrick Mayhew, was quick to issue a statement in which he reinforced Britain's policy that Northern Ireland's future would continue to be decided by the majority of its population. Known as the 'Principle of Consent', Mayhew told the House of Commons that, 'It is with the IRA, and with it alone, that the key to peace lies. It is for the IRA to decide whether it brings to an end violence that should never have been begun'.[49] One of those who understood the challenges now facing the IRA at the time was Sinn Féin strategist Danny Morrison. Despite being imprisoned in 1990, he remained a key architect of the strategy that enabled the IRA to end its armed campaign. In a letter to Gerry Adams, Morrison warned that they should not 'confuse our base' by giving the impression that 'we are indecisive and weaker than we are and give the opposition an opportunity to hammer us'.[50] Morrison later claimed, he had 'never heard the word decommissioning until the IRA ceasefire … The idea that they would have to go through a period of decontamination before they can enter talks. Who the fuck do these people think they are? Up to their necks in blood, and assassination, and "dirty war", and they're lecturing to us'.[51] As Rogelio Alonso has so effectively observed, on the IRA's own terms, 'it is hard to agree with those who claim that there was no military defeat. In short, because the IRA failed to attain its political aims, military action must also be seen to have failed.'[52]

## Britain's intelligence-led security strategy

The British government's stated aim throughout the Troubles was to defeat terrorism.[53] With IRA leaders articulating their desire for 'an honourable peace' by the early 1990s, the British were in a strategically advantageous position to meet this policy. Yet, ironically, the single greatest obstacle to administering a *coup de grace* on the IRA was the competition within Britain's Security and Intelligence Agencies. Even though the Army and RUC interpreted a decisive military defeat as the only realistic outcome, MI5 believed that only a political solution that would guarantee the long-term defence of the realm.[54] Mrs Thatcher herself took a longer-term view and had even attempted to lance the boil of inter-agency rivalry by dispatching retired MI6 Chief Sir Maurice Oldfield to Belfast in the autumn of 1979 to coordinate London's secret intelligence struggle against the IRA.[55] Oldfield's appointment followed in the wake of the IRA's assassination of Lord Mountbatten, a member of the Royal Family, and a deadly bomb attack at Warrenpoint, which killed 18 soldiers. Oldfield was largely successful at laying the foundations of an intelligence-led response, which saw 'almost all the military structures which eventually defeated PIRA … [put] in place'.[56] These structures were coordinated under the emerging Tasking and Coordination Group (TCG), which acted as a clearing house for all security forces operations.[57] As one former Special Branch officer recalled:

> Now, the common denominator was the TCG. But, don't forget, all these things worked on a 'need-to-know principle.' There was very little that I wouldn't have been told, for instance. But you were working through a system where some of the key agent handlers would have held back … Now, the whole theory was that the TCG was the whole centre of intelligence and I would say it was probably getting 85 percent of it. Even the army would have played silly buggers and held back. And that was the biggest mistake they all made.[58]

Under the leadership of RUC Chief Constable John Hermon and successive GOCs Richard Lawson, Robert Richardson and Robert Pascoe, police–army relations improved in the 1980s. The military, in particular, were well-positioned to meet the political demands on them, as they later noted in a doctrine note:

> The three key tenets of Army policy in the 1980s were reassurance, deterrence and attrition. Most of the attrition took place through arrest and conviction. Overt operations probably killed at most a dozen terrorists during the 1980s. The number killed in covert operations was in absolute terms not much greater. In any case attrition of individual terrorists of itself had little effect on the outcome of the campaign. However, PIRA seem to have been brought to believe that there was no answer to Army covert operations, and that they would not win through violence. That was probably a key factor.[59]

The pressure exerted by covert action eventually brought the IRA to what one former Military Intelligence Officer called a 'culminating point'.[60] However, it is important to recognise that such tactical responses were devoid of strategic meaning. As Colin S. Gray points out in his study of Special Operations Forces, in order 'to realise their strategic potential, the tactical behaviour of SOF requires conversion into the coin of strategy', which only makes sense if they are put to 'specific purposes'.[61]

The 'specific purposes' would soon be identified in a major review of Counter Terrorism operations by Permanent Under Secretary at the Northern Ireland Office, Sir John Blelloch, in January 1990. In his 'Defeating Terrorism' papers (DTP), he outlined a comprehensive political-led strategy for combatting the IRA:

> HMG's security policy is based on a firm commitment to defeat terrorism. This translates into resolute action, both to prevent or interdict terrorist activity and to react quickly in pursuit of the terrorists (and/or to protect the local community) if terrorist activity does take place. It is therefore both proactive and reactive. It consists (or should consist) at all times of the even-handed application of the law. This means that, as necessary, there will be vigorous action against loyalist paramilitaries, and, indeed, against members of the Security Forces themselves, if they fail to meet the high standards expected of them. It is a policy which in its implementation has the police in the lead role – even in the 3 Brigade area, where the Army's operational independence in counter-terrorist activity sits within a police-led general framework.[62]

By setting out a goal-orientated approach a decade after the establishment of the 'Oldfield System', Blelloch moved to exert greater political control over the security forces. In a sense, Blelloch was resorting to strategy – in Michael Howard's classic definition is 'the use of available resources to gain any objective'[63] – which would give the British a winning edge over the IRA. In Blelloch's view:

> The PIRA campaign of violence will end when PIRA and/or its supporters are brought to the point when even they recognise its futility. They must be 'motivated' to end it. Continuing pressure by the security forces, equipped with the necessary resources, will be a necessary condition for achieving that objective; but it will not itself be a sufficient one. It must therefore be reinforced by effective Government action in the political social and economic fields. All the Government's policies seek to take account of the wider geographical, social and political contexts.[64]

The DTP review not only took stock of Britain's successes against the IRA but essentially mapped out how the security instrument might be utilised in a strategic fashion that would yield a decisive blow against their terrorist opponents.

Blelloch's replacement at the NIO, Sir John Chilcot, was an even greater proponent of strategy. He interpreted strategy according to the precepts of grand strategy, which was a somewhat more flexible process that could be adjusted to 'shifting realities, means, resources and personalities',[65] thereby yielding greater benefit to London. As he told respected academic Graham Spencer:

> I think there was a short-term perception on the right side of British politics that the PIRA could be beaten, but there was little advantage in that perception. I think that kind of thinking may have been more prevalent in the mid- to late 1980s after the hunger strikes, but that was also true on the other side of the political divide. However, as time went on the possibility and expectation of military victory certainly lost appeal.[66]

In December 1992, Chilcot presided over the decision to form a Province Executive Committee (PEC), which was 'created as a result of a desire by the Army to have a formal strategic body looking at security issues across Northern Ireland'.[67] The PEC met at RUC Headquarters and its attendees included senior RUC officers, Army commanders and MI5 personnel. Its purpose was 'to ensure that all of the key players had full cooperation and an understanding of all overt and covert intelligence activities being conducted. It also gave senior personnel from the Security Service and the Army the opportunity to consider the division of labour in terms of the intelligence requirements throughout Northern Ireland'.[68] Intelligence was the lifeblood of the PEC, which, as far as British government ministers were concerned, had always been:

> The key to the fight against terrorism, both in Northern Ireland and elsewhere; without good intelligence we are be [sic.] unable to target our limited resources to best effect. It is the single most potent force in countering the activities of all paramilitary groups. The information which we obtain from agents and informers is critical to our assessments of their policy, plans and psychology.[69]

Although it took the British three decades to perfect a successful strategy for defeating the IRA's violent campaign, it would not amount to much until it was translated into the accomplishment of policy aims. With the competition between the rival Security and Intelligence agencies only really resolved after the formation of the PEC, the British soon realised the merits of the 'whole of government' approach[70] first articulated by Blelloch and implemented under Chilcot's stewardship.[71]

## Conclusion: Who won and who lost?

In light of this appraisal of the strategic contest between the IRA and Britain's security forces, is it possible to say who won and who lost the war? Respected journalist Henry McDonald observes that Britain 'pushed the Provisionals into a position where they had accepted the principle of consent. In doing so they had lost the Long Game'.[72] This was a view echoed by one of the IRA's leading members Tommy Gorman:

> The IRA at the end was like the Sword of Damocles. Adams was definitely speaking to – we know he was speaking to – these other people and we know Martin McGuinness was speaking to [them too], without the okay of anybody else. He done it himself. And I was told it was going on from 84. I heard in 84 it was going on. And so you had this thing where these people were talking to the secret service and all sorts of people. And other volunteers would maybe speak to the Special Branch in Castlereagh and were getting shot dead for it. So, you had this contradiction. And if it's run down, then stop it now. I say because ... we spoke about it and said, 'Listen, what's going on here is just gratuitous violence. Just going out for killing's sake. Sending wee lads out to kill and be killed when they know that, very soon, that there's going to be some sort of halt to this.' And we spoke out against this.[73]

Gorman was soon joined by other dissenting voices who doubted the strategic sense of the 'armed struggle'. Anthony McIntyre had been a leading member of the IRA in Belfast prior to his arrest in the mid-1970s. Emerging from prison after an 18-year prison sentence, he argued that, 'British state strategy in terms of the IRA was remarkably successful'.[74]

Unsurprisingly, perhaps, there remains no academic consensus on the question of who won and who lost the war. For John Bew and Martyn Frampton, 'the highly effective unofficial war against the republican movement... had a decisive impact on what occurred subsequently'.[75] In contrast, Paul Dixon argues that the British government 'did not dictate terms to republicans during the peace process because there is little evidence that the IRA was either "defeated" or had "won"'. Seeking refuge in the myth of 'stalemate', Dixon denies the ensuing saw 'the victors imposing their will on republicans but rather a complex process of negotiation – choreography and co-operation among elites as well as conflict – which involved uncertainty, contingency and characterised by a considerable degree of pragmatism, compromise and deception'.[76] More recently, Thomas Leahy has concluded that Britain 'did not "win" the intelligence war'; rather, in line with Danny Morrison, he too believes the conflict 'ended in a stalemate situation'.[77] Although it suits political actors, including Sinn Féin, to talk about the outcome of the war in Northern Ireland as if the concepts of victory and defeat are somehow meaningless, this chapter has evaluated the contest in broadly strategic, rather than narrowly parochial terms.

## *Notes*

1 This chapter is dedicated to the memory of Professor Colin S. Gray (1943–2020), formerly of the University of Reading, who was one of the leading torch bearers of Strategy Studies. Professor Gray was gracious in his comments and encouragement of my application of the lens of Strategic Theory to the Northern Ireland case.

2 On the 'peace strategy' followed by the Provisional IRA and its political associates in Sinn Féin, see Kevin Bean, *The New Politics of Sinn Féin* (Liverpool: Liverpool University Press, 2007) and Tommy McKearney, *Provisional IRA: from Insurrection to Parliament* (London: Pluto Press, 2011).

3 Carl von Clausewitz, *On War*, edited by Michael Howard and Peter Paret (Princeton, NJ: Princeton University Press, 1976), 75.
4 Ibid., 75.
5 Ibid., 529.
6 Ibid., 91.
7 MoD, *British Army Field Manual, Volume 1, Part 10: Countering Insurgency*, Army Code 71876 (London: MoD, October 2009), 1-1.
8 Hew Strachan 'British Counter-Insurgency from Malaya to Iraq', *RUSI Journal*, Vol. 152, No. 6 (December 2007), 10; see also Aaron Edwards, 'Misapplying Lessons Learned? Analysing the Utility of British Counter-insurgency Strategy in Northern Ireland, 1971–76', *Small Wars and Insurgencies*, Vol. 21, No. 2 (June 2010): 303–330.
9 MoD, *British Army Field Manual, Volume 1, Part 10: Countering Insurgency*, 1–5.
10 General Sir Richard Dannatt, Address to the International Institute of Strategic Studies, 21 September 2007.
11 Aaron Edwards, 'Beating the Retreat on a Contested Past? The British Army and the Politics of Commemoration in Northern Ireland' in Jim Smyth (ed), *Remembering the Troubles: Contesting the Recent Past in Northern Ireland* (Notre Dame: University of Notre Dame Press, 2017), 77–95.
12 For survey research undertaken by the Army's Officer Tutors, see Claudia Harvey and Mark Wilkinson 'The Value of Doctrine', *RUSI Journal*, Vol. 154, No. 6 (2009): 26–31.
13 Colin S. Gray, 'Irregular Warfare: One Nature, Many Characters', *Strategic Studies Quarterly*, Vol. 1, No. 2 (Winter 2007): 36.
14 See Rod Thornton, *Asymmetric Warfare: Threat and Response in the Twenty-First Century* (Cambridge: Polity Press, 2007). Thornton defines asymmetry as primarily being a position where one can exploit the weaknesses of one's opponent by stretching their resources. See also Ivan Arrenguín-Toft, 'How the Weak Win Wars: A Theory of Asymmetric Conflict', *International Security*, Vol. 26, No. 1 (Summer 2001): 93–128.
15 Gray, 'Irregular Warfare', 37.
16 Ibid., 39.
17 Michael Howard, 'When Are Wars Decisive?', *Survival*, Vol. 41, No. 1 (Spring 1999): 126.
18 Ibid., 132.
19 Aaron Edwards, 'Deterrence, Coercion and Brute Force in Asymmetric Conflict: The Role of the Military Instrument in Resolving the Northern Ireland 'Troubles'', *Dynamics of Asymmetric Conflict*, Vol. 4, No. 3 (December 2011): 238.
20 McGuinness gave this speech at the Sinn Féin *Ard Fheis* in 1991. Cited in Liam Clarke and Kathryn Johnston, *Martin McGuinness: From Guns to Government* (Edinburgh: Mainstream, 2001), 188.
21 For more on this point see Graham Spencer, *From Armed Struggle to Political Struggle: Republican Tradition and Transformation in Northern Ireland* (London: Bloomsbury, 2015); and Aaron Edwards, *Agents of Influence: Britain's Secret Intelligence War against the IRA* (Newbridge: Merrion Press, 2021).
22 See Gerry Adams, *Selected Writings* (Dingle: Brandon, [1994], 1997), 278; it was also emphasised by Danny Morrison during an interview with the author on 23 November 2010.
23 MoD, *Operation Banner: An Analysis of Military Operations in Northern Ireland*, Army Code No. 71842, (London: MoD, July 2006), 8-3. Archived at: https://www.vilaweb.cat/media/attach/vwedts/docs/op_banner_analysis_released.pdf. Accessed: 13 May 2020.
24 MoD, *Operation Banner*, 8-3.
25 Kevin Bean, *The New Departure: Recent Developments in Irish Republican Ideology and Strategy*, Occasional Papers in Irish Studies, 6 (Liverpool: Institute of Irish Studies, October 1994), 5; James W. McAuley, Catherine McGlynn and Jonathan Tonge, 'Conflict Resolution in Asymmetric and Symmetric Situations: Northern Ireland as a Case Study', *Dynamics of Asymmetric Conflict*,

Vol. 1, No. 1 (2008): 88–102; Jonathan Peter Shirlow Tonge and James McAuley 'So Why Did the Guns Fall Silent? How Interplay, not Stalemate, Explains the Northern Ireland Peace Process', *Irish Political Studies*, Vol. 26, No. 1 (2011): 1–18. For an alternative, strategic perspective see Edwards, 'Deterrence, Coercion and Brute Force in Asymmetric Conflict'.

26 Paul Dixon, 'Guns First, Talks Later: Neoconservatives and the Northern Ireland Peace Process', *The Journal of Imperial and Commonwealth History*, Vol. 39, No. 4 (November 2011): 668.

27 Alex S. Wilner, 'Deterring the Undeterrable: Coercion, Denial, and Delegitimization in Counterterrorism', *Journal of Strategic Studies*, Vol. 34, No. 1 (2011): 4–5.

28 In arrangements made during the partition of Ireland in the early 1920s, the Unionist government acquired considerable authority to make decisions on matters including policing and justice but had to ask London for Military Aid to the Civil Power. For more on this point see Robin Evelegh, *Peace-keeping in a Democratic Society: The Lessons of Northern Ireland* (London: Hurst, 1978).

29 On the IRA and its split in 1969 see Brian Hanley and Scott Millar, *The Lost Revolution: The Story of the Official IRA and the Workers' Party* (Dublin: Penguin Ireland, 2009).

30 Author interview with Tommy Gorman, 23 June 2010.

31 Robert White, *Out of the Ashes: An Oral History of the Provisional Irish Republican Movement* (Newbridge: Merrion Press, 2017), 150.

32 Author interview with Danny Morrison, 23 November 2010.

33 White, *Out of the Ashes*, 150.

34 Gary A. Ackerman, 'The Provisional Irish Republican Army and the Development of Mortars', *Journal of Strategic Security*, Vol. 9, No. 1 (2016): 12–34.

35 The National Archives (TNA), WO 305/5815, 1 KOSB Post Tour Report, dated 22 July 1985. The 1985 South Armagh tour was 1 KOSB's sixth operational deployment in Op Banner by that date. Its principal operation was Op Cara, force protection for the rebuilding of security forces bases in the region.

36 Foreword to the 1 Royal Anglian Post Tour Commemorative Booklet (1990).

37 John F. Morrison, 'The Provisional Irish Republican Army' in Andrew Silke (ed.), *Routledge Handbook on Terrorism and Counterterrorism* (Abingdon: Routledge, 2017), 327–328.

38 Thomas Leahy, *The Intelligence War Against the IRA* (Cambridge: Cambridge University Press, 2020), 5, 176.

39 Author interview with a former British agent inside the IRA, October 2018.

40 Author interview with a former senior RUC Special Branch officer, October 2011.

41 Rogelio Alonso, *The IRA and Armed Struggle* (Abingdon: Routledge, 2007), 145.

42 Aaron Edwards, *UVF: Behind the Mask* (Dublin: Merrion Press, 2017), 202–203.

43 Moloney, *Secret History*, Chapter 13.

44 Niall O'Dochartaigh and Isak Svensson 'The Exit Option: Mediation and the Termination of Negotiations in the Northern Ireland Conflict', *International Journal of Conflict Management*, Vol. 24, No. 1 (2013): 44.

45 Edwards, *Agents of Influence*.

46 Peter Taylor, *Provos: The IRA and Sinn Féin* (London: Bloomsbury Publishing, 1998), 318.

47 McGuinness gave this speech at the Sinn Féin Ard Fheis in 1991. Cited in Liam Clarke and Kathryn Johnston *Martin McGuinness: From Guns to Government* (Edinburgh: Mainstream, 2001), 188. Emphasis added.

48 Eamonn O'Kane, 'Anglo-Irish Relations and the Northern Ireland Peace Process: From Exclusion to Inclusion', *Contemporary British History*, Vol. 18, No. 1 (2004): 85.

49 House of Commons Debates (Hansard), 29 November 1993, Vol. 233, Col. 788.

50 Danny Morrison, *Then the Walls Came Down: A Prison Journal* (Cork: Mercier Press, 1999), 240.

51 Author interview with Danny Morrison, 23 November 2010.

52 Alonso, *The IRA and Armed Struggle*, 151–152.

53 Aaron Edwards, *Defending the Realm? The Politics of Britain's Small Wars since 1945* (Manchester: Manchester University Press, 2012), 210.
54 Christopher Andrew, *Defending the Realm: The Official History of MI5* (London: Allen Lane, 2009); see also Tony Craig, 'From Countersubversion to Counter-Insurgency: Comparing MI5's Role in British Guiana, Aden and the Northern Ireland Civil Rights Crisis', *Journal of Intelligence History*, Vol. 14, No. 1 (2015): 38–53.
55 Aaron Edwards, '"A Whipping Boy If Ever There was One"? The British Army and the Politics of Civil-Military Relations in Northern Ireland, 1969–79', *Contemporary British History*, Vol. 28, No. 2, (2014): 182.
56 MoD, *Operation Banner*, Para 812.
57 Mark Urban, *Big Boys' Rules: The Secret Struggle Against the IRA* (London: Faber, 1992), 95.
58 Author interview with a former RUC Special Branch officer, 26 October 2011.
59 MoD, *Operation Banner*, 2–15. In fact, the Army never set its own policy–the three key tenets were actually an operational framework for security forces response to terrorism generally. For a strategic-based analysis of the Army's operations in the post-ceasefire years see Bass, Christopher and MLR Smith 'The Dynamic of Irwin's Forgotten Army: A Strategic Understanding of the British Army's Role in Northern Ireland after 1998', *Small Wars and Insurgencies*, Vol. 15, No. 3 (Winter 2004): 1–27.
60 Author interview with a former Military Intelligence Officer, June 2019.
61 Colin S. Gray, *Tactical Operations for Strategic Effect: The Challenge of Currency Conversion*, Special Report for the Joint Special Operations University, (Tampa, FL: JSOU, November 2015), 44.
62 John Blelloch, 'Defeating Terrorism', PRONI, NIO 10/9/13A, 23 January 1990.
63 Michael Howard, *The Causes of War* (London: Counterpoint, 1983), 36. As Neumann and Smith have argued, 'clearly, and logically, yields to our understandings of war as a goal-orientated enterprise'.
64 Blelloch, *Defeating Terrorism.*
65 Graham Spencer (ed) *The British and Peace in Northern Ireland The Process and Practice of Reaching Agreement* (Cambridge: Cambridge University Press, 2015), 98.
66 Ibid., 87–88.
67 Sir Michael Morland, Dame Valerie Strachan and Sir Anthony Burden, *The Rosemary Nelson Inquiry Report* (London: HMSO, 23 May 2011), 168.
68 Ibid., 168.
69 Sir Desmond de Silva, *The Report of the Patrick Finucane Review, Volume II* (London: TSO, 2012), Top Secret correspondence from Tom King to the Attorney General, dated 19 March 1991.
70 There is evidence that the RUC Special Branch believed this broader approach effectively 'won' the war. See William Matchett, *Secret Victory: The Intelligence War that Beat the IRA* (Lisburn: Hiskey Ltd, 2016), 234.
71 For a more detailed analysis of the use of the military instrument in ending the Northern Ireland 'troubles', see Edwards, 'Deterrence, Coercion and Brute Force in Asymmetric Conflict'.
72 Henry McDonald, *Gunsmoke and Mirrors: How Sinn Féin Dress Up Defeat as Victory* (Dublin: Gill and Macmillan, 2008), 100.
73 Author interview with Tommy Gorman, June 2010.
74 Author interview with Anthony McIntyre, 20 October 2019.
75 John Bew, Martyn Frampton, and Inigo Gurruchaga, *Talking to Terrorists: Making Peace in Northern Ireland and the Basque Country* (London: Hurst, 2009), 247.
76 Dixon, 'Guns First, Talks Later', 653.
77 Leahy, *The Intelligence War Against the IRA*, 5.

CHAPTER 15

# Russian Views of 'Winning': '*Velikaya Pobeda*' and '*pobedonosnaya voina*'

Andrew Monaghan

2020 saw the 75th anniversary of victory in Europe, and in Russia it has been treated as a 'Year of Remembrance and Glory'. As Putin said in late 2019, 'We will mark the 75th anniversary of the Great Victory with grand celebrations on a national scale. This day is sacred for all citizens of Russia, a symbol of courage and heroism of our Army, Navy and all Russian people'.[1] The importance of the occasion in Russia is such that the marking of the anniversary is being overseen by the Russian Victory Organising Committee, which includes very senior officials from all the major parts of the political and bureaucratic leadership as well as public organisations. Plans for marking the anniversary nationally focus on the annual gathering and military parade on 9 May ('Victory Day'). But they also include a range of other events, including the 'Echelon of Victory',[2] and the opening of the main cathedral of the Armed Forces, the Church of the Resurrection, and, as Putin, who is taking a very active personal role in the discussion about the war, put it, 'a number of other grandiose events'. Russia's armed forces, he felt sure, will celebrate the anniversary 'by showing even higher performance results in ensuring the nation's defence capability and security'.[3] In the event, these plans were altered by Covid-19, though this largely served to increase the number of shows and parades. The planned 9 May parade in Moscow was replaced by an air military parade and a small parade of the Kremlin regiment. The main parade itself was held on 24 June (the anniversary of the original Victory Day parade of 1945).

The Great Patriotic War, as World War II is known in Russia, is the focal point of contemporary Russian thinking about war and Victory. Victory (*Pobeda*) in that war is not considered merely 'victory', it is capitalised and raised to a higher status as '*Velikaya Pobeda*' ('The Great Victory'). Indeed, as Putin suggests, for many Russians the Victory is a 'sacred issue', and seen as Russia's most important contribution to world history.[4] Other wars feature, of course, in the Russian debate about winning, but all roads lead to the Great Victory, both in the public-political debate and in the defence community's thinking about and preparation for contemporary war.

The Euro-Atlantic community's debate about Russian warfare and victory is rather different. Some point to the greater prevalence of defeat than victory. The historian Stephen Kotkin argues that Russia's moments of victory have been 'fleeting' and only come against a declining Sweden, a 'wildly overstretched' Napoleon or the 'maniacal gambler' Adolf Hitler.[5] And there are persistent stereotypes about how Russia has won its victories, most often that they have been achieved due to geography and climate (Generals 'Winter' and 'Mud') and the advantage of overwhelming mass of troops.[6]

Equally, the eruption of war in Ukraine in 2014 and then the Russian intervention in the Syrian civil war in 2015 have made these questions ones of contemporary importance. Russia's role in the new era of 'Great Power Competition' has generated concern in Euro-Atlantic capitals about the Russian leadership using armed force as a tool of policy to challenge the international order, and a range of views about why and how Moscow has gone to war, how it seeks to win, and its chances of success.

Some assert that Moscow has gambled on launching unprovoked wars to boost declining domestic popularity. This echoes an approach attributed to Vyacheslav Plehve, Russian Interior Minister in the early 20th century, who is said to have advocated such a policy to avert domestic revolution. This, it is often argued, underpinned the Russo-Japanese war of 1904–05 (which Russia lost).[7] Equally, however, Euro-Atlantic officials and observers have asserted that by intervening in Ukraine and – even more so – in Syria, Russia would find itself inextricably stuck in a costly quagmire; essentially that Moscow faced inevitable failure in war from the outset. But as one observer has suggested, 'since Washington had judged that force could not be used to achieve political ends in Syria, it assumed the same would be true for Russia' and that 'the Russian air campaign was insufficient to change the balance of forces'.[8]

Another view addresses Russia's perceived actions in the Euro-Atlantic area. It is an orthodoxy that because Moscow could not hope to defeat the west in a conventional war, it is obliged to avoid direct confrontation and resort to indirect measures – measures short of war. These include hostile measures such as propaganda, disinformation and cyber attacks, the use of economic pressure and other tools to destabilise western states and societies. In a similar vein, some suggest that the Russian military now seeks to win wars without using violence.[9]

Such views do not correlate smoothly with Russian views or activity. The Russian leadership has conducted a sustained and substantial modernisation and development of conventional military capability, and has deployed it, including in combat. In Ukraine, Russia deployed substantial conventional firepower to achieve its ends, fighting several battles, including at Debaltseve which led directly to the signing of the Minsk II agreement in early 2015. In Syria, the Russian leadership has explicitly emphasised its purposeful deployment of limited but overwhelming firepower. Clearly Moscow sees intervention in Syria in a different way and has been employing an evolving strategy to end the war on its terms. Though the war in Syria continues at the time of writing, Russia's intervention in late 2015 turned the tide of the war, and Russia is now firmly established in the region. Thus, senior Russian officials already seek to portray success. Putin stated in late 2019 that Russia has 'defeated the terrorist international that was winning on Syrian territory, and prevented the return, the infiltration of hundreds and later perhaps thousands of armed cutthroats into our country', as well as launching a Syrian and international political process. This 'diplomatic alignment' is now a '*fait accompli*', he suggested, one that can become a model for regional crisis resolution elsewhere.[10] In other words, although it is too early to be conclusive, Putin was asserting victory.

Such differences suggest that there is both a poor understanding of (and even some confusion about) Russian thinking about warfighting and war winning in the Euro-Atlantic discussion, and a need to reflect again on how the Russian leadership understands war and victory. Two intimately related features of Russian views of 'Victory' stand out, the spiritual and the scientific, and this chapter addresses each in turn. The first part looks at the ambiguity between victory and defeat, how victory is sometimes snatched from the jaws of defeat, or defeat is masked and redefined as victory as a result of mythmaking. This is where the moral or spiritual feature becomes clear, in terms of patriotism and courage, and the emergence of elements of both holy and civic religion. The link between the two features, spiritual and scientific, is shown by the nature of the Soviet victory in the Great Patriotic War. The primary example of disaster being converted to Victory, and the ultimate sacred victory, also reveals the importance of more profane features, including Russian and Soviet military science.

The second part looks at the more scientific approach to achieving victory, and how this has evolved since the interwar period. This part looks at how the Russian defence community defines and discusses victory in scientific terms, looking at the tools and approaches to winning in war, and the purpose of war. This is reflected in the term '*pobedonosnaya voina*' – literally translated as 'victory bringing war'.

Before turning to these main points, though, it is worth acknowledging some limitation to the extent of what can be addressed here. The potential scope of even the specific question 'what are Russian views of victory?' is vast. There is so much that could be included, from Russia's imperial wars and territorial expansion, to discussion of some of their most successful generals: General Alexander Suvorov, for example, is widely considered to be one of the great commanders in military history. Undefeated in major battles, he wrote a book entitled *Nauka pobezhdat* (*The Science of Victory*), and his maxims continue to be influential in contemporary Russian thought. There are many other potential examples in the 18th century alone. Peter the Great's victory in the Great Northern War (1700–21), including the decisive battle of Poltava, established Russia as a major power in north-eastern Europe. Russian victories in the Seven Years' War, in which Russia occupied Eastern Prussia and defeated Frederick II's forces on numerous occasions, and Catherine the

Great's territorial expansion and victories over the Ottoman Empire also stand out as clear Russian victories and wars won because of the enduring nature of the results in terms of territorial expansion and the establishment of Russia as a Great Power in Europe. Their results meant that, as Sergei Lavrov, Russia's long-serving Foreign Minister has put it (quoting then-Russian Chancellor Alexander Bezborodko), 'not a single cannon in Europe could be fired without our consent'.[11] Here, though, the focus is largely on the two historical pillars of Russian 'Victory', the Napoleonic Wars and the Great Patriotic War, with some other illustrative examples drawn from the 19th and 20th centuries. Though the question of nuclear war and whether and how one could be 'won' has been long debated in Russia, official Russian military definitions of victory in war treat the question of nuclear war separately, and note that the unacceptable scale of damage and loss excludes the possibility of talking of 'victory'. Consequently, the chapter focuses on victory in conventional war.

## Sacred victory: '*Velikaya Pobeda*'

There is considerable ambiguity in Russian thinking about warfighting, and the close intertwining of near total defeat and complete victory is evident, even in the same wars. One way that victory and defeat are interwoven is the nature of the victory: the war may be won militarily, but through such a poor performance or at such a high cost that victory appears hollow or worthless. A recent example of this might be the short Russo-Georgia war of 2008. The performance of the Russian armed forces was chequered, with poor coordination between the services, and rather high casualties due to friendly fire and accidents.[12] There was some debate, including in Russia, about whether Russia had sacrificed so much international political capital that the victory was hollow, if not tantamount to a defeat since it worked against Russian strategic interests.[13] Nevertheless, Russian forces ejected Georgian forces from South Ossetia and Abkhazia, recognised them as states and established military bases there, effectively resolving the political question that underpinned the war, and the overall cost in international political capital was limited and temporary.

The USSR's Winter War with Finland is a more obvious example. Launched by the USSR in November 1939, the Soviets sought to acquire portions of territory considered essential for the defence of Leningrad. But the poor performance of the numerically considerably superior Soviet forces, and the very heavy casualties they suffered have blurred – if not completely hidden – the point that the USSR, having reorganised its forces, finally overwhelmed Finnish forces in March 1940, achieving Moscow's initial policy aims. With the Moscow Peace Treaty of March 1940, Finland ceded 11% of its territory – more than Moscow's pre-war demands. Even so, this did not represent complete victory for Moscow. But not only was the damage done to the reputation of Soviet forces seen by many to have encouraged both German and Japanese leaders in the view that their own attacks on the USSR would be successful, but it did not resolve the Finnish question. Finland invaded the USSR in 1941, initiating the Continuation War, a move commonly explained as an attempt to reverse the results of the Winter War. The USSR's war with Finland was not fully resolved, therefore, until the signing in 1948 of the Agreement of Friendship, Cooperation and of Mutual Assistance, which was both much less favourable to Finland, and which the Soviets regarded as satisfactory guarantee against another war of revenge by Finland.

An important question here, though, is the recasting of defeat into success. There have been occasions when other factors have mitigated the defeat or even overcome it, rendering the consequences only short term or comparatively marginal. Sergei Lavrov often draws on history, for example, to show how diplomacy can serve to neutralise the impact of, or even reverse military defeats. He argues that Russia overcame the consequences of military defeat in the Crimean War in 1856 relatively quickly because of the skilful policies of Prince Gorchakov, who quickly restored Russian influence through diplomacy.[14] Another example in which battlefield defeat has been more immediately turned into a satisfactory settlement through negotiation is the Treaty of Tilsit in 1807. Though the Russian army under von Bennigsen was decisively defeated at the battle of Friedland, Alexander I negotiated a settlement with generous peace terms that made little substantial demand of Russia.

The intimate connection between defeat and victory, though, is best illustrated by the conversion of defeat into victory in the creation of myths. These serve to adjust the points of focus, whether to the specific units that achieve smaller triumphs while wider defeat or disaster is taking place all around them, or to a more spiritual level, mitigating the defeat transforming it into a different form of victory. A recent example of the former is the film *Panfilov's 28*, about a unit defending the approaches to Moscow in 1941. There are numerous other cultural representations of Russian troops facing overwhelming numerical and technical odds, but fulfilling their duty by engaging the enemy and not surrendering, regardless of the cost, and in which defeats and retreat are translated into tactical level or personal victories.[15] This approach frames an understanding of many battles: from the battles of Eylau in 1807 and Borodino in 1812, and the battle of Chemulpo Bay in 1904, when the cruiser *Varyag* and the gunboat *Koreets* refused to surrender and sallied out of the harbour to engage a much superior Japanese force, to the defence of Brest Fortress and other defences in the Great Patriotic War.

The mention of Borodino, of course, brings us to the pre-eminent exponent of the latter, more spiritual approach, illustrated by Leo Tolstoy most famously in his novel *War and Peace*. Whether Borodino was a defeat or victory for Russia posed some military ambiguity. But as Gregory Carleton, a cultural historian has noted, Tolstoy's answer was to recast it. How the battle is described, Tolstoy suggested, did not need to derive from its tactical outcome, or measurements of short- or long-term gain. Instead, in certain cases there are different criteria, such as the demonstration of force against an invading and seemingly invincible enemy. Regardless of the casualty toll or punishment suffered, if the army did not break then the steadfastness set the standard creating a moral victory over the enemy. This moral victory reflected the steadfastness of the Russian army – Napoleon's army had thrown everything at it, inflicting unprecedented damage, but without consequence. In turn, this was said to send a jolt of terror through the French army.[16]

Tolstoy's depiction of victory at Borodino, and even the Russian success in the wider campaign of 1812, is undoubtedly very influential, including in Russia. Tolstoy emphasises the moral strength, courage and patriotism of ordinary Russians, and is scathing about military professionals and the strategic competence of the leadership. But as the historian Dominic Lieven has pointed out, Tolstoy greatly underestimates the Russian achievement in defeating Napoleon, and ignores entirely Russian campaigning in 1813–14. Lieven notes both the significant improvement in Russian tactics, logistics and staff work, and the overarching point that Russian leaders out-thought Napoleon, planning and then imposing a kind of war on him that he was ill-equipped to wage. Russian policy was thus 'far removed from Tolstoyan mythology'.[17] Indeed, as discussed below, this more practical approach to winning wars is an important feature of Russian military science.

Nevertheless, the Tolstoyan point of patriotism and courage has a significant role to play in understanding Russian notions of victory. An important concept here is '*stoikost*'. Stemming from the verb '*stoyat*', to stand, *stoikost* is a term which encapsulates courage, resilience and defiance. *Stoikost* is, as Carleton notes, the 'favoured term for when a certain fanaticism takes over', a power that 'allows the Russians to absorb casualties that would cause others to flee'. The root is an important one in Russian military thinking, depicting the prolonged (and ongoing) debate about the comparative value of technology or a soldier's spirit in war, and the view that spirit is what is decisive on the battlefield. Thus, General Mikhail Dragomirov could advocate the view that if Russian soldiers suffered more it was a good sign, one that pointed to Russian superiority: more dead on the battlefield meant more *stoikost* and less retreat.[18] It also features in other Russian military terminology, such as '*boevaya ustoichivost*' – roughly translatable as 'combat stability': the ability to fulfil a task while absorbing substantial punishment. This is how the results of battles such as Eylau and Borodino, though perhaps tactically ambiguous, can be understood as victories. Victory in this sense, therefore, is both the refusal to surrender, sometimes combined with the stopping of a previously undefeated, even apparently unstoppable and invincible enemy. This is the spiritual or moral aspect of victory.

The Napoleonic wars are one of the central pillars of Russian thinking about victory, particularly in spiritual terms. But it is the Great Patriotic War that is the most obvious example of spiritual victory and *stoikost*. As Carleton suggests, the Great Patriotic War offers Russia the greatest of war stories: a 'seemingly invincible demon invades from the west … again the fate of a nation is decided in a titanic struggle on its own soil, and again the war ends when Russia, at the head of a coalition of liberators, enter the enemy's capital'. But he depicts it as the 'burden of victory', since it is rare that a nation's greatest triumph and its greatest tragedy constitute the same chapter as they do in Russia's case. The fighting (and destruction) was on an unprecedented scale, with approximately 25–28 million people killed (some 12% of the pre-war population), millions more rendered homeless and disabled, 1,710 towns and cities destroyed and 70,000 villages burned and bombed. The war has become a symbol of an 'epic struggle', showing Russia's capacity to withstand anything at any cost, and its 'tenacious refusal to surrender'.[19]

This scale of conflict is what has rendered it a sacred issue. As one prominent observer has suggested, since the 1990s a 'divine interpretation' of the history of the Great Patriotic War has emerged, with the assertion of miracles and portents anticipating Soviet victories, and particularly of a causal link between faith, the Russian Orthodox Church's spiritual and material contributions and Soviet battlefield successes. Senior Soviet military leaders including Boris Shaposhnikov, Alexander Vasilevsky, Georgi Zhukov and even Stalin are portrayed as believers who prayed for victory, and Zhukov is equated with other faithful Russian military heroes, including Suvorov and Kutuzov.[20]

The Great Patriotic War has also become the basis of a civic religion. As Putin has put it, because it has left a 'deep trace in the lives of peoples of the USSR and the Russian federation', it is 'inseparable from the history of every Russian family'. Thus, where war is concerned, preserving the memory of *Velikaya Pobeda* dominates the public-political horizon, reflected in the recently emerged – and very popular – tradition of the Immortal Regiment, and various other organisations, such as the fallen heroes search units.[21] Consequently, the Great Patriotic War plays a special role in the development of a patriotic educational agenda to strengthen society and reflect Russian traditions and national identity, as well as the entire history of Russia with its trials and triumphs. Putin continued, 'Memory and pride should unite us, make us stronger, and help young people fully recognise their engagement with the Motherland, the great deeds of their ancestors and responsibility for the future of Russia. This is what should become the main spiritual meaning of the Year of Memory and Glory.'[22]

And much effort is put in by the Russian state to emphasise the Soviet victory, including multivolume histories of the war published by both the Ministry of Defence and the Ministry of Foreign Affairs, as well as those supported by the Ministry of Culture. Here it is important to note, though, that Russian military history, and particularly its role in World War II, is currently very contentious, including the much-debated elements of 'memory wars', the accusations of the hijacking by the Russian state of the Immortal Regiment, differing interpretations for the responsibility for World War II, and even representation of specific battles such as Kursk. As one Russian commentator recently noted, the 'war for history is ongoing'.[23]

Many of these questions lie outside the immediate scope of a discussion of Russian victory,[24] but one point deserves mention here: the debate about the *cost* of victory. At different times in both the Soviet and post-Soviet eras, the sense of triumph in war, particularly in the Great Patriotic War, but also Afghanistan and Chechnya, has been questioned and even challenged. There are numerous Russian cultural representations, for instance, of courageous Russian (or Soviet) soldiers being unnecessarily sacrificed because of inattentive or incompetent leadership. Some have argued that the victory was so expensive in terms of lives that pride in sacrifice should become shame in carnage, that this represented the point where pyrrhic victory was no longer victory.[25]

This leads to consideration of the military aspect of the war: if the huge societal and economic effort and sacrifice underpinned the Soviet war effort, the military effort was on a similar scale. In keeping with other questions about Russian military history and the war, there is some debate about the Soviet contribution to victory. If for years the Soviet view was that the west did not sufficiently acknowledge the Soviet role (even when senior western historians were emphasising just that), there has been what historian Mark Edele has called an emergent 'amnesia' about this, and the war is increasingly seen again by some as an 'avoidable German defeat' rather than a Soviet victory.[26]

Broadly, the war, and the transition from disaster to victory can be divided into three parts – first, from 22 June 1941 to November 1942, the period of German strategic initiative, second, from mid-November 1942 to December 1943, when the strategic initiative was moving towards the USSR, and third, 1944–45, when the USSR held the strategic initiative. With the Red Army caught mid reform and effectively taken by surprise, the first period was catastrophic for the USSR: some 200,000 tanks, guns and mortars, 12,000 combat aircraft and 6.4 million permanent casualties were lost as well as vast swathes of territory. But by 1944, the Red Army was executing increasingly deep, continuous high-speed offensives, often without pause. Soviet offensives encompassed the whole combat front from the Baltics region to Belgrade, and the Red Army undertook simultaneous and often consecutive offensives across the entire front. As the military historian Charles Dick notes, therefore, the explanation for the transition from disaster to success is to be found in these offensives – by waging war at the operational level, the Soviets generated strategic victory.[27]

There is another echo here of the Napoleonic wars: not only the transition from costly retreat to strategic victory, but how the military success is portrayed. As noted above, it is often asserted that Soviet victory in the Great Patriotic War was due to Russian geography/climate or the Russian 'steamroller' of hordes of soldiers. Of course, the scale of the effort suggests huge resources and overwhelming force. But victory was achieved more through significant improvements in military capability. Just as the Russian military's performance became much more skilled and sophisticated from 1812–14, learning lessons, improving logistics, tactics and command and control, the Red Army also improved as the war wore on. As Dick has put it, 'too many historians have taken the complete and relatively rapid reversal of fortunes for granted, believing it to be an inevitable result of inexhaustible Soviet numbers'. Instead, the transformation of the Red Army into a 'war winning sword of vengeance', was not a matter of 'brute strength, though that had its importance, but a question of operational cleverness stemming from the development of a fundamentally sound doctrine. The Wehrmacht was to be beaten at its own game'.[28]

Victory is also to be found, therefore, in how the Soviets developed sophisticated theoretical approaches to warfighting. These built on Soviet theoretical developments of the interwar period that sought to examine the influence of the much greater spatial and temporal scale in massed force warfare, and the idea of achieving victory by conducting a series of major offensive operations each stepping towards the overall strategic goal. This was the basis of operational art: the organisation of tactical actions, both simultaneous and sequential, such that their sum is greater than the individual action. Though tactical steps were important therefore, it was the operational level that counted, intended to destroy a major enemy grouping and seize strategically important terrain. The accumulation of operational successes led to the achievement of strategic aims.

Much of this theory was lost as a result of the purging in 1937–38 of its authors and those able to put it into practice. But it was re-learnt in part during the battles at Khalkin Gol in 1939,[29] and then in the hard conditions of combat against the Wehrmacht. By 1943 and particularly into 1944, the Red Army was implementing operational art and reaping strategic successes as they employed a clear chain of command to achieve high-tempo manoeuvre to penetrate deep into enemy defences along specifically chosen axes, and encircle and destroy major German force groupings. This destruction of enemy forces was seen as the primary aim, since it would break enemy resistance, and thus deliver territory. Mass was a feature, but the essential elements of this were deception to allow the build-up of carefully worked out superiority in numbers and weight of fire, and momentum to prevent the enemy reforming coherent defence.[30]

Zhukov, one of the USSR's most successful commanders, summed it up thus: for the Soviet people, the road to victory was a difficult one, costing the lives of millions, but the Soviet armed forces kept to their agreement to force Germany into complete defeat and unconditional surrender. The aim was to dig out Nazism in German society and government completely and severely punish Nazi leaders. The aim for the complete defeat of Nazi Germany was a life and death struggle, with the intention in preparing the assault on Berlin to 'stun and smash [*podavit*] the enemy completely', swamping them in a 'sea of fire and metal'.[31] This reflected the attainment of victory through a crushing blow – '*sokrushitelny udar*' – to overwhelm the adversary through the deployment of greatly superior forces and devastating power.

If the moral-spiritual aspect is one pillar of Russian thinking about victory, therefore, military science is the essential other: for the Russians, *Velikaya Pobeda* was won through the combination of resilience but also the development of successful military operational concepts leading to strategic victory. Echoing the point about the Russian leadership out-thinking Napoleon in 1812, the Great Patriotic War was won conceptually at the operational and strategic levels. These points reverberate through contemporary discussion: Sergei Karaganov, a prominent Russian commentator, evokes Tolstoy when he notes that victories are won by those firmly resolved to win, and asserts that Russia can claim victory today by holding out and consolidating.[32] And the contemporary relevance of the practicalities of victory in 1945 are repeatedly evoked by the Russian Chief of General Staff Valery Gerasimov.[33] This leads us to the more specific question of how victory is to be achieved, in a more technical sense. This is the remit of Russian military science to which we now turn to examine in more depth.

## Russian military science: '*pobedonosnaya voina*'

The Russian defence community puts much effort into thinking about and debating how to achieve victory in war, both in terms of the broader changing character of war and in terms of specific questions, such as the correlation of forces and ratio of force to space and the factors that determine tempo. To examine and understand these questions is considered to be an essential ingredient in successful warfighting.

In some ways, the Russian understanding of victory is related to western views. The many shared experiences of war, the echoing of requirements of different 'eras' of warfare (the so-called 'Cabinet wars', followed by the Napoleonic Wars, and then 20th-century wars of mass mobilisation), and the influence of thinkers such as Carl von Clausewitz all connect Russian and western understandings of war and victory. Similarly, long-running theoretical debates about the superiority of the offensive or defensive, or of the advantages in battle of technology or the human spirit, all find their echo in the Russian discussion about how to win wars, as do arguments about the advantages of long, attritional wars or quick, decisive strikes.

Despite this reciprocal relationship, Russian military science is distinct. Russia's military historical experience differs from that of European states, and consequently Russian war fighting concepts have generated different points of focus and emphasis. Western observers during the Cold War contrasted the Soviet intention to knock the opponent out, emphasising the *permanence* of the victory, with western warfare, where shifting alliances meant that the adversary's permanent defeat could be detrimental in future. Another contrast was the Western European focus on winning decisive battles, while the Russian – and Soviet – focus was always on winning the war as a whole. Though decisive battle plays an important role in Russian thinking, winning battles or successful tactical outcomes are far less important in Russian thinking than achieving lasting strategic success.[34]

An important feature of Russian military science is the attempt to define terminology. Igor Popov and Musa Khamzatov, two prominent Russian specialists, suggest that while understanding 'victory' is one of the most important categories of military science, there is no 'concise and unambiguous' definition of the term, or, 'more accurately, there are many definitions, but they are all in one way or another unsatisfactory'. They offer, therefore, a sliding scale of stages of success and failure, in which 'political victory' is the ultimate success, followed by 'military victory' and finally 'the avoidance of military defeat' (at the other end of the scale are 'no military victory', 'military defeat', and 'complete destruction'). Developing this further, they argue that defining military victory is 'always concrete: to smash, to destroy, to annihilate, or to capture the enemy, to occupy his positions and seize military trophies'. But all wars are different, and to open up this question is to lead to many questions and 'ambiguities: victory can be military, political or moral; full or partial; real or perceived; accepted or unaccepted and, finally, pyrrhic'.[35] Much of this we have already discussed above: political or moral victories, real or perceived, and even pyrrhic.

Given these ambiguities, Popov and Khamzatov thus seek to define further a spectrum for illustrating victory in which the successful interplay of four factors – material, moral, information and cognitive (decision-making) – creates a 'zone of victory' for each side in a military conflict. This is defined as being between the minimum of one side's 'achievement of superiority' (*prevoskhodstvo*) in strength and means (both in quantity and quality) and the maximum result of 'full and conclusive victory' (*polnaya i okonchatel'naya pobeda*).[36]

The Russian Ministry of Defence offers an official definition. Tellingly, it is categorised in its military encyclopaedia specifically as '*pobeda v voine*' (victory in war). It is worth quoting at length, partly because it makes explicit some of the points noted above, and because it offers some clarity of focus in terms of the sources and measures of victory. The broad base of the definition is simply put: that 'victory in war' is 'the bringing of defeat to the enemy and the achievement of the political and military-strategic aims of one of the warring parties, be it a state, a coalition or a political movement'. The result of victory is the conclusion of hostilities on the terms of the victor in the form of an international legal document, such as an act of surrender/capitulation, or the signing of a truce.

Importantly, the main means of achieving victory in war is defined as success in the conduct of the armed struggle – the combat activities of the armed forces lead to the 'destruction of the enemy's armed forces, and the occupation of the most important parts, or all, of the enemy's territory'. Here we see some overlap with Popov and Khamzatov. Other aspects of the struggle, though, are also mentioned, including the collapse of the enemy's economic potential, the disruption of the structures and activities of enemy political power and political-military alliances, and the exhaustion of the enemy's spiritual ability to continue the fight. The MoD definition then goes on to identify also the *sources* of victory. These are to be found in bringing together of all the material and spiritual capabilities of the state and the effectiveness of their implementation in the generation of war power.[37] Here, therefore, we see made explicit the link between the moral resilience of the state and the effective generation of military power. But the definition also emphasises the broader horizon in Russian thinking, of what comes after 'winning'.

Again, the Russian military lexicon illuminates this. The term '*pobedonosnaya voina*' can be translated literally as a 'victory bringing war'. This point emphasises that in Soviet and Russian doctrine, war is understood as a tool of policy: the Soviets started a war to achieve specific policy ends. This reflects a Clausewitzian influence but has two important implications: war is thought of in terms of resolving the conflict and taking sufficiently thorough measures to make it impossible for the defeated adversary to conduct a war of revenge. The policy conflict is only resolved by a peace treaty or unconditional surrender. We have seen this illustrated in the discussion above, not least in the examples of the Russo-Finnish war and the Great Patriotic War.

Another important point in the Soviet and Russian approach to the decision to go to war is caution. As western observers noted during the Cold War, when the USSR 'did decide to use force, it always did so against a country whose military capacity was much less than its own', or 'only when the Bolsheviks were quite certain of an easy and overwhelming victory did they permit themselves to attack openly with guns'.[38] Two factors here shaped this. First, the Soviets sought to use a range of measures to prepare the context to minimise adversary capability, both beaming propaganda directly at its citizens and conducting intensive diplomacy. Indeed, considerable attention was devoted to diplomacy to shape the theatre, to ensure that Moscow did not face a two front war itself, but also to isolate the adversary and ensure the neutrality of others. Second, much attention was paid to preparing the military situation: encirclement of the adversary as far as possible and achieving strategic surprise while delivering a sufficiently heavy blow by well-co-ordinated, pre-mobilised forces to achieve initiative from the start.

Such an approach underpinned the central idea of the initial period of war (*nachalny period voiny*) to achieve a short, decisive victory. As the senior Soviet commander Boris Shaposhnikov suggested, the intention was for the enemy to be defeated by political means before a shot was fired, so that 'when troops were finally deployed all they were required to do was to push over a tottering edifice'.[39] Two illustrations of this approach can be seen in the Soviet Manchurian offensive in 1945 and the invasion of Czechoslovakia in 1968. Both were preceded by intense diplomatic activity, followed by a surprise attack, and the delivery of sufficiently weighty blows from several points on the map to bring a victory in short order at comparatively little cost. Much of this thinking reverberates through Russian military thinking today: the emphasis on diplomacy and propaganda to shape the conditions, and on preparing militarily before the war starts, including completing mobilisation preparation before conflict starts are evident.[40]

How Moscow understands the focus of the campaigning, though, has evolved during the last century. Initially, the Red Army sought the destruction of the enemy as its main aim: as discussed above, this involved seeking deep battle, encirclement and the destruction of large groupings to generate operational momentum for strategic effect. The capture and holding of ground was seen as a good consequence, but the main aim was to conduct deep operations across enemy territory and disrupt cohesion on a large scale, depriving the enemy of the chance to respond to a rapidly evolving situation. The successful – if extremely costly – offensives of 1944 and 1945 restored the primacy of the offensive to Russian thinking.

But by the mid-1960s, evolving technology shifted the focus: the development of long-range precision guided munitions added a new dimension to the deep operations. Now the adversary's economic and military potential could be attacked. The Gulf War only confirmed the importance of this development: a high tempo, high intensity land-air operation in which front lines disappeared and the side with the preponderance of weaponry lost. If previously victory was achieved by the destruction of the enemy's military forces, the destruction of their economy and the elimination of their political system, by the 1990s this view had changed. Instead, victory in the war of the future was seen to be achieved by destroying enemy means of counterattack and strikes by precision-guided munitions against armed forces, economic targets, command and communications infrastructure and national leadership.[41]

Much of this thinking continues to be the primary concern of the Russian leadership about how advanced states achieve victory in war. As Gerasimov stated in 2013, 'long distance, contactless actions against the enemy are becoming the main means of achieving combat and operational goals', and the 'defeat of the enemy's objects is conducted throughout the entire depth of his territory.' More recently, he has pointed to Russia's growing combat might and the ability of the Russian armed forces to 'resolve strategic missions in remote theatres', emphasising Russia's strategic reach.[42]

Though the main focus of the Russian official definition is about victory in what is understood as traditional war, a coda is necessary: there is also extensive discussion in Russian military science about non-traditional wars. If victory in traditional or classical wars can be defined, Popov and Khamzatov note that victory in non-traditional wars – intrastate wars, civil wars, partisan/guerrilla or terrorist wars – is very difficult to fit into the framework of a concrete definition. They point out that victory in civil wars cannot be defined by numbers of citizens killed, and that regime change or 'democratisation' is also an ambiguous measure. Understanding of victory in such cases is subjective, generating different images and associations.[43]

This is an important point, though, because of how the Russian defence community views 'colour revolutions'. This is Russian shorthand for what Moscow sees as western-created uprisings through pro-democracy protest movements aided and abetted by the western information war, and diplomatic and financial support: effectively, externally imposed *coups d'état*. As Gerasimov stated in 2013, a 'perfectly thriving state can, in a matter of months and even days, be transformed into an arena of fierce armed conflict, the victim of foreign interventions and sink into a web of chaos, humanitarian catastrophe and civil war.'[44]

Other senior figures suggest that this points to an important change. Yuri Baluyevski, formerly Chief of General Staff and now an advisor to the director of the Russian National Guard, suggests that the aim of war in the 21st century will not be to occupy territory, but the subordination of the governing apparatus by the population of that state. The first targets are the leadership and political elite, then the economy and infrastructure, and, last of all, the armed forces.[45]

The Russian answer is twofold. The first part is the prioritisation of territorial defence, which is to be achieved by enhanced coordination of all the arms of the state, combined with moral and education and organisation. This is one of the reasons behind the attempts to encourage patriotism, and has also had practical consequences, including the establishment of the Main Directorate of Political-Military Affairs in the Ministry of Defence and Yunarmiya, a national cadet movement. The focus here is on socio-political resilience. The second part is enhanced military capability. Although other measures are understood to be increasingly important, they are considered to be effective only on the basis of military force. Indeed, the Russian leadership often emphasises how the use of force, including against the adversary's forces remains a necessary feature of deterring and countering such warfare. This is a theme to which Gerasimov has often returned, not least with regard to the war in Syria.

## Russian views of victory

Victory in Russian thinking is based on two aspects, the sacred and the profane: victory as sacrifice and victory as science. These are often intertwined across wars of necessity and wars of choice. There are numerous tensions within this, to be sure, shown above. The sacred aspects of Russian views of victory pervade the contemporary Russian socio-political environment. The idea of resilience and sacrifice in the defence of Russia is a prominent and persistent feature of contemporary culture and political discourse, with consequences for the political and defence landscape.

The profane aspects are perhaps more revealing. First, to the Russian leadership, winning looks like the result of an unpredictable and almost certainly expensive effort: with vanishingly few exceptions, Russia's experience of war and achieving victory is that it is very costly. Even 'small' wars of choice have tended to see high casualty rates; indeed, this is envisaged by Russian military science. This has engendered a persistent sense of caution in Moscow's views of war as a political tool, a caution that has both led to a strong emphasis on diplomacy and context shaping prior to the war, and even constrained Russian military efforts during the war. Moreover, while military theory may reflect the search for the ideal one blow, 'knockout' war, the practicality of experience is quite different. When the Russian leadership chooses to go to war, it is on the basis of prior preparation simultaneously to achieve surprise and overwhelming superiority. It is not the stereotypical steamroller, but the Russian military is more club than rapier. The evolution of Russian military science has meant that this club is at times artfully wielded, but it is a club. Finally, a hierarchy is clear: war – and victory – is a tool of policy: war as '*pobedonosnaya*' – 'victory bringing'. It is intended to achieve policy aims, and victory is seen on these terms: the adversary's unconditional surrender or a peace treaty that puts a final end to the policy dispute. Victories in battle and in campaigns are important (particularly as a feature of the sacred victory), but these are only of value if they bring strategic goals by ending the political question.

## *Notes*

1 'Zasedaniye kollegii Ministerstvo oborony', Website of the Presidential Administration, 24 December 2019, http://kremlin.ru/events/president/news/62401.

2 '"Eshelon pobedy" prosleduet po territorii YuBO s ostaovkami na 30 stantsiyakh', Website of the Ministry of Defence, 11 March 2020, https://function.mil.ru/news_page/country/more.htm?id=12281605@egNews.

3 'Zasedaniye kollegii Ministerstvo oborony'.

4 D. Trenin, *Post-Imperium. A Eurasian Story* (Moscow: Carnegie, 2007), 215–216.

5 S. Kotkin, 'Russia's Perpetual Geopolitics. Putin Returns to the Historical Pattern', *Foreign Affairs* (May–June 2016).

6 R. Pennington, 'Was the Russian Military a Steamroller? From World War II to Today', *War on the Rocks*, 6 July 2016, https://warontherocks.com/2016/07/was-the-russian-military-a-steamroller-from-world-war-ii-to-today/; D. Lieven, *Russia Against Napoleon. The Battle for Europe, 1907–1814* (London: Allen Lane, 2009).

7 Although the phrase is commonly translated into English as a 'short victorious war', the Russian word used is 'malen'ki' which, though it does not rule out 'short', is closer to a '*small* victorious war'. There is ongoing debate about this question in Russia about whether Plehve himself said this, or whether it was posthumously attributed to him, or even whether the phrase is merely a 'legend'. M. Saevskaya, 'K voprosu o proiskhozhdeniye phrasy o 'malen'nkoi pobedonosnoi voine', https://www.portal-slovo.ru/history/46352.php

8 M. Kofman, 'The Russian Quagmire in Syria and Other Washington Fairy Tales', *War on the Rocks*, 16 February 2016, https://warontherocks.com/2016/02/the-russian-quagmire-in-syria-and-other-washington-fairy-tales/.

9 R. Thornton, 'The Russian Military's New 'Main Emphasis': Asymmetric Warfare', *RUSI Journal*, Vol. 162, No. 4 (October 2017): 18–28.

10 'Zasedaniye diskussionnovo kluba "Valdai"', Website of the Presidential Administration, 3 October 2019, http://kremlin.ru/events/president/news/61719.

11 S. Lavrov, 'Russia's Foreign Policy in a Historical Perspective', *Russia in Global Affairs*, No. 2 (April–June 2016), https://eng.globalaffairs.ru/articles/russias-foreign-policy-in-a-historical-perspective/.
12 R. Pukhov (ed) *The Tanks of August* (Moscow: CAST, 2010).
13 S. Blank, 'Georgia: The War Russia Lost', *Military Review* (November–December 2008); Trenin, *Post-Imperium*, 78.
14 'Minister No', *Foreign Policy*, 29 April 2013, https://foreignpolicy.com/2013/04/29/minister-no/.
15 Of course, such myth-making methods are not limited to Russia: this was reversed completely later in the war when German units pointed to their own tactical victories while much greater disaster was engulfing them on the larger operational and even strategic levels.
16 G. Carleton, *Russia. The Story of War* (London: The Belknap Press of Harvard, 2017), 126–127.
17 Lieven, *Russia Against Napoleon.*
18 Carleton, 122, 128–30.
19 Carleton, 80; M. Edele, 'Who Won the Second World War and Why Should You Care? Reassessing Stalin's War 75 Years After Victory', *Journal of Strategic Studies* (December 2019).
20 D. Adamsky, *Russian Nuclear Orthodoxy. Religion, Politics and Strategy* (Stanford: Stanford University Press, 2019), 70–1.
21 The Immortal Regiment is a parade that started in 2012. The parade, in which Russian citizens march carrying photos of family members who served during the war, has grown substantially in scope–including internationally–often with hundreds of thousands or millions taking part. The fallen heroes search units are volunteer search groups that find lost remains of the war dead and rebury them.
22 'Zasedaniye Rossiiskovo organizatsiyonnova komiteta "Pobeda"', Website of the Presidential Administration, 11 December 2019, http://kremlin.ru/events/president/news/62293.
23 '"Pol'sha prava, kompromiss nerealen". Voina za istoriyu ne konchaetsa', 29 January 2020, *Svoboda*, https://www.svoboda.org/a/30404527.html.
24 See M. Edele, 'Fighting Russia's History Wars: Vladimir Putin and the Codification of World War II', *History and Memory*, Vol. 29, No. 2 (Fall/Winter 2017); 'Kursk WW2: Why Russia is Still Fighting World's Biggest Tank Battle', BBC News, 13 July 2019, https://www.bbc.co.uk/news/world-europe-48963295; 'US Ambassador Joins Row Between Poland and Russia Over World War II', Forbes, 2 January 2020, https://www.forbes.com/sites/jamesrodgerseurope/2020/01/02/us-ambassador-joins-row-between-poland-and-russia-over-world-war-ii/.
25 For further discussion, see Carleton, 84.
26 Edele reflects on this and confirms the consensus view that without the Soviet effort, Hitler would not have been defeated. 75% of German equipment was knocked out by the Red Army and nobody killed as many German soldiers as the Red Army. Edele, 'Who won', 5.
27 D. Glantz, *The Soviet-German War, 1941–45. Myths and Realities*, Lecture to Clemson University, 11 October 2001, https://www.youtube.com/watch?v=5qkmO7tm8AU; C. Dick, *From Defeat to Victory: The Eastern Front, Summer 1944* (Kansas: University Press of Kansas, 2016). There is a vast and evolving literature on the Great Patriotic War/Eastern Front, for overviews in English and Russian, see D. Glantz and J. House, *When Titans Clashed: How the Red Army Stopped Hitler* (Kansas: University Press of Kansas, 2015); S. Tyushkevich (ed), *Vtoraya mirovaya voina: itogi i uroki* (Moscow: Voennoe Izdatelstvo, 1985); M. Gareev, *Velikaya Otechestvennaya Voina 1941–1945: v dvenadtsati tomakh* (Moscow: Kuchkovo Pole, 2012).
28 Dick, 40, 159, 247.
29 These took place in the undeclared Soviet-Japanese border conflict.
30 For concise examinations of Soviet Operational Art, see Dick, *From Defeat* on which this brief overview draws, and D. Glantz, *Soviet Military Operational Art: In Pursuit of Deep Battle* (London: Frank Cass, 1991).

31 'Den, kogda konchilas voina', 9th May Project, *Kommersant*, https://www.kommersant.ru/projects/9may.
32 S. Karaganov, 'God pobed. Shto dalshe?', *Rossiiskaya Gazeta*, 17 January 2017, https://rg.ru/2017/01/15/sergej-karaganov-vozobnovlenie-dialoga-rossiia-nato-oshibka.html.
33 V. Gerasimov, 'Sila Velikoi Pobedy', *Voenno-Promyshlenny Kurier*, 11 May 2015, https://vpk-news.ru/articles/25167.
34 P. H. Vigor, *Soviet Blitzkrieg Theory* (London: Palgrave MacMillan, 1983); Dick, *From Defeat to Victory*.
35 I. M. Popov, and M. M. Khamzatov, *Voina budushchevo. Kontseptualnie osnovy i prakticheskie vyvody* (Moscow: Kuchkovo Pole, 2016), 444–445.
36 Popov and Khamzatov, 448–463.
37 'Encyclopaedia', Russian Ministry of Defence Website, https://encyclopedia.mil.ru/encyclopedia/dictionary/details.htm?id=8964@morfDictionary.
38 Vigor, *Soviet Blitzkrieg Theory*, 60; E. Crankshaw, *Putting up with the Russians* (London: Penguin Books, 1985), 20.
39 Cited in Crankshaw, 20.
40 See Gerasimov's often referred to, but much less often read and understood 2013 article, V. Gerasimov, 'Tsennost nauki v predvidenii', *Voenno-Promyshlenny Kurier*, 26 February 2013, https://www.vpk-news.ru/articles/14632.
41 L. Grau and T. Thomas, 'A Russian View of Future War. Theory and Direction', *Journal of Slavic Military Studies*, Vol. 9, No. 3 (September 1996): 510.
42 Gerasimov, 'Tsennost nauki'; V. Gerasimov, 'Mir na granyakh voiny', *Voenno-Politichesky Kurier*, 13 March 2017, https://www.vpk-news.ru/articles/35591.
43 Popov and Khamzatov, 446.
44 Gerasimov, 'Tsennost nauki'.
45 Y. Baluyevsky, 'Voina ne konchaetsa, ona–zamiraet', *Nezavisimoe Voennoe Obozrenie*, 26 May 2017, http://nvo.ng.ru/realty/2017-05-26/1_949_rosgvardia.html.

CHAPTER 16

# 'Winning the Peace': The Peacebuilding Paradigm and its Implications for Peacekeepers in the 21st Century

Nicholas Rees

## Introduction

Winning wars and achieving victory has traditionally been one of the core objectives of military forces, alongside providing national defence and supporting international security; although engaging in armed conflict is normally a last resort and most political leaders will try to avoid embarking on such a dangerous route. In the latter part of the 20th and early 21st centuries, there has been a decline in conventional inter-state wars, although there has been an increase in intra-state conflict, including civil wars and ethnic conflict. The challenge today has shifted from winning wars to ending intra-state violence and ensuring a sustainable peace. There is a recognition amongst those involved in helping to bring such conflicts to a close, including diplomats, political and military leaders, that ensuring a lasting peace and rebuilding fragile states is difficult to do. There is little point in investing resources in bringing a conflict to a closure, unless there is a way of ensuring that longer-term peace can be sustained. This chapter argues that this is an inherently political process and one that can only be achieved through local engagement and support for the peacebuilding process, with the support of international actors such the UN, regional organisations, NGOs and supportive states. There needs to be significant financial investment in the peacebuilding process, ensuring a recognition of the causes of conflict and a sustainable peace settlement.

The idea of winning the peace needs to be at the heart of any strategy aimed at ensuring a lasting and sustainable peace. In too many cases, military intervention, whatever the intention, has not ensured a lasting peace because insufficient reflection and planning has gone into thinking about what happens next. What does the peace look like when there is an absence of violence? Winning the peace is increasingly associated with peacebuilding, the term first popularised by the UN Secretary-General Boutros-Ghali in *An Agenda for Peace*[1] although originally associated with the work of Johan Galtung.[2] The term peacebuilding is often conceptualised as being about creating a positive peace, which the Global Peace Index defines as 'the attitudes, institutions, and structures that create and sustain peaceful societies.'[3] Much of the recent literature on peacebuilding and conflict transformation informs military thinking about its role in helping to win the peace. Notably, the 2016 *Understand to Prevent* (*U2P*) report highlights the importance of key principles, such as 'do no harm' and 'conflict sensitivity', whilst reflecting on what role the military could play in supporting efforts to counter violent conflict.[4] The report emphasises the need for effective coordination between comprehensive players, including all those involved in the peacebuilding process. The report also highlights the types of skills sets that military officers would need to be 'credible' and 'contribute' in such situations.

This chapter considers the UN's role in international peacekeeping and peacebuilding, how it has evolved and the growing role of regional organisations, including focusing on the development of peacekeeping doctrine and winning the peace. It considers the UN and British contributions to the development of peacekeeping doctrine, as well as the challenges this brings in working with other regional actors. It also examines what role Britain's military has played in these areas and what are the prospects for its future involvement in peacebuilding. The UK has been a significant supporter and financial contributor to UN peacekeeping and peacebuilding initiatives, as well as an active supporter of UN peacekeeping reform.[5] The UK as pen-holding member of the UN Security Council has been a supporter and proposer of Security Council resolutions, including as a lead nation in developing UN peace mission mandates and new initiatives around gender and sexual violence in peacekeeping. However, UN peacekeeping has not featured prominently in UK security policy and various governments have not been willing to commit large numbers of UK troops and police to peacekeeping missions.[6]

## Moving from peacekeeping to peacebuilding: winning the peace

The UN's early involvement in peacekeeping between 1945 and 1989, possibly with the exception of the Congo, was largely based on the peacekeeping principles adopted in the 1950s. Peacekeeping was not a high-profile activity during the Cold War period, and the UN was largely involved in peripheral conflicts, many of which were not seen as important to the P5.[7] These traditional peacekeeping operations aimed at separating the warring parties and supporting the implementation of a peace agreement. The focus was on the conclusion of a peace agreement and not on winning the peace through a process of peacebuilding, thus largely neglecting the long-term vision of 'winning the peace'. The move from peacekeeping to peacebuilding, and thus the real acceptance of the idea that 'winning a peace' is an on-going process of long duration, began with the end of the Cold War and an expansion in the number of UN peacekeeping missions at the beginning of the

1990s.[8] The then Secretary-General, Boutros Boutros-Ghali, in *An Agenda for Peace,* considered the challenges of the post-Cold War period, arguing that the growth in intra-state conflicts led to a demand for more peacekeeping operations, including peacebuilding to support peace agreements. There was also an expansion in the tasks outlined in the Security Council mandates, with a move from tasks associated with traditional peacekeeping to multi-facetted peacebuilding. The UN sought to develop mechanisms around early warning, preventive diplomacy, peacemaking, peacekeeping and peacebuilding.

The focus on peacebuilding was further intensified following the disasters of the 1990s, where the UN and other actors with limited troops and resources were unable to prevent violent conflicts in Rwanda, Bosnia-Herzegovina and Somalia. There followed a process of reflection and debate in the UN, including around how best to prevent such violent conflicts in the future and how to respond to them when they did occur. *The Brahimi Report*, published in 2000, proposed a number of significant reforms in the way peace operations should to be undertaken. Brahimi called for an integrated approach to peace support operations, bringing together all UN actors to maximise their impact. The renewed interest in conflict prevention, including the establishment of the UN Peacebuilding Fund and the reorganisation of peacekeeping led to a commitment to engage in multi-dimensional peacekeeping operations and support for humanitarian intervention. This was first put into practice in Kosovo, where the post of Deputy Special Representative of the Secretary-General was created, bringing together the functions of the Resident Coordinator and the Humanitarian Coordinator.[9]

The changing UN approach to peacekeeping, reflected in the use of the term peace support operations, highlights the growing recognition that peacekeeping is in itself usually insufficient to ensure a lasting peace. In post-conflict situations, the progression from conflict and violence to peace is seldom straightforward and the risk of violence reoccurring is high in many cases. There needs to be a strong commitment to disarmament, demobilisation and reintegration (DDR), which has become a key feature of UN peace support operations in places such as the Central African Republic, Darfur, the Democratic Republic of Congo, Liberia, Mali and South Sudan. UN peacekeepers have become used to supporting DDR processes, including working with other partner organisations, but can find post-conflict situations challenging when keeping the peace is fragile. Peacebuilding is dependent on the existence of security, but, if the root causes of the conflict have not been addressed and resolved, then lasting peace is more difficult to achieve. The challenges faced by those involved in such situations include supporting democratisation, human rights, civil society, the rule of law and economic liberalisation. In some cases, the priority has been on regional stability and security, especially where such conflicts have involved intervention by the UK, USA, France or Russia, with further state-building processes slower to develop. The main concern has been to return power to favoured political and economic elites, preserving the status quo and the territorial integrity of the state.

The focus today in peacebuilding is increasingly on supporting the emergence of authentic voices and a credible political structure. As Richmond and Franks note, 'Liberal peace transitions effectively appear to be more about "transition" to statehood and regional stability than about the norms and institutions of the liberal peace and their implications for the post-conflict individual; they are indeed far from producing an emancipatory form of peace.'[10] The focus is now much more on bottom up initiatives and participatory approaches which include local actors and civil society. De Coning refers to this change in focus as adaptive peacebuilding, whereby 'The focus is on the means or process, and the end-state is open to context-specific interpretations of peace.'[11] Peacebuilding has moved from being considered as a programmatic activity to being understood as a political one where problems need to be solved.[12] This change in thinking reflects the reality that attempts at peacebuilding in Afghanistan and Iraq, as well as in the Balkans, Middle East and parts of Africa were not effective, not least because these attempts had primarily focused on aspects of nation- and state building, as, for example, the Bonn Agreement of 2001, which aimed at the establishment of a secure and democratic Afghanistan. This agreement could, in retrospect, not achieve its aim for a number of reasons: the concentration on state-building meant that the emphasis was originally placed on the capital, Kabul, and neglected the rest of the country. The process was extended into the rest of the country once it was realised that a concentration on Kabul could not achieve what the UN had set out to achieve, because it ignored the political, tribal and social realities of Afghanistan. In addition, the mission was made more complicated by the fact that, more or less parallel to the UN's initiatives, the anti-terrorist and military-focussed, US-led operation of the 'global war on terror' was conducted in Afghanistan. The UN now recognises that peacebuilding is an essentially local, context specific and political process, with a growing focus on supporting the work of the UN Peacebuilding Commission and UN special political missions.[13]

Beyond New York, the UN has sought in the field to focus on transitions as peace operations come to an end, because these operations are only the first step towards a long-term peace in war-ridden countries and regions. There is a recognition that such transitions can prove problematic, especially as funding declines and political attention shifts. This is best exemplified in the changing requirements being made by the UN on missions, beginning with the UN Mission in Liberia (UNMIL), where the Security Council requested that the Secretary-General prepare a peacebuilding plan for Liberia, as a part of its final 15 month drawdown plan, as it transitioned to the UN country team. This national peacebuilding plan was to be led by national stakeholders and supported by the UN, the EU and the World Bank. Similar activities were planned around the UN missions in Sudan, Haiti and Guinea-Bissau.[14] The

Peacebuilding Commission also focused on cross-border and regional issues in the Sahel and the Great Lakes Region, aiming to address the issues crossing borders in these regions.

Developments in the field are closely related to the organisational reforms that the UN Secretary-General has implemented in order to stream-line the UN's efforts and make the UN as an organisation more effective and efficient. In the UN headquarters, this has included merging several departments in order to avoid duplication of effort. The work of the UN Peacebuilding Commission has been prioritised, with aim of supporting the transition process in a number of states: DRC, Guinea-Bissau, Haiti, Liberia, Mali and Sudan. The focus has been on supporting the UN resident coordinator's offices and ensuring early planning, financing, operational assistance and staffing. The UN Secretary-General has sought to shift resources from peacekeeping to prevention, mediation and peacebuilding, which will require support from the member states.[15]

Most recently, the UN Secretary-General António Guterres launched 'Action for Peacekeeping' plan or A4P (March 2018), which was aimed at ensuring renewed political commitment to peacekeeping operations. He called on all member states, the Security Council, troop and police contributing countries, regional partners and financial supporters to renew their engagement with UN peacekeeping and a commitment to a set of agreed principles. In August the Secretary General published a *Declaration of Shared Commitments on UN Peacekeeping Operations*, which to date has been supported by 152 states and four regional organisations.[16] The commitments made in the declaration have been subject to regular monitoring to ensure that implementation is taking place and the UN is achieving its aim to be more effective in peacekeeping.[17] This included supporting political solutions and sustainable peace (peace agreements); performance, safety and security; women, peace and security; conduct and discipline; and partnerships.

## UN and regional organisations

The UN's move from traditional peacekeeping operations to supporting peacebuilding has been led by the UN and its agencies work and supported by regional organisations, NGOS and host countries. The UN's engagement with regional actors has become increasingly formalised and institutionalised, although the manner in which this works can be complex and competitive.[18] The UN Charter had always envisaged that the UN would work with regional organisations in support of peace and security. In Europe this included working with the European Union, NATO and the OSCE, while elsewhere in the world, this included the African Union, the Economic Community of West African States (ECOWAS), the Arab League and ASEAN amongst others.[19] In light of its own financial and organisational constraints, the UN viewed such organisations as partners and enablers, as well as resource providers and sub-contractors who could undertake peace operations (including contracting out).[20] As early as 2004, during the Irish EU presidency, there were requests from the UN for greater cooperation between the EU and the UN in crisis management operations.[21] This was based on a call from the UN Secretary-General for the EU to support the UN in its peace support operations.[22]

In 2015 the High-Level Panel Report on Peace Operations (HIPPO) highlighted the need for 'stronger, more inclusive peace and security partnerships'.[23] A call echoed by the then UN Secretary-General, Ban Ki-moon in his response to the panel's report.[24] There has, since the 2000s, been an increasing diversification of regional organisations and state actors involved in supporting and leading peace support operations, including NATO in Bosnia-Herzegovina and Kosovo (IFOR, SFOR and KFOR), the NATO-led Mission in Afghanistan (ISAF), the AU-UN in Darfur, Nigerian peacekeeping forces in Liberia (ECOMOG/ECOWAS), the UK-led International Military Advisory Team (IMAT) in Sierra Leone and Australia in East Timor (INTERFET/UNTAET).[25]

The changing nature of peace support operations means that the military remit now includes a diverse array of tasks, including restoring the host state's ability to ensure security and maintain order, strengthen the rule of law and judicial institutions, respect for human rights, supporting the development of legitimate political institutions and popular participation, and promoting social and economic recovery. The degree to which regional organisations and individual states are able to support such activities depends on their own ability and capacity to support such measures. In terms of peacebuilding, the EU, NATO and OSCE already have the capacity to support such activities. A good illustration of the changing role played by such organisations is that of NATO in Afghanistan, where ISAF was replaced in 2015 by a new NATO-led operation, Resolute Support Mission (RSM). RSM's remit was to train, advise and assist the Afghan security forces and institutions. These regional organisations have also sought to enhance the capacity of other regional organisations, such as the African Union, as well as individual states, to undertake such tasks, although this still remains challenging.[26] These activities are also inter-connected, with high degrees of inter-organisation dependence on each other and a division of labour, as well as some competition.[27] This is evident in the support provided by individual states as supporters of peacebuilding, including Sweden, the UK, France, Germany and the USA. A good example of this has been in Kosovo where the EU has been the main international (political) actor (EU Special Representative and EULEX), but NATO (KFOR) has provided troops, maintained stability and advised the Kosovo security forces. The UN, through UNMIK and the work on the UNHCR and the UNDP, supports projects around democratic governance and peacebuilding.[28]

There is cooperation between the UN, EU and NATO in the way they approach peace support operations, including having adopted the UN's integrated mission method.[29] This is hardly surprising given the commitment to the UN Charter, the parallel development of peacekeeping doctrine in the organisations, and reality that many of the same military troops, police and civilians are deployed on operations under different 'hats'. However, there is a divergence in strategic visions, given how these organisations view the world, their priorities and assessment of threats.[30] There are, of course, differences in the ways in which missions are formed and undertaken, which is reflected in whether missions are undertaken on the basis of a UN Security Council Resolution or at the invitation of a host state, such as in Mali or the Central African Republic. There is also competition between the organisations, especially in relation to the EU and NATO, with each organisation developing its own vision, interests and strategies. This was evident in recent EU and NATO missions in Libya (EUFOR) and Afghanistan (EUPOL), where the EU found itself playing a secondary role to NATO and struggling to coordinate with it. NATO has also been far more likely to engage in peace enforcement operations, as evidenced in its involvement in the Balkans and Libya, where it has engaged in active military responses. In contrast, the EU through military and civilian operations, has been able to exercise more soft power diplomacy and adopted a more comprehensive country-level approach closer to that of the UN. There has also been competition between the UN and EU in operations, such as in Chad, Mali, DRC and Kosovo, where each actor has sought to preserve its independence.[31]

## UN and UK doctrine on peacekeeping

The UN's approach to peacekeeping has significantly changed over time, mostly notably since the Brahimi Report, which proposed that a robust posture might be taken by UN peacekeepers in future operations.[32] The review panel, led by Lakhdar Brahimi, the former Algerian Foreign Minister, suggested that the future of UN peacekeeping would necessitate an increase in the organisational capacity of the UN to undertake peacekeeping operations. It also suggested that the UN should only deploy a peacekeeping operation where there was a realistic and achievable mandate, backed with sufficient resources to undertake the operation. Such forces should be able to engage in robust peacekeeping, provide protection of civilians, and support peacebuilding initiatives. The report, and subsequent actions by the UN Secretary-General, including the deployment of stronger regional peacekeeping forces, were considered to be more culturally sensitive and supportive of local populations.

The UK doctrine on peacekeeping followed much of what had been developed within the UN but there have been differences. The UK approach to peacekeeping was predicated on the assumption that a peace enforcement posture might need to be adopted from the outset of a mission, given the consent of those involved in the conflict might change over time, and a robust approach might prove necessary.[33]

The UK doctrine was based on the *Army Field Manual: Wider Peacekeeping* (1994), which at the time sparked a debate over whether peacekeeping could ever be impartial and what the distinction was between peacekeeping and peace enforcement. UK doctrine was later updated in the *UK Joint Warfare Publication 3-01* (1997), and the follow-on Joint Warfare Publication *The Military Contribution to Peace Support Operations 3-50* (1999 and 2004).[34] The UK was willing to deploy military force to undertake robust peace enforcement, as exemplified in Sierra Leone, building on the lessons learnt from UNPROFOR during the 1990s. The UK was still committed to supporting the underlying principles of consent, impartiality and the use of minimal force, but there was a recognition that in some situations this was likely to vary over time. There were similar concerns around the idea of impartiality, ensuring that a peace support operation should treat all parties in the same manner, and be prepared to use force in a similar manner towards all parties. The UK's military focus in the 2000s was on Afghanistan and Iraq, where more traditional military tactics were employed, although not exclusively, given the nature of the security threats and other demands on the military. There was limited appetite for involvement in UN peacekeeping beyond the UK's ongoing commitments in Cyprus (UNFICYP), Bosnia-Herzegovina (UNPROFOR), Kosovo (UNMIK) and Sierra Leone (UNSIMIL). The UK did go onto contribute troops to UN operations in Somalia (UNSOS) and South Sudan (UNMISS).

In this period the UN was itself rethinking its approach to peace support and peacebuilding operations. In 2004 the UN Secretary-General's High-Level Panel on Threats, Challenges and Change, which assessed the threats to international peace and security in the 21st century, published its report, *A More Secure World: Our Shared Responsibility*. The report examined the new global security landscape, highlighting the indivisibility of security, economic development and human freedom, and the relationship between states, sovereignty and responsibility, including the right to intervene to protect citizens. The UN Secretary-General used the themes in the report as the basis of his recommendations *In Larger Freedom: Towards Development, Security and Human Rights for All* (March 2005), which endorsed the Responsibility to Protect principle and provided the basis for further discussions by member states on the 60th anniversary of the UN. In December 2005, a new *Note on Guidance on Integrated Missions* was published, which outlined a shared vision of the UN's strategic objectives, including the need for integrated planning and missions, as well as a set of agreed results, timelines and responsibilities of tasks critical to consolidating peace. There were also to be agreed mechanisms for monitoring and evaluation. The notion of integrated missions was closely linked to the idea of multi-dimensional peacekeeping, thereby recognising that such missions could not deliver peace on their own. This represented a change in UN peacekeeping doctrine, with an expansion of the underlying principles from consent, impartiality and minimum force to include credibility, legitimacy and the promotion of national and local ownership. A number of states, however found these changes to be unacceptable and the new principles were downgraded to success factors.[35]

Following on these attempts at reform, the UK and France published a joint non-paper (2009) in which they argued for better preparation, planning, monitoring and evaluation of missions. This informed a wider debate in the UN around reform, with the Departments of Peacekeeping Operations and Field Support publishing *A New Partnership Agenda: Charting the New Horizon for UN Peacekeeping* (2009). The aim of the New Horizons initiative was to maintain and intensify the dialogue among peacekeeping stakeholders on options for improving UN peacekeeping. This included improving consultation with those member states contributing troops and police contingents to missions. The proposals also included wider consultation with Military Force commanders, which addressed a long-standing issue whereby force commanders had found it difficult to communicate with those working at higher levels in the UN system.

The UK approach was, however, shaped by its experiences in Afghanistan and Iraq. As Curran and Williams note, the UK was intent on developing a military doctrine based on stabilisation.[36] This was described in its report on *Security and Military Stabilisation: The Military Contribution* (2009), which defined stabilisation operations as being long-term, multi-dimensional and largely political affairs.[37] In such operations the military's role was to support the stabilisation process based on nine security principles, including serving in combat, undertaking peacekeeping and providing humanitarian assistance. The overall aim was to support a political settlement, working closely with the host government, local elites and population, whist discouraging spoilers.[38] This approach went beyond the traditional UN approach to peacekeeping, suggesting a more interventionist approach where peace settlements might not be in place and in which the intervening force would be acting in a far more political manner than traditionally envisaged in UN peacekeeping. The UK's limited commitment to becoming more involved in peacekeeping, was evident in the Conservative–Liberal Coalition Government *Strategic and Defence Review,* which focused on making cuts and reducing the size of the UK's armed forces.[39] The UN did not feature prominently in the review, but the government did commit as one of its goals to work with the UN Secretariat, regional organisations and member states to ensure that conflict prevention played a central role in UN efforts to foster global peace and security alongside peacekeeping and peacebuilding.[40] This was linked closely with the *Building Stability Overseas Strategy* (FCO, DFID and MOD, July 2011), which aimed to promote stability and prosperity in countries and regions where it was morally right and UK interests were at stake. The UK was supportive of UN peacekeeping, calling on the UN to improve its efficiency and effectiveness and support political processes (prevention) aimed at providing long term stability.

In 2011 the UK published a new Joint Doctrine Note on *Peacekeeping: An Evolving Role for Military Forces* (July 2011), replacing the earlier JWP 3-50 *The Military*

*Contribution to Peace Support Operations* (1998 and 2004). Unlike the 2009 focus on Stabilisation, this new doctrine assumed that a peace agreement and ceasefire would be in place, thereby more closely mirroring UN doctrine, and suggesting that the role of military forces were to support the implementation of the peace agreement and maintain the ceasefire. This did not preclude the use of force against spoilers but did not suggest such missions be undertaken for particular political purposes. It was suggested that the UK might focus on niche areas in UN peacekeeping – high readiness capability, support helicopters, intelligence, logistics, field hospitals, etc.[41] The UK was also willing to support training initiatives aimed at building peacekeeping capacity and provided support mechanisms, such as the Military Stabilisation Support Group.[42] The UK also led on the 2012 Prevention of Sexual Violence Initiative (May), which was linked closely with the government's commitment to protecting civilians, aimed at both preventing sexual and gender-based violence and ensuring peacekeepers were also held to account for such crimes when on missions.

The UK's approach to peacekeeping has also been shaped by its involvement in NATO, as well as with the US Military.[43] The Ministry of Defence follows the guidance provided under the NATO Allied Joint Publications, including NATO's *Allied Joint Doctrine for the Military Contribution to Peace Support.*[44] This doctrine defines peace support as 'Efforts conducted impartially to restore or maintain peace. Peace supports efforts can include conflict prevention, peacemaking, peace enforcement, peacekeeping and peacebuilding'.[45] It is, however, predicated on the belief that there should be an overarching political strategy, which include a mix of peace and security tools. It requires a comprehensive approach, including the combined efforts of civilian and military actors.[46] The UK has also worked closely with the EU, supporting the development of the EU's global strategy, but still preferring to commit forces through NATO.[47]

The UK has continued to make commitments in support of international peacekeeping and peacebuilding. In January 2015 the Ministry of Defence signed a new MOU with the Irish Department of Defence, aimed at enhancing cooperation, especially around training and potentially joint Irish/UK contributions to peace support operations.[48] The Conservative government also made further commitments to the UN in the 2015 *Strategic Defence and Security Review*, including doubling the number of UK personnel on UN missions, building on an earlier commitment made by the Prime Minister at the Peacekeepers Leaders' Summit in September. The government also agreed to provide more law enforcement and civilian experts and to train foreign peacekeepers. It created cross-Whitehall joint UN Peacekeeping Policy Unit to maximise UK military and civilian impact. Underlying this approach was the view that conflict prevention and de-escalation were more cost-effective than having to respond to conflicts when they occur.[49] There was a significant commitment to supporting peacebuilding initiatives, which culminated in 2015 in the formal establishment of the Conflict, Stability and Security Fund, which provided greater opportunities for cross-departmental collaboration. Underlying all of this has been Britain's commitment to stabilisation, aligned with broader foreign policy objectives and overseas interests. The UK has been one of the largest financial contributors ($483 m in 2016) to civilian peacebuilding, conflict prevention and resolution over the period 2007–16.[50]

The UK's commitment to supporting peacebuilding is evident in *Shaping a Stable World: The Military Contribution* JDP-05 (March 2016). JDP-05 'outline[s] how the UK seeks to help shape a more stable world as part of our national strategy and examines the military role within.'[51] The emphasis being on a more 'forward leaning and engaging approach' based on cross- government cooperation and inter-agency collaboration. The armed forces would be required to respond to crises, including supporting humanitarian assistance and disaster response, and conduct rescue operations; conduct strike operations; conduct operations to restore peace and stability; and conduct major combat operations through NATO (Article 5). The continuing focus on stabilisation, including on the role of the military in supporting such a process, is evident throughout the document.[52] There is a commitment in the *UK National Action Plan on Women, Peace and Security, 2014–17* (FCO) to ensuring women's participation as critical to peace and preventing conflict.

The UK has also supported capacity building via the British Military Advisory Training Teams (Sierra Leone, the Balkans, Jordan, Ghana, Nigeria), British Peace Support Teams (South Africa and Kenya) and the British Army Training Unit (Kenya).[53] Other examples of projects include support for the AU in enhancing its capacity to prevent, manage and respond to conflicts in Africa, as well as supporting the development of troop contributing countries for participation in AU and UN missions (Africa Peace and Security Programme). Other UK-funded projects in the recent past have focused on Somalia (IED diffusion), South Sudan (conflict reduction), Sudan (conflict reduction and stabilisation), Afghanistan (security sector and peacebuilding support; peace process), Nepal (peacebuilding), Western Balkans Stability and the Good Governance Fund.

## Conclusion

Looking to the future, the UN will face significant challenges, including calls for reform, but it is likely to survive, as long as states see it as vital to international peace and security.[54] However, the growing influence of China and Brazil, as major financial contributors, as well as India, Nigeria, Brazil and South Africa as major UN troop contributors, will lead to more calls for change and reform of the UN system. There is, then, a shift in the balance of power from West to East, with intra-state and non-state conflicts likely to grow. The rise of China, the resurgence of Russia and American unilateralism are all major issues of concern. There are significant global challenges around inequality and social justice, religious intolerance, nationalism and immigration, which may fuel further conflicts. There is also likely to be a further blurring of the boundaries between war and peace, with a growth in hybrid approaches to warfare. By implication modern military forces will need to be prepared to respond to such changes and challenges across the globe, where political instability remains a problem, whether in parts of Africa, the Middle East and South Asia, whether unilaterally or as part of coalitions with other states and international organisations.

What role might the UK play in supporting the peacebuilding paradigm? If the UK is to play a more prominent role in UN peace operations, then a number of barriers will need to be overcome. First and foremost, there needs to be the political will to play such a role in UN peace support operations and peacebuilding. Second, there is also a financial consideration, given the competing priorities in the UK's defence and security budget and the continuing pressure to reduce costs, which have impacted the organisation of UK military forces.[55] If the UK is to contribute to peacekeeping support missions and peacebuilding, there needs to be sufficient funds to do so. Finally, the other key factor is UK military attitudes towards the UN and its peace support and peacebuilding. In the past there has been some scepticism within the military of the value of the UN and its competency in peacekeeping. There is also some concern that military personnel have little to gain from participating in peace support operations, if the experience is not valued by senior military leaders.

There are new opportunities for the UK military to gain experience and play a role in peacekeeping and peacebuilding aimed at winning the peace. The roles that the UK chooses to play will depend on the context, but the UK has the ability to deploy at high readiness, possibly in advance of a UN missions, or play a more traditional role with a UN peacekeeping support operation, including providing niche expertise which the UN lacks. This does mean that UK troops and police will face the challenges of becoming involved in peacebuilding processes, supporting demobilisation, disarmaments and reintegration programmes, as well as security sector reform, policing, rule of law and justice programmes. The UK military will need to further its understandings and the skills required in working with all the players involved in peacebuilding. There will be a continuing need for training, education and support for those involved in such peace operations, including context specific knowledge of the UN and local political environment and an understanding of the mandate and the rules of engagement. As the 2016 *U2P* report noted military officers will need to develop their skills sets to be 'credible' and 'contribute' to peacebuilding, which will require them to understand the theories of conflict, peace and prevention, as well as having the skills to support negotiation, mediation and dialogue.[56]

## *Notes*

1 Boutros Boutros-Ghali, *An Agenda for Peace* (New York: United Nations S/24111, 1992).

2 J. Johan Galtung, 'Three Realistic Approaches to Peace: peacekeeping, peacemaking and peacebuilding', *Impact of Science on Society*, Vol. 26 (1976): 103–115.

3 Institute for Economics and Peace, *Global Peace Index 2019: Measuring Peace in a Complex World* (Sydney: Institute for Economics and Peace, June 2019), 66. Available from http://visionofhumanity.org/reports, accessed 8 May 2020.

4 Multinational Capability Development Campaign, *Understand to Prevent: The Military Contribution to the Prevention of Violent Conflict* (London: Ministry of Defence, 2016), 116. Available from https://www.gov.uk/government/collections/multinational-capability-development-campaign-mcdc, accessed 8 May 2020.

5 In 2016 the UK was the 6th largest provider of assessed contribution to the UN peacekeeping budget, behind the USA, Japan, France, Germany and China.

6 A. Johnson, 'After Afghanistan: A British Military Return to Peacekeeping', New Trends in Peacekeeping: In Search for a New Direction (2014): 86. In 2020 the UK deployed some 8,000 troops overseas, but the contributions to UN-led peacekeeping were modest, including 292 troops in South Sudan (UNMISS), 13 troops to the logistical support to the UN Support Office in Somalia (UNSOS) and a smaller number of military staff to other UN peace operations.

7 T. O'Neill and Nick Rees, *United Nations Peacekeeping in the Post-Cold War Era* (London: Routledge 2005).

8 S. Ryan, 'The Evolution of Peacebuilding' in R. McGinty (ed), *Routledge Handbook of Peacebuilding* (London: Routledge, 2013), 25–35.

9 L. Debuysere, and S. Blockmans, 'Crisis Responders: Comparing Policy Approaches of the EU, the UN, NATO and OSCE with Experiences in the Field', *European Foreign Affairs Review*, Vol. 24, No. 3 (2019): 243–264.

10 O. Richmond and J. Franks, *Liberal Peace Transitions: Between State Building and Peacebuilding* (Edinburgh: Edinburgh University Press, 2009), 205.

11 C. De Coning, 'Adaptive Peacebuilding', *International Affairs*, Vol. 4, No. 2 (2018): 301–317, here 301.

12 Ibid., 303.

13 It was also reflected in the work of the UN Advisory Group of Experts, *The Challenge of Sustaining Peace; Report of the Advisory Group of Experts for the 2015 Review of the UN Peacebuilding Architecture* (New York: United Nations, 2015), A/69/968-S/2015/490. Also, see General Assembly Resolution 70/262 and Security Council Resolution 2282 (2016) on peacebuilding and sustaining peace.

14 UN Security Council Report, *Peacebuilding and Sustaining Peace* (New York: United Nations: July 2019). Available from https://www.securitycouncilreport.org/monthly-forecast/2019 07/peacebuilding-and-sustaining-peace 2.php, accessed 25 April 2020.

15 De Coning, 'Adaptive Peacebuilding', 308.

16 https://peacekeeping.un.org/sites/default/files/a4p-declaration-en.pdf, accessed 26 April 2020.

17 https://peacekeeping.un.org/en/a4p-highlights-of-key-achievements, accessed 26 April 2020.

18 J. A. Koops and T. Tardy, 'The United Nations' Interorganisational Relations in Peacekeeping' in J. A. Koops et al. (eds), *The Oxford Handbook on UN Peacekeeping Operations* (Oxford: Oxford University Press, 2014).

19 The EU, Germany, the Netherlands, the UK, Norway and Sweden have pooled funding to support the AU's peace and security facility.

20 The EU's early missions included Artemis and EUFOR RD Congo (supporting MONUC 2003 and 2006) and EUFOR Tchad/RCA (supporting MINURCAT 2008-09). The UN also created a UN Liaison Office for Peace and Security in Brussels in 2011 (UN DPKO and Department of Political Affairs). NATO and the UN signed a 'Joint Declaration on UN-NATO Secretariat Cooperation' in September 2008.

21 Note that the EU-UN signed 2 formal declarations in 2003 and 2007 and then a formal agreement in 2014 and are also committed to a Strategic Partnership on Peace Operations and Crisis Management (2015).

22 The degree to which the organisations can work together also depends on what is technically possible, given the various militaries involved, levels of training and the types of equipment and communications being used.
23 United Nations, *The Challenge of Sustaining Peace; Report of the Advisory Group of Experts for the 2015 Review of the UN Peacebuilding Architecture* (New York: United Nations A/69/968-S/2015/490, 2015), viii.
24 T. Tardy, 'The European Union and UN Peace Operations: What Global-Regional Peace and Security Partnership' in C. de Coning and M. Peters (eds) *United Nations Peace Operations in a Changing Global Order* (London: Palgrave Macmillan, 2019), 231–251.
25 T. Woodhouse and O. Rambotham, 'Peace Operations' in M. D. Cavelty and V. Mauer (eds), *The Routledge Handbook of Security Studies* (London: Routledge, 2009).
26 The UN and AU have cooperated on peace operations, such as in Burundi, Darfur, Somalia, Mali, and Central African Republic. This has also included hybrid operations, such as the UN-AU Mission in Darfur (UNAMID), as well as AU Mission in Somalia (AMISOM).
27 P. Petrov et al., 'All Hands on Deck: Levels of dependence between EU and other international organisations in peacebuilding', *Journal of European Integration,* Vol. 41, No. 8 (2019): 1027–1043, here 1031.
28 P. Petrov et al., 'All Hands on Deck', 1032.
29 T. Tardy, 'The United Nations' Interorganisational Relations in Peacekeeping', 237.
30 The different UN and NATO understandings of peacekeeping and peace enforcement was summarised in the *UK Joint Doctrine,* Note JDN 5/11. See *UN Peacekeeping Operation: Principles and Guidelines* (London: Ministry of Defence, 2008).
31 T. Tardy, 'The United Nations' Interorganisational Relations in Peacekeeping', 243.
32 L. Brahimi, *Report of the Panel on United Nations Peace Operations* (New York: United Nations S/2000/809, August 2000).
33 D. Curran and P. D. Williams, *Contributor Profile: The United Kingdom* (International Peace Institute, 16 August 2016), 75. Available from http://www.providingforpeacekeeping.org/wp-content/uploads/2016/08/UK-Curran-Williams-15Aug2016-3.pdf, accessed 26 April 2020.
34 Also see Joint Doctrine Note 5/11, *Peacekeeping: An Evolving Role for Military Forces* (London: Ministry of Defence, 2011). This was replaced by. *Peace Support Operations,* AJP 3.4.1 (NATO).
35 United Nations, *UN Peacekeeping Operations: Principles and Guidelines,* Department of Peacekeeping Operations (New York: United Nations, 2008).
36 Curran and Williams, *Contributor Profile,* 79.
37 See *Security and Military Stabilisation: The Military Contribution,* JDP 3-40 (London: Ministry of Defence, 2009). The overarching policy is the *UK Defence Doctrine*, JDP 0-01 and the *UK Joint Operations Doctrine,* JDP 01.
38 In 2009 the UK chaired the informal UN Security Council Expert Group on the Protection of Civilians.
39 C. Thomson and D. Blagden, 'A Very British National Security State: Formal and institutions in the design of UK security policy', *British Journal of Politics and International Relations,* Vol. 20, No. 3 (2018): 573–593.
40 A. Johnson, 'After Afghanistan: A British Military Return to Peacekeeping Conference on New Trends in Peacekeeping: In Search for a New Direction', paper presented at the NIDS International Symposium on Security Affairs, 5–6 November 2014, 84. Available from www.nids.mod.go.jp/english/event/symposium/pdf/2014/E-04.pdf, accessed 24 April 2020.
41 The challenge being to deploy UK military forces that were appropriate in a peacekeeping context, although this might depend on the security need, and whether they were deployed as an advance force prior to a UN deployment.

42 This unit is jointly owned by the FCO, DfID and the MOD and aims to provide civil military cooperation in crisis operations. It also identifies staff to be deployed on UN missions and provides a repository for lessons learned.

43 In 2019 UK support for NATO operations included circa 800 troops in Europe, the Royal Navy support for NATO Standing Naval Forces in the Baltic and Black Sea and some 150 personnel as part of the NATO Enhanced Air Policing (London: Ministry of Defence, 2019).

44 North Atlantic Treaty Organisation, *Allied Joint Doctrine for the Military Contribution to Peace Support*, AJP 3.4.1 (December 2014). Available from http://nso.nato.int/nso/zPublic/ap/.

45 NATO, 2014, 1–2.

46 NATO, 2014.

47 The UK has also contributed to a range of EU missions over the years, especially in the Balkans, including the EU Mission in Bosnia Herzegovina (EUFOR ALTHEA) and KFOR (2007-), as well as EU missions in Africa.

48 Ministry of Defence, 'UK and Ireland sign a historic defence agreement' (19 January 2015). Available from www.gov.uk/government/news/uk-and-ireland-sign-historic-defence-agreement, accessed 20 April 2020.

49 The UK annual review of *Countries at Risk Instability* report led by the Cabinet Office provides a watch list of fragile states.

50 A. Sherriff et al., *Supporting Peacebuilding in Times of Change* (Brussels: European Centre for Development Policy Management, 2018), 6.

51 Ministry of Defence, *Shaping a Stable World: The Military Contribution,* JDP-05 (March 2016), iii.

52 Note the importance of the Stabilisation Unit, as an integrated civilian-military operational unit, whose purpose is to help the government respond to a crisis and consider the causes of instability.

53 http://www.providingforpeacekeeping.org/wp-content/uploads/2016/08/UK-Curran-Williams-15Aug2016-3.pdf, accessed 26 April 2020.

54 Ministry of Defence, *Global Strategic Trends: The Future Starts Today*, sixth edition (2018), 106.

55 P. Antill and J. Smith, 'The British Army in Transition', *The RUSI Journal,* Vol. 162, No. 3 (June/July 2017): 56.

56 Multinational Capability Development Campaign, *Understand to Prevent: The Military Contribution to the Prevention of Violent Conflict* (London: Ministry of Defence, November 2014), 117. Available from https://www.gov.uk/government/collections/multinational-capability-development-campaign mcdc, accessed 8 May 2020.

# Conclusion: So, What is 'Winning'?

Andrew Sharpe

Sir Hew Strachan opens this book by observing that doubts over the efficacy of force in providing clear strategic or political results – of providing national political leaders with a 'win' – is not a new concern, but, despite that, it has become a particular obsession born of recent experience, in Afghanistan, Iraq and Libya. Recent apparently indeterminate results have cast re-doubled doubt on whether military action is wise or useful in the attaining of national or international strategic ends. Tactical success is achievable, and even campaign success is seen as a realisable goal in UK doctrine, but above that level, in the strategic corridors of Washington and Whitehall, Paris or The Pentagon, success is a much more nebulous concept, and, regardless of a political desire to talk of demonstrable 'wins' and 'victories', life must inevitably cease to be about success and failure *per se*, but rather about 'maintaining a position of advantage'. For much of history, however, the philosophy and doctrine of war has been much less muddy, grey or doubtful about the utility and purpose of force in the pursuit of strategic ends. To paraphrase both Sir Hew and several of the other authors of this book, and the passage of Clausewitz that they quote, war used to be about beating the enemy sufficiently soundly that you could demand of them whatever you wished once the fighting was over. Simple. Niccolò Machiavelli would have agreed.

Yet as we have seen, very few wars or conflicts since World War II ended with decisive outcomes as a result of swift or direct military action. The Falklands model is not a typical model replicated widely in the second half of the 20th century or the first decades of the 21st. Since World War II a number of things have combined to change the calculus previously so clear for war philosophers from Sun Tzu to Liddell Hart, for example: the replacement of centuries of regnal, imperial or dictatorial models of governance by various models of democracy; international law (not least Article 51 of the UN Charter); complexity; interconnectivity; new views, amalgamations and perceptions of risk and threat and of security and strategy; a 'global public voice'; and mankind's ability, for the first time in its history, to destroy itself completely if war is not controlled (and thus a switch of emphasis from a 'War

Office' to a 'Ministry of Defence' and from an imperative to be able to defeat the enemy to an imperative to deter the enemy); to name but a few. In short, there has been a relatively recent (in terms of the last two thousand years of history) switch of emphasis from winning wars to avoiding wars.

The nuanced nature of the question of winning is not, by any means, unique to the world of wars and statecraft. It applies, at its most basic level, in any walk of life. In business, for example, there is a marked difference between those who conduct their affairs for tactical gains alone or who make their profit through tactical action, and those who conduct business with strategic aims. A venture capitalist, for example, may be uninterested in the long-term sustainability of a business; the goal may be to buy up a failing business, strip from it any assets that can be sold on for advantage, and then turn the business around, in the short term, so that its reinvigorated and streamlined new form can be rapidly sold on for a profit. Long-term success for such a business model depends entirely upon the ruthless pursuit of tactical goals and the handing over of long-term strategic issues to others. For this sort of business the only long-term goal is to keep finding short-term opportunities to turn a profit. By contrast, the founder of a family business may lay out plans that revolve around making the business sustainable and profitable in the long term, such that the priority becomes ensuring that a going concern is passed on to the next generation, or the generations thereafter. For such a business, issues such as client satisfaction, reserves, reputation, contingencies, staff wellbeing and so on have a different importance from the quick-win-and-walk-away venture capitalist. The venture capitalist can, and must, see life in terms of a series of 'wins'; the family business owner can, and must, see life in terms of a series of moderated actions that keep them 'winning'.

Furthermore, the successive chapters of this book have very clearly illustrated that the answer to the question posed by the book's title is always going to be relative. There would appear to be no absolutes; despite centuries of hyperbole in the language of both victors and the defeated, outcomes, consequences, repercussions, rewards, costs, returns, and after-effects are very rarely clear-cut. The answers to the question, both general and specific, will always, it seems, be sensibly prefaced with 'well, it depends …'. Few things illustrate this imbalance of calculus as well as this short vignette:

On the evening of 13 August 1704, as the late sunshine pushed its way through the scattered clouds and the fast-dispersing fog of war in the aftermath of the battle of Blenheim, two men stopped to exchange comments. Both were dressed in the same red coats and black tricorn hats that set them out for all to see as English fighting men. No doubt one red coat was rather better tailored than the other, rather less stained with sweat, dirt, and blood and rather more braid-bedecked. And, no doubt, one tricorn hat was perfectly shaped, while the other was sagging and drooping, one hat was rather more ornamentally fringed and be-feathered, and rather less dusty and grimy than the other. For one of the men was the Duke of Marlborough, and the other a nameless English Musketeer. Both men were 'victors' in the historic struggle of the day; both had played their part and given their best to the labours of the day. They were both 'winners'.

The Duke had paused as he rode by, however, because he detected no excitement in the soldier, who he expected to be filled with the same elation in this moment of victory as he himself was, but who was instead leaning in a melancholy way on his musket, deep in thought. Their conversation, as reported by the then Colonel Lord de Ros[1], went as follows:

'Why so sad, my friend, after so glorious a victory?'

'It may be glorious, My Lord, but I am thinking how much blood I have spilled this day, for fourpence.'

Thus perspectives of winning are not just about the relative costs and gains between nations or alliances on a grand scale, are not just about the relative weight of strategic outcomes or the flow of campaigns and of so-called operational art, are not just about the consequential politics and economics of nations and international systems, but they are also and at the same time an intensely small-scale, personal calculation. For one, the battle of Blenheim brought eternal fame, an eponymous palace and grounds from a 'grateful nation', wealth and reward, and a line of successors that would go on to play significant roles in the history and fate of their country. For the other, no doubt, just more dusty marches, more blood, sweat and labour (for fourpence a day), and the right, should he survive to old age, to sit in a quiet corner of a tavern and say 'I was there with The Duke'.

One of those iconic moments of the early 21st century was the vision of George W. Bush standing on the flight deck of the USS *Abraham Lincoln* at the start of May 2003, with the 'Mission Accomplished' banner behind him, receiving the rapturous applause of a US Service audience. Iconic now because it has become an enduring symbol of political hubris. And iconic too because it has come also to symbolise so much of the strategic doubt that Sir Hew highlights in his opening chapter. (The fact that, during his address, Bush made it clear that he was talking of 'Combat Operations' and that 'our mission continues' and that 'we still have difficult work to do in Iraq' was almost instantly forgotten or conveniently ignored.) Yet, at the time, it made perfect sense to a savvy political operator, to transmit to the world a double-sided message: for an internal US audience – 'I'm a strong and successful president, and we should be proud of what our servicemen and women have achieved'; and for an international audience – 'don't mess with the might of US arms, or even risk messing with them, because you'll lose'. And, indeed, at that particular moment in time, if you had stopped any US soldier or marine, from private first-class to four-star general, (or, indeed, any Coalition soldier or marine for that matter) and asked them if they had 'won' the war against Saddam, they would have given you a quizzical look and a (rightly) emphatically positive answer.

In just a month of warfare the region's most armed and apparently war-ready nation had been undeniably defeated. (The story is familiar to us all, and is explored in detail in this book by Dan Marston and Carter Malkasian: the airstrikes began and the Coalition crossed the berms into Iraq on the night of 19/20 March 2003, Baghdad airport fell on the 3rd of April, the 'Thunder Runs' into Baghdad proper began the next day, and Baghdad fell on 10 April. By 1 May, the date of Bush's appearance on the flight deck, combat operations against the Iraqi armed forces, throughout Iraq, had, indeed, come to an end.) This was an enemy that Coalition troops had been briefed to consider as a real threat, as an opponent not to be underestimated, as an opponent that (probably) held and would (probably) use weapons of mass destruction, and as an opponent that would put up a real fight. Yet, in a blitzkrieg that surpassed the German drive to the Channel Ports in 1940, that enemy had been swept aside through a combination of superior firepower, technology, applied tactics and fighting morale. Saddam Hussein had been toppled from power (and was subsequently dug out of a grubby burrow in the ground and hanged). To those who had swept up the Tigris and Euphrates valleys into Baghdad, to be met by cheering crowds of Iraqi civilians, it felt very much as if the mission had undeniably been accomplished. And with some style. It felt very much like 'winning'.

Indeed, some 18 months later, as the US and the world struggled with the results of the Coalition Provisional Authority's attempts to restore some form of order and governance in an Iraq writhing in the throes of a multifaceted insurgency and post-war chaos, those same sweating and dusty PFCs and square-jawed generals (now on their second or even third tours of duty) were still emphatic in their view that they had, irrefutably, won the war against Saddam. It was, still, self-evident. They were soldiers; they had been asked to defeat an enemy army; they had done so; convincingly. Period. The fact, they argued with real self-conviction, that they were now, post-victory, being asked to police a state in chaos, while whoever was responsible for getting its economy and governance back on the rails got their act together, should not be allowed to detract from the victory that they had achieved. Journalists who made clever references to the unfortunate (and very Iraqi-un-empathetic) naming of the vast base outside Baghdad International Airport – Camp Victory – were invariably met with a steely glare and an answer that ran something along the lines of 'it's named to celebrate our victory in the war. It's about what *happened*, not about what's happening.' In a very different way from the Duke of Marlborough and his melancholy musketeer, the view of 'Victory' was still just a matter of personal perspective. For the person paid to plan, deliver and sweat their way through the firefight, tactical success (and even short-term operational success) had been achieved, and should be recognised. For the person paid for their ability to catch the reader's or viewer's eye, tactical success (and even short-term operational success) gone strategically wrong made a much better story. Again, the delivery of 'victory', or 'winning' was just a matter of perspective.

The perspectives evident of the month-long battle for Iraq in 2003 are typical of the characterisation of those wider Coalition campaigns in Afghanistan and Iraq that dominated the opening decade of the 21st century. Talk to any military veteran of those campaigns and they will describe countless tactical successes in the face of a clever enemy that used tactics designed to avoid direct confrontation wherever possible. Talk to any journalist, or academic, and they will describe pointless tactical engagements resulting in no identifiable strategic success, as we have found reflected in the relevant chapters of this book. Very few allied belligerents from those campaigns would admit to having ever even lost a firefight. Very few allied strategists, practitioner or academic, would claim that the coalition of those involved had ultimately 'won'. There is a paradox in these views.

The historical analysis chapters of this book abound with examples of great tactical successes that have not resulted in great operational or strategic outcomes. Hannibal's failure in 216 BC to follow up his unprecedented tactical success at the battle of Cannae by deciding not to advance on Rome, allowing himself instead to be outmanoeuvred and distracted by Fabius Cunctator, 'The Delayer', is an object lesson in a failure to understand that campaigning is not a matter of simply winning a series of battles or tactical engagements such that they accumulate some kind of a score that ultimately results in a strategic 'win'. As both Hannibal's cavalry commander Maharbal and, seemingly, his brother Hasdrubal, are reputed to have said to him, ruing the fact that he understood tactics but not strategy: 'You know how to win a victory, Hannibal, but you do not know how to use it.'[2] Thus 'winning' can occur, or not, on a number of different levels simultaneously.

By contrast, as we have seen in Ali Parchami's chapter on classical antiquity, Rome saw war as a key tool in a constantly rolling strategy. It saw neither victories nor defeats as finite, but rather as cumulative. So long as the wins outweighed the losses Rome would continue to grow and prosper. 'Peace' for Rome, *Pax Romana*, was about a state of subjugation (and assimilation) of its enemies or opponents. At the same time, Rome's understanding of strategic depth meant that they structured themselves to be able to absorb successive military defeats, and to continue to pursue strategic goals, simply by raising more legions to replace those defeated in battle. In so doing Rome repeatedly confounded opponents who were operating on the basis that they were fighting to defeat *the* Roman army rather than understanding that they were in fact facing just *a* Roman army that operated within a Roman construct and mindset. In short, Roman armies were conducting wars not as an end in themselves, but as part of an overarching strategic continuum.

Importantly, in antiquity, although there were conventions and expectations about inter-state conduct, there were no rules – there was no international law. 'Just War Theory' was in its infancy. Thus a survival-of-the-fittest or devil-take-the-hindmost state of affairs existed. It was possible to win simply by defeating the armed force of others and taking from them whatever one wished. Winning was therefore a mixture of cumulative tactical successes, a refusal to admit defeat in the face of reverses or to submit to subjugation, a survival or expansion of one's own cultural norms and mores, and the retention of influence or control. Laced into the ancient world is also a theme that has been reflected throughout much of military history, exemplified perhaps in Alexander the Great, which is the pursuit of victory as an end in itself, or rather in producing such things as fame and honour for the victor as an end in itself. In parallel to the accumulation of power, therefore, is the accumulation of victories for victory's sake – not unlike the accumulation of sporting trophies, kudos and celebrity in a modern sense. When a combination of these two states of affairs exists (i.e. an absence of regulated constraint and a desire for power, fame or glory), war becomes a more likely state of affairs, and both the effort to deliver victory itself and the impetus to impose harsh absolutes on those defeated become magnified. 'Winning' becomes a more likely and a more desirable state of affairs – both in its actuality and in its intensity.

Furthermore, as Sir Michael Howard reminds us in his book *The Invention of Peace,*[3] the notion that peace is the normal state of affairs for human interaction, interrupted only infrequently and undesirably by short and intense bouts of war, is a relatively modern mindset. John France's chapter on the Middle Ages reminded us that for a large part of human history a state of war (that modern doctrine dissects into all sorts of sub-states of conflict, confrontation, insurgency and counter-insurgency, and so on) has been the norm. The clashes of armies in decisive engagements in which one side or the other produced a win with dramatic and instant strategic consequences are rare (and, as a norm, were avoided). In the 'Hundred Years War' between England and France (which lasted 116 years) one struggles to identify many major battles – perhaps only Crecy, Poitiers and Agincourt on land and Sluys at sea. That represents four days' worth of activity in nearly forty-two and a half thousand days. Even with a list of sieges and chevauchees, this long war was as much about a state of bellicose inactivity as it was about the business of attempting to 'win' through serious armed conflict. And those battles named above each resulted in English victories, yet even the most partisan of historians would be hard pushed to describe that war as an English strategic victory in its final outcome. Whether between crowns, as in the 'Hundred Years War', or coalitions of interest, as in the Crusades, or simply in the constant state of uneasy co-existence between holders of power and influence throughout medieval Europe, a state of so-called war was as commonplace as tactical engagements and decisive strategic outcomes were rare. In short, war was as much (or more) about a factional or national strategic state of being as it was about the use of force *per se,* and, certainly, was rarely about winning in any final or decisive sense.

And little in European, or global, history changes as we enter the early modern period. David Parrott opens his chapter on Europe between 1500 and 1715 with the statement: 'The waging of war was the principal occupation of European rulers and their states throughout this 200-year period.' In this chapter we considered that the most frequent motive for warfare, and the usual determinant of what constituted winning, was based upon the accumulation of power by successful (or aspirant) families or factions. War was about the securing of the interests, and the longevity, of dynasties. Both within states and beyond their borders, internal and external power-broking was conducted against the backdrop of the use of military force to 'win' an accumulation or maintenance of dynastic power. Territories (or states in the current sense) were therefore a much more nebulous concept than they are now, expanding and contracting as families expanded or contracted, supplemented by marital successes. Subsequently, even as Just War Theory began to come of age during this period, war and warfare were not seen as unnatural events, but rather as a normal state-of-being as powerful entities sought short-term wins to strengthen their long-term positions. The practical and pragmatic view espoused by Niccolò Machiavelli in *The Prince* made perfect sense. War was about winning tactical engagements, certainly, but it was not about bringing about a state of peace through a decisive conclusion to those engagements. Winning was just tactical; and tactical wins were just so much coinage in the accumulation and retention of dynastic power, which was, by design, a constant competition.

In Jonathan Riley's chapter on The Age of Reason we found that new factors became important in the late 18th century: the notion of statehood took on new meaning with the Treaty of Westphalia; the fight for political ideas took on new significance and was added to dynastic and religious motivations for warfare; and technology brought the importance of science in the conduct of war increasingly to the fore. Importantly, the science of fortification brought in a new emphasis on attrition in warfare, rather than just upon conclusive battles. Once span or scale was added to the equation – politics, economics, strategy and war with continental, and even global, reach – the likelihood of a short and vigorous campaigning season culminating in a final and conclusive battle were greatly reduced. So, as Riley indicates, the sort of outcomes generated by Gettysburgs or Waterloos had become increasingly rare. Furthermore, success and failure in various campaigns rarely brought about finite changes in the strategic configuration of the belligerents involved – merely in changes to the balance which allowed those various participants to engage in their wider global ambitions with more or less power and influence. 'Winning' in war, in any decisive sense, was becoming harder.

Yet, as both Riley and Clausewitz remind us, this is perhaps to focus too closely on the character of war, rather than upon its nature. Regardless of these changes in emphasis, an enduring theme remains intact: tactical successes continue to be achievable and tactical actions are thus finite; battles and engagements can still be won. And campaigns can still be brought to a successful conclusion; but the outcomes of wars remain merely part of the continuum of strategic inter-relationships. We have seen that Napoleon could bring campaigns to a successful conclusion (as in the 1806 Prussian campaign), but still find that an aggregation of tactical and operational success did not ultimately culminate in unchallenged and unchallengeable lasting strategic success – parallel strategic and operational miscalculations in Spain and Russia, and then a final throw of the tactical and operational dice in the Low Countries in 1815, put an end to the Napoleonic dynasty. But not, of course, to France (albeit greatly weakened) as a strategic player in Europe. The fighting may be suspended, and may result in advantage to one side or the other, but the ebb and flow of strategy continues. As Riley observes, warfare for Napoleon was a tool for gaining and maintaining the strategic initiative, and in this light it should always be seen. It is this observation that is much more useful in this debate than an unproductive search for a chimerical conclusive and finite 'win' as a result of war.

The journey through the history of European warfare was then brought into the 20th century by the chapters provided by Lothar Höbelt and Rob Johnson. To pick up where we left off with the Age of Reason, it is interesting to note that Lothar Höbelt takes us through a series of winners from the outcomes of World War I, but he is using the term 'winner' not in a finite sense, but in a sense of ending up strategically ahead of the competitors. His lens is the same as the lens in the foregoing chapter. It is telling therefore, as our examination of the concept continues to develop, that Höbelt's chapter is an exploration of relative advantage, not an exploration of conclusive outcome. Rob Johnson hammers the point home; in his opening paragraph he makes it clear that his chapter: 'examines the character of *military* [his italics] victory, rather than the full spectrum of winning in diplomacy, economics, or other factors, although these elements cannot be separated entirely, not least because, as Basil Liddell Hart put it, the purpose of war is a 'better peace'.

In his examination of Lloyd George's concept of victory for Britain in World War I, Johnson explains that there were very clear military goals: the liberation of France and Belgium, and the administration of a crushing military defeat on Germany. And that the achievement of those military goals should also have subsequent military effects in terms of capability and doctrine: to dethrone the military hierarchy; the erasure of Prussian militarism from Germany; and the removal of Germany's ability to make and wage war. For Churchill, twenty-two years later, it was also a fight that was both substantial and conceptual. He wanted a tangible 'victory at all costs', by which he meant the physical destruction of the German war machine (and, indeed those of Japan and Italy and their allies) and the restoration of conquered territories; and he wanted the eradication of the Nazi ideology. In other words, both British war leaders wished to inflict that set of circumstances that Strachan reminds us in the opening chapter were those that Clausewitz considered the whole point of using the military instrument – to bring your competitor to a point where they have to accept whatever terms you offer them for continued existence in the strategic context thereafter.

So, Johnson explains how victory, winning, could be achieved with the military instrument, but, vitally, he also concludes by reminding us what that it is important to understand what that victory was achieved *for*. Military force had utility, he tells us, in securing ownership, in intervention for effect and in fighting for ideas. These three purposes echo those put forward in the previous chapters covering two thousand years of military endeavour. The immediate achievement of each of these things represented a military win. But only in the exploitation of those things for advantage, after the win, did the efforts have meaning. In other words, only by using a win is the win useful!

Our review of 2,000 years of European history became global as it ended with a chapter on the Cold War. Early in this chapter Jonathon Riley clearly articulates the initially very bounded military aim for the west: 'To halt the forward progress of communist expansion, but not necessarily to push it back or to win the battle of ideas.' Yet, very quickly, not least because of the unconscionable consequences of the Cold War going hot (in other words, Mutually Assured Destruction that would ensure that no-one could win), it soon became clear that the war could only ever be won, without going hot, by the triumph of one ideology over the other, either through pragmatic circumstance (such as through economic effect) or through conceptual circumstance (people simply no longer accepting one doctrine over another), or, more likely, though a combination of both. Wars, such as they were within that superpower context, were about fights for relative advantage through the use of proxies within carefully limited boundaries. As we have seen elsewhere, such constrained wars could be 'won', but these limited victories (or, indeed, stalemates) were of equally limited value in their own right – their utility came in how they were used to influence the overall superpower balance. Importantly, as our theme continues to develop, Riley concludes this chapter by pointing out that, despite the fact that the west would claim to have 'won' the Cold War, current international relations circumstance would suggest that Putin's Russia would not necessarily share that view, and thus, as he says, that nothing – including winning – is finite.

What must we therefore conclude from this illuminating gallop through the history of European warfare? It would seem that, by way of the briefest of summaries of over 2,000 years of the use of force in conflict and competition, up until the 20th century war was seen as a normal state of being. The notion that such a thing as a 'war to end all wars' could even be conceptualised would have been anathema. The events of the first half of the 20th century, specifically the two world wars, followed by a Cold War for most of the rest of the century in which the notion of war was so apocalyptic that the preservation of peace became the core purpose of military force, turned the traditional calculus on its head. As Sir Michael Howard would have it, the modern characterisation of peace was invented. It became more important that wars could be 'won' so that they could be finished; so that peace could be restored.

And, in parallel, it became equally important that wars could also be 'limited' so that they happened elsewhere and to others so that their effects did not spill over into the realm of warfare with global impact. The fact, as we have seen, that conflict that our forbearers would have unhesitatingly called war, but that modern doctrines (both military and academic) have sought to codify in various different ways other than war, has raged on all over the world to resolve power issues either directly or through proxies seems to have become lost in what Sir Hew Strachan refers to in the introductory chapter of this book as fashionable pigeon-holing. This study of history has shown us that, by stripping away a fascination with the Clausewitzean character of relatively recent warfare, and studying instead the more reliable enduring Clausewitzean nature of warfare over twenty centuries, this particular form of human interactivity is essentially about achieving success within short-term bounded goals (winning) in order to maintain an advantage in the constant state of competition (conflict, war, tension, statecraft, strategy, international politics, call-it-what-you-will) between human interest groups. Rather depressingly it suggests that the relatively recent notion that these are abnormal violent clashes that can be brought to neat conclusions with a 'win' so as to restore an (elusive) normal state of peace is a flawed one.

Perhaps the big message here is that 'winning' is only relevant if you understand what you are winning *for*. In other words, tactical 'winning' can rarely be a goal in itself. It can rarely deliver a clear-cut end state. Conflict may not be as simple as Clausewitz's characterisation of a game of cards – which is a pastime with a definable goal and a finite end.

The chapters of more distant historical analysis are complimented by a series of relatively recent studies that serve, together, to illustrate in even more stark focus, this clear theme that has emerged. These examples are all examinations of the specific character of war through the lens of a series of more recent events. The previous chapters revealed an enduring truth about the nature of war. These more recent character-of-warfare studies re-emphasise this as the theme recurs despite the very different character of the campaigns themselves. As Marston and Malkasian observe in the opening paragraph of their study of the Iraq and Afghan campaigns of the early 21st century: 'What emerges is that clearly defined policy goals are a helpful component of strategy but perhaps less so than recognising that strategy can be a process of muddling through.' Thus they start with the enduring theme of this book. Strategy is not finite. Strategy is a dynamic management of events, as they unfold, endlessly, seeking ways to seize, maintain and exploit the initiative over rivals (be they friendly competitors or outright enemies). And again, shortly thereafter: 'Although the traditional interpretation of winning can still pertain at the operational and tactical levels of war, it has become much less helpful to think about strategy in terms of "winning", rather than in terms of attaining policy goals.' Or, as we have seen from our historical chapters, of providing those who formulate policy goals with the initiative over their rivals and competitors such that they can exploit the context to stay ahead on the unremitting international chess board of statecraft.

It is telling, therefore, that the majority of the chapter on Afghanistan and Iraq concentrates not just on the actions of those in-country, but also on the shifting sands of the policy goals that framed the efforts of the contributing countries (and in particular the US and UK). This chapter therefore suggests to us that the problem here was not so much a failure of the actions in-country to meet the remit set, but rather an opaqueness and confusion of fluctuating remit, alongside a fractured politico-military dialogue, such that there was a separation of action from purpose. Marston and Malkasian quote both Strachan and Freedman in pointing out that policy goals will inevitably change in the ebb and flow of strategic statecraft – but they can only do so to advantage if they do so at the measured and considered design of both those delivering outcomes and those subsequently left holding and exploiting the initiative.

In Nick Rees's chapter on peacekeeping we explored the difficulty in conceiving what 'winning' may look like in a context that is as much about stopping things from happening as it is about making things happen. He looks at UN operations across their breadth (Congo, Rwanda, Bosnia, Somalia, Kosovo, Liberia, Mali et al). Importantly he also examines both the strategic goals of the UN itself and those of a range of contributing nations – sometimes in concert sometimes not. Inevitably he also mentions Cyprus and it is that example alone that perhaps serves as the starkest illustration of the difficulty of imagining what 'winning' may look like in this environment. When the Turkish army landed on Cyprus in 1974 UNFICYP had already been in existence for ten years. The mission continues and, if one attends an UNFICYP briefing in the Green Line buffer zone, one would be forgiven for believing that the mission in Cyprus was an outstanding example of the UN at its best; and so it may be – yet nothing is resolved. The whole point of the mission is to enforce stalemate. By design, the point is to prevent anyone from 'winning'. So it must almost inevitably be with peacekeeping – unless one were to make Sir Michael Howard revolve with ire in his grave by suggesting, rather tritely, that 'peace itself must be the winner'! Yet this awkward conclusion does not contradict our theme – for if the stalemate can be used to steady the context in which those imposing the state of affairs may profit then a 'win' may be seen to have been achieved. Even the outcome itself does not need to be conclusive if the set of circumstances that the outcome delivers gives advantage, and the initiative-holders (and there could be many of them) subsequently exploit that advantage.

In Chris Tuck's chapter on the Indonesian Confrontation of 1963 to 1966 the paradox of outcomes is explored further. One side 'won' yet the other side also profited from the outcome. Malaysia and Britain won in the operational and campaign sense, yet Indonesia emerged much stronger in the strategic sense. Both sides were able to portray their engagement and outcomes in a positive light. As Tuck says: 'In the final analysis, though, Confrontation illustrates that victory is not necessarily a zero-sum condition. Both the British and the Indonesian governments in the end were able to convince their domestic constituencies that the outcomes of the Confrontation had been favourable to them. Whatever the actual complexities, these perceptions trumped reality. In many respects, therefore, victory can be construed as an imagined condition.' Exactly so. And therefore, by extension, yet again, we see that outcomes of military action are important in their own right, but only in the light of how they are used, portrayed and exploited post-outcome.

Aaron Edwards, using the IRA as his vehicle, gives us a searching insight into the complexity of 'winning and losing' in a terrorist campaign. Very much an ongoing feature in the 21st century, the so-called 'war on terror', in all of its forms, is a recurring feature of violent human interaction and is thus likely to continue for the foreseeable future as a major defence and security concern for nations across the globe. Edwards emphasises an interesting extra ingredient to our debate, an angle that has appeared throughout this book, but to which he adds explicit clarity. He says: 'The important thing to note is that one's enemy can only be said to be truly defeated if their will to fight has been completely eroded. The outcome of the engagement, therefore, plays a far-reaching role in creating the kind of peace that follows war.' Or to paraphrase an exhortational (and rather tautological) speech I once heard from an American company commander to his troops as they prepared to go into action: 'You are going to beat these guys. No doubt about it. They will be beat. But you gotta understand that they won't be beat, even when you have beaten them, until they *decide* they're beat. So, not only have you gotta beat them, you gotta make sure that they *know* they're beat.' Quite.

Edwards makes another telling point, in reminding us that regardless of whether you are using the military instrument in a national war of survival, a war of limited goals and boundaries, a peacekeeping operation or counter-insurgency, you are still using force, or the threat of force, or the capabilities that military force bring with them, as an instrument of coercion or persuasion. So, the Clausewitzean principles of the use and utility of force as an extension of policy goals remain sound regardless. Vitally for our discussion in this book, Edwards's analysis and narrative conclude with the insight that the British government could consider the campaign to have been 'won' because force had been used by both sides as an instrument of policy, but had failed for Sinn Fein to bring about the policy changes desired, and succeeded for the British government in maintaining the governance status quo which was their policy aim. (Yet, still, perhaps, Sinn Fein argue that their position in the governance of the province now demonstrates that campaign outcomes only matter in the way that you use those outcomes in the strategic continuum over time.)

With this flow of logic behind us, Richard Kuno's thought-provoking and wide-ranging analysis of 'Assad's Hollow Victory' in Syria adds real focus to the theme. In this comprehensive overview Kuno sets the Middle Eastern context with real clarity, explains the system of power and governance that Assad has used, and examines the sequence of events that unfolded surrounding the destructive civil war in Syria. This war is filled with minor 'wins' and 'counter-wins' and, as Kuno points out, has drawn in the strategic interests of friends and enemies, close and distant neighbours alike, all of whom have used those 'wins' and 'losses' with varying degrees of effectiveness. At the same time, those minor wins in a wider context are seen to be almost irrelevant when set against the backdrop of the heart of the conflict itself, and how it has been so ruthlessly used – that is to shore up both Assad and his Alawite clansmen. In a neatly circular fashion we could return to the anecdote from the battle of Blenheim at the start of this summary chapter because in this Syrian chapter we find a 'victory' that, aside from its bloody cost, is at one and the same time both national-strategic (and regional) in its implications, but is also intensely personal, on an individual scale, not just for the unfortunate people of Syria, but also for both Assad and those closest to him.

In Richard Reid's exploration of the spectrum of 'winning' in African history he opens his analysis by framing the chapter with the two fundamental considerations of this concluding chapter. He says: 'Our assessment of what constitutes victory depends on two important considerations. The first relates to the actual *objectives* of the respective combatants, and an analysis of whether or not these were achieved. This is (arguably) the more straightforward of the two. The second concerns the timeframe within which we are positioning our analysis. In other words, what often looks like a victory (or a defeat) *in the moment* – that is, in the immediate context of the armed encounter itself – may not actually prove to be such over a more extended timeframe.' And we have found it to be so in every aspect of the flow of analysis throughout this book. The core of his message, as he guided us through centuries of African conflict and competition, is that short-term wins, whether peacefully or violently achieved, only have utility in how they are exploited in the long term, aggregated together with other factors and events, be they military or non-military.

The remaining chapters of the book widen our horizons to a global level. The study without them would be very western-centric (despite the odd reference to Sun Tsu and others) in its viewpoint. Russia, China and Iran are not just competitors to the west in the ebb and flow of international relations, but are also entities that approach strategy, philosophy and statecraft through a variety of lenses that are very different from those of the western tradition. Ali Parchami guides us through an Iran struggling with contrasting identities and commensurately divergent views of winning. The goals of Iranian nationalism and Islamism sit uncomfortably alongside each other, maintaining a constant state of uneasy tension that plays out on both the domestic and the international stages. Kerry Brown shows us a China that, deeply imbued with an underdog feeling, may talk of 'win–win' situations, but is very clear indeed upon its endgame: to be as far ahead of its rivals as possible in the long game of international relations. To win. But, as he reminds us, the world has no actual experience of what that might mean; as he says: 'In this context, we do have a history of knowing what major powers like the US look like as winners. But we don't have any modern experience of what this means for China, in terms of geopolitical role and status. China as the PRC has been involved since 1949 in a long process of perpetual limbering up, getting ready for the final stages of a competition to prove what it is really capable of.' So for China, right now, the wins are bound to be incremental as it manages them, as wisely as it can, to build a position of increasing ascendency and advantage. Andrew Monaghan steers us through a characteristically insightful exploration of Russian history, military doctrine, politics, philosophy and mindsets. He explores the nature and notion of winning in and through war, to the Russian mind, and, importantly he explores how such 'wins' are portrayed and used (even when others may not see them as a 'win' for the Russians!).

These three excellent chapters, when taken together and at the end of our flow of analysis, add real impact to the conclusions that have already been drawn. Impact because, despite their widely divergent contexts and philosophical approaches to our question, their answers, when unpacked, have huge resonance with our other collective western-centric answers to the question. For example: use war with caution and never by default; understand what you are trying to achieve by the use of the military instrument before you resort to its use; set clear goals, but be flexible in their pursuit as their context changes; be pragmatic; do things that surprise and unsettle your competitors; be careful, deliberate and imaginative in how you portray outcomes (both domestically and internationally); and use incremental small wins to build momentum and initiative. And ultimately, if you have to use war at scale, then make sure that you use it in a way that gives you the very best chance of putting your opponent in a position where the terms post-war will be set by you. Andrew Monaghan's closing words offering a Russian view echo the central conclusion of every other chapter of the book: 'Victories in battle and in campaigns are important (particularly as a feature of the sacred victory), but these are only of value if they bring strategic goals by ending the political question.'

Hew Strachan concludes the opening chapter of this book by stating: 'Going to war requires of those who do it a commitment to win – and a sense of what victory would look like. Without that, war lacks purpose and the strategy to shape it is robbed of coherence.' Indeed. It is hard to understand how to *win* a game if you do not know the *object* of the game. Those who design games, whether board games, wargames or card games, once they have settled upon a general idea, start their game design process by setting down the 'aim of the game', or, for wargamers, 'the victory conditions'. They understand full well that it is very difficult indeed to design a playable game (let alone to play it) if you have little clear understanding from the outset of the object of the game. So, whether designing your game of strategic international relations or playing in it, knowing what you are trying to achieve before you start is not a bad idea. Equally, however, we have concluded from this analysis that war may or may not be, as Clausewitz described it, like a game of cards (with clear object and outcomes), but what is for certain is that war's context will be malleable, changeable, unpredictable and ongoing.

Those who study strategy disagree on much, but, usefully, they also agree on several things. One of which, as we have seen throughout this book, is that strategy and strategic outcomes are iterative. Theorists and practitioners alike recognise that, while a simple description of the balancing of ends, ways and means may be a fair characterisation of strategy, it lacks a key ingredient – which is to say that strategy must be a dynamic process. From Heuser to Liddell Hart and from Sun Tzu to Freedman observations about the need for dynamism, for recalculation, for readjustment, for post-goal-setting and post-planning continuous management abound. It is about the balancing of those three things, while at the same time accepting that once 'a strategy' has been formed the job is not done. In other words, strategy does not finish when the plan is written. Strategy must be a constantly adapting activity, balancing the desired strategic path against the combination of logic, chance and unpredictable human interactions that will drive the ebb and flow of events. And here we find a key point in this discussion. 'Victory' is a finite notion. It is something to be achieved or that has been achieved. It is a noun, not a verb. It is therefore a thing, not an action. In statecraft and strategy the practitioner, because they understand that their art is iterative and never-ending, seeks to retain the initiative, to achieve goals that keep them ahead, in short to *be winning*, because the job will never be complete – so the strategist cannot 'win', they can only ever 'be winning' – or, perhaps more usefully, 'be ahead'.

And it is here, at this late stage, that the discussion finds utility in the unlikely realm of grammar. The trouble identified by Hew Strachan at the outset comes when people think that strategy is finite and can be 'won'. Analysts of the conflicts of the first decade of the 21st century are fond of asking: 'We may have succeeded in every battle that we fought, but does it really look as if we have won?' And: 'Are we really arguing that we are the winners as a result of this international interventionism?' This muddled thinking is to suppose that strategy has fixed outcomes, that the strategy of a few years ago is now somehow over; that strategy and statecraft, far from being interwoven and continuous, is somehow separated out into finite breaks and starts. The fact that 'winning' is a verb matters – this is because verbs are about doing things, not about final outcomes; and the tense also really matters. The title of this book is *Winning Wars*. Not *What is won?* or *What is having won?* And this is the key distinction that troubles those who are looking for simple answers. The notion that international strategic affairs can be won or lost is, as we have seen in every one of the foregoing chapters of this book, a very modern European and North American unsophisticated view of events that does not reflect experience. Despite the comments in the opening paragraphs of this chapter about the clear-cut definitions of past theorists and philosophers, both Clausewitz and Machiavelli would have understood this completely. The use of force, and the cessation of its use, are merely a part of the continuum of strategic events. War may well be a continuation of politics or policy by other means, but it is also about the politics and policies to which one returns post-war (imagined and aimed for pre-war by any self-respecting statesperson or strategist, one would hope). Each Machiavellian short-term win only has value if the Prince ruthlessly exploits the outcome of that win thereafter to ensure that he stays ahead of his rivals.

Thus for the strategist the question can never be 'have we won?', but rather should always be 'are we winning?'. Strategic goals cannot be finite, but rather should be iterative. On the other hand, tactical actions (be they military, national or even business) should only be undertaken for clearly identifiable reasons, outcomes or goals. In military mission terms, tactical actions are often described before they are undertaken in terms of achievable goals: 'you are to do x and y in order to bring about the following state of affairs'. The x and y in this case should be clear, finite and achievable. Thus, when x and y have been achieved the tactical actor can be said to have 'won', to have achieved a 'victory'. Those actions, however, only have real meaning or real value if they are linked to the 'in order to', to the reason for carrying out those actions, or in other words to the higher cause, be it on a small scale, or on a large one. So, for force to have utility, whether in a purely military micro-tactical sense, or in a grand-strategic sense, one should decide what one wants to use force for in advance. War therefore cannot only be an extension of politics or policy by other means, but it must also be the interlude between the politics and policies, and thus it is the policies to which one wishes to return after the 'win' is obtained by the use of force that make the use of force apposite. For it to be otherwise is both pointless in a utilitarian sense and immoral in the sense of human cost for frivolous or ill-considered ends.

So, what then is 'winning' in war? This book has taken us through 2,000 years of European warfare, has taken a guided tour through conflicts in the last 50 to 60 years in which armed force has been used as an instrument of policy imposition, and it has taken a glimpse at the concept from a range of global lenses. This is what that journey has led us to conclude:

It is possible to win, in war, providing clear and achievable goals are set. Whether those goals are micro-tactical (seize and hold the bridge at Ponte Grande) or grand-strategic (force unconditional surrender upon the Axis powers) the setting and subsequent achievement of those goals does amount to a win. To victory. But the accomplishment of those goals only has relevance if they are achieved for a reason. If those successes are used. If they continue, after their accomplishment, to keep those who benefit from their achievement 'winning'. Holding the initiative is only relevant if that initiative is used. Achieving a 'win' in war is done in order to seize, re-gain or maintain the initiative. But *having* the initiative is not enough; *exploiting* the initiative is how one stays ahead – how one remains in a state of 'winning'.

One thing is for certain. If you use force as an instrument of policy, on the micro-scale, there are *always* going to be 'winners' and 'losers'. Lives and property will be casualties. Perspectives will matter – especially where they are of the personal and visceral kind that war evokes. Whether in the lives and experiences of the combatants, or those of non-combatants caught up in the violence of conflict, war remains an unpleasant, vicious, primitive, passionate, hard-to-control and messy instrument of policy. There is always a cost, from the intensely personal to the national-strategic level, when it is embarked upon. And, in a world of the increasing influence of individual personal public voices, those micro-outcomes, and the way that they are portrayed to the world, are likely to be increasingly seen and felt by others and thus to have increasing influence on many levels from the individual human level the macro-international stage.

Winning *in* war is bound-able, definable and achievable. Winning *from* war is a very much more demanding and long-term prospect. Winning *in* war is the job of those who conduct wars. Winning *from* war is the job of those involved in statecraft who made the decision to use war to serve their ends.

## *Notes*

1 Major General Lord William de Ros, *The Young Officer's Companion or Essays on Military Duties and Qualities* (London: J Parker and Son, 1857).
2 Livy, *History of Rome*, Vol III, 22:51.
3 Michael Howard, *The Invention of Peace* (London and New Haven: Profile Books and Yale University Press, 2020).

# Select Bibliography


Jonathan Bailey, Richard Iron and Hew Strachan (eds), *British Generals in Blair's Wars* (Farnham: Ashgate Publishing Ltd., 2013)

Gregory Carleton, *Russia. The Story of War* (London: The Belknap Press of Harvard, 2017)

John Childs, *Warfare in the Seventeenth Century* (London: Cassell, 2003)

Peter Clarke, *The Last Thousand Days of the British Empire: Churchill, Roosevelt, and the Birth of the Pax Americana* (New York: Bloomsbury Press, 2008)

Carl von Clausewitz, *On War* (ed and translated by Michael Howard and Peter Paret) (Princeton: Princeton University Press, 1976)

Philippe Contamine, *War in the Middle Ages*, translated by M. Jones (Oxford: Blackwell, 1984)

Harold Crouch, *The Army and Politics in Indonesia* (Ithaca: Cornell University Press, 1988)

Cedric de Coning and Mateja Peters (eds), *United Nations Peace Operations in a Changing Global Order* (London: Palgrave Macmillan, 2019)

Philip de Souza and John France (eds), *War and Peace in Ancient and Medieval History* (Cambridge: Cambridge University Press, 2008)

Avinash Dixit and Barry Nalebuff, *Thinking Strategically: the Competitive Edge in Business, Politics and Everyday Life* (New York: W. W. Norton, 2008)

David Easter, *Britain and the Confrontation with Indonesia, 1960–66* (London: Taurus Academic Studies, 2004)

Niall Ferguson, *The Pity of War* (London: The Penguin Press, 1998)

John France, *Victory in the East. A Military History of the First Crusade* (Cambridge: Cambridge University Press, 1994)

Lawrence Freedman, *Strategy* (Oxford: Oxford University Press, 2013)

David French, *British Strategy and War Aims, 1914–1916* (London: Allen and Unwin, 1986)

David Glantz and Jonathan House, *When Titans Clashed: How the Red Army Stopped Hitler* (Kansas: University Press of Kansas, 2015)

Adrian Goldsworthy, *The Pax Romana: War, Peace and Conquest in the Roman World* (London: Weidenfeld & Nicolson, 2017)

Stephen Gunn, *The English People at War in the Age of Henry VIII* (Oxford: Oxford University Press, 2018)

Alistair Horne, *A Savage War of Peace: Algeria 1954–1962* (London: MacMillan, 1977)

Michael Howard, *The Invention of Peace* (London and New Haven: Profile Books and Yale University Press, 2020)

Robert Rhodes James, *Churchill: A Study in Failure, 1900–1939* (London: Weidenfeld & Nicolson, 1970)

Matthew Jones, *Conflict and Confrontation in Southeast Asia, 1961–65: Britain, the United States, Indonesia and the Creation of Malaysia* (Cambridge: Cambridge University Press, 2002)

Francois Jullien, *A Treatise on Efficacy: Between Western and Chinese Thinking,* translated by Janet Lloyd (Honolulu: University of Hawaii Press, 2004)

Ross A. Kennedy, *The Will to Believe: Woodrow Wilson, World War I, and America's Strategy for Peace and Security* (Kent: Kent State University Press, 2009)

Henry Kissinger, *A World Restored: Metternich, Castlereagh and the Problems of Peace 1812–1822* (Boston: Houghton Mifflin, 1957)

Thomas Leahy, *The Intelligence War Against the IRA* (Cambridge: Cambridge University Press, 2020)

Sandra Mack, *The Iranians: Persia, Islam and the Soul of a Nation* (New York: Plum, 1998)

Margaret MacMillan, *Peacemakers: The Paris Conference of 1919 and its Attempt to End War* (London: Murray, 2001)

Noel Malcolm, *Useful Enemies. Islam and the Ottoman Empire in Western Political Thought 1450–1750* (Oxford: Oxford University Press, 2019)

Robert Mandel, *The Meaning of Military Victory* (Boulder, CO: Lynne Rienner Publishers, 2006)

William C. Martel, *Victory in War: Foundations of Modern Military Policy* (Cambridge: Cambridge University Press, 2007; revised ed., 2011)

Ewan Mawdsley, *The Russian Civil War* (Edinburgh: Birlinn 2nd ed., 2008)

Roger McGinty (ed), *Routledge Handbook of Peacebuilding* (London: Routledge, 2013)

John McHugo, *A Concise History of Sunnis & Shi'is* (London: Saqi Books, 2017)

Martin J. Osborne and Ariel Rubenstein, *A Course in Game Theory* (Cambridge, MA: MIT Press, 1994)

Emanuele Ottolenghi, *The Pasdaran: Inside Iran's Islamic Revolutionary Guard Corps* (Washington DC: FDD Press, 2011)

David Parrott, *The Business of War: Military Enterprise the Military Revolution in Early Modern Europe* (Cambridge: Cambridge University Press, 2012)

Thomas Philipp and Christoph Schumann, *From Syrian Land to the State of Syria and Lebanon* (Beirut: Ergon Verlag, 2004)

Kurt Raaflaub (ed), *War and Peace in the Ancient World* (London: Blackwell, 2007)

Richard J. Reid, *Warfare in African History* (New York: Cambridge University Press, 2012)

William Reno, *Warfare in Independent Africa* (New York: Cambridge University Press, 2011)

Oliver Richmond and Jason Franks, *Liberal Peace Transitions: Between State Building and Peacebuilding* (Edinburgh: Edinburgh University Press, 2009)

Jonathon Riley, *Napoleon and the World War of 1813: Lessons in Coalition Warfighting* (London: Routledge and Keegan Paul, 2000)

Victor Rothwell, *War Aims in the Second World War: The War Aims of the Belligerents, 1939–45* (Edinburgh: Edinburgh University Press, 2005)

Benjamin M. Rowland (ed), *Balance of Power or Hegemony: The Interwar Monetary System* (New York: New York University Press, 1976)

Graham Spencer, *From Armed Struggle to Political Struggle: Republican Tradition and Transformation in Northern Ireland* (London: Bloomsbury, 2015)

Roel Sterckx, *Chinese Thought from Confucius to Cook Ding* (London: Pelican Books, 2019)

Hew Strachan, *Directions of War* (Cambridge: Cambridge University Press, 2013)

Hew Strachan, *European Armies and the Conduct of War* (London: Routledge, 2005)

Jacques Waardenburg, *Muslims and Others. Relations in Context* (Berlin: De Gruyter, 2003)

Robert White, *Out of the Ashes: An Oral History of the Provisional Irish Republican Movement* (Newbridge: Merrion Press, 2017)

James Q. Whitman, *The Verdict of Battle: The Law of Victory and Modern War* (Cambridge, MA: Harvard University Press, 2012)

Peter Wilson, *Europe's Tragedy: A History of the Thirty Years War* (London: Harvard University Press, 2009)

# Index